Essential
Electrical S
for HVACR
Theory and Labs

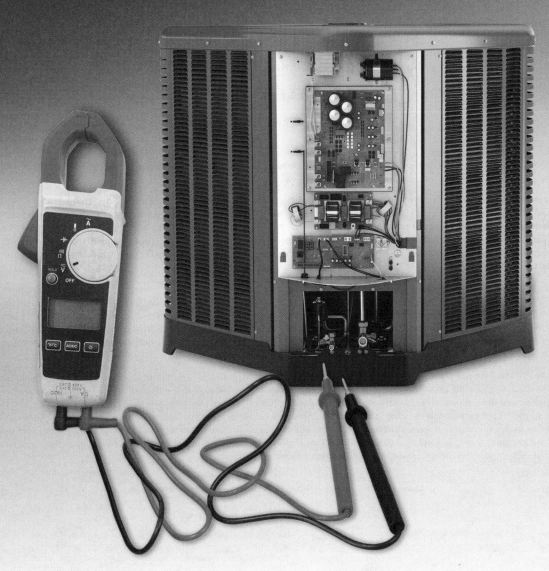

Ernesto Reina

Certified Master HVACR Educator (CMHE)
Lincoln Technical Institute
Union, NJ

Publisher
The Goodheart-Willcox Company, Inc.
Tinley Park, IL
www.g-w.com

Preface

Essential Electrical Skills for HVACR: Theory and Labs addresses today's need to produce HVACR technicians that are proficient in electrical fundamentals and troubleshooting modern HVACR systems. This is a combined electrical text and lab manual that provides a practical approach to core electrical theory supported by integrated, hands-on labs. With concise coverage of the material required to find success in the field, each chapter is presented in a logical, linear order where learning is reinforced by the previous chapter and built upon in the next one. This text serves as supplemental electrical material for those currently using *Modern Refrigeration and Air Conditioning* as part of their HVACR curriculum.

Essential Electrical Skills for HVACR provides numerous detailed illustrations and examples to visually depict component operation and support electrical theory. Forty-five hands-on labs are included in the text, following each chapter as opportunities to apply learning into trade-related practices. End-of-chapter review questions also provide a final reinforcement to evaluate student comprehension.

This text has a strong emphasis on electrical troubleshooting, which begins with a solid understanding of electrical fundamentals. These fundamentals, such as constructing a simple circuit, applying Ohm's law, application of alternating and direct current, and reading the various types of electrical diagrams are presented concisely in early chapters for students to master. Students will practice using electrical hand tools, preparing wire terminations, and working with various components. They will also be instructed on how to assemble and operate a universal lab board that is required for most labs assignments in later chapters.

Once these basic proficiencies are established, the text will cover special features, including multimeter operation, troubleshooting microprocessor-based printed circuit boards, and exploring HVACR systems. A technician must know how a multimeter works in order to troubleshoot successfully and provide accurate analysis. Lacking this understanding can lead to inaccurate diagnoses of system operation. To maximize troubleshooting skills necessary for modern electrical systems, this text covers in detail how a multimeter works and expectations about what the meter can do.

Students will then transition to microprocessor-based printed circuit boards and learn how to troubleshoot them. These are found in most systems today and, thus, the HVACR technician must be prepared to troubleshoot these systems proficiently. These labs will use a real circuit board, an integrated furnace control, for evaluation by the student. The student will provide the necessary input signals sequentially to produce the desired operation outcome. The lab will also guide the student to analyze failure modes. Students will wrap up the text with practice on troubleshooting HVACR systems that include the use of wiring diagrams.

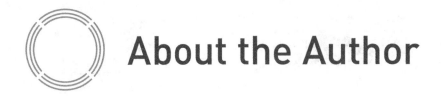

About the Author

Ernesto Reina is an instructor at Lincoln Technical Institute, where he has taught courses in HVACR and for electrical-electronics service technician and customized training programs. He also is an alumnus from Lincoln Technical Institute, where he developed a special interest in troubleshooting electrical systems. He then moved on to receive a diploma in Industrial Electronics from RETS Electronics Institute and pursued an A.A.S in Electronics Engineering Technology from Union County College, New Jersey.

Before becoming an instructor, Ernesto worked at General Motors (GM), where he expanded his knowledge in mechanic, electrical, and manufacturing engineering through corporate training and work experience. His responsibilities included HVACR and electric system quality assurance. He had the opportunity to learn about the design and manufacturing process for electromechanical components and computerized control systems. He designed and developed specialized test equipment for root cause analysis of failed components. During his career, he also provided electrical training to the in-house repair personnel.

Mr. Reina has received the Certified Master HVACR Educator (CMHE) certificate from HVAC Excellence. The CMHE certificate requires passing six subject matter expert exams. He is a Specialist Member (electrical) of the Refrigeration Service Engineers Society (RSES) and received the Imperial award for scoring the highest test grade in the country. He is a licensed master HVACR contractor in New Jersey.

Reviewers

The author and publisher wish to thank the following industry and teaching professionals for their valuable input into the development of *Essential Electrical Skills for HVACR*:

Terry Carmouche
River Parishes Community College
Gonzales, LA

Samuel A. Heath
Southern Technical College
Brandon, FL

John Labriola
West Virginia Northern Community College
Wheeling, WV

Charles William (Bill) Ledford
Pikes Peak Community College
Colorado Springs, CO

Richard L. Pasznik
Lincoln Technical Institute
Union, NJ

Richard J. Ruscigno
Lincoln Technical Institute
Union, NJ

Frank Sloan
Lincoln Technical Institute
New Britain, CT

Charles Smith
Lincoln Technical Institute
Cobb, GA

Acknowledgments

The author and publisher would like to thank the following companies, organizations, and individuals for their contribution of resource material, images, or other support in the development of *Essential Electrical Skills for HVACR*:

Carel Industries
Carrier Corporation, Subsidiary of
  United Technologies Corp.
Danfoss
Dial Manufacturing, Inc.
DiversiTech Corporation
Emerson Climate Technologies
Hubbell Inc.

Ideal Industries, Inc.
Lennox Industries Inc.
Parker Hannifin Corporation
Rheem Manufacturing Company
Sporlan Division, Parker Hannifin Corporation
Tempstar
Uline
White-Rodgers Division, Emerson Climate Technologies

TOOLS FOR STUDENT AND INSTRUCTOR SUCCESS

Student Tools

Student Text

Essential Electrical Skills for HVACR: Theory and Labs delivers focused instruction and hands-on application of electricity and controls to prepare students for a successful career in HVACR. This first edition is a combined essential electrical text and lab manual that provides a review of core electrical concepts and a deeper exploration into the concepts through integrated hands-on labs to master application and comprehension. This text serves as supplemental electrical material for those currently using *Modern Refrigeration and Air Conditioning* as part of their HVACR curriculum.

Instructor Tools

LMS Integration

Integrate Goodheart-Willcox content within your Learning Management System for a seamless user experience for both you and your students. LMS-ready content in Common Cartridge® format facilitates single sign-on integration and gives you control of student enrollment and data. With a Common Cartridge integration, you can access the LMS features and tools you are accustomed to using and G-W course resources in one convenient location—your LMS.

G-W Common Cartridge provides a complete learning package for you and your students. The included digital resources help your students remain engaged and learn effectively:

- **Online Textbook.** All content from the print textbook is included with exact page reproductions to ensure that students do not miss any important information or illustrations. Navigation is easy with a linked table of contents, and the search function helps students to easily find information.
- **Drill and Practice.** Learning new vocabulary is critical to student success. These vocabulary activities, which are provided for all key terms in each chapter, provide an active, engaging, and effective way for students to learn the required terminology.

When you incorporate G-W content into your courses via Common Cartridge, you have the flexibility to customize and structure the content to meet the educational needs of your students. You may also choose to add your own content to the course.

For instructors, the Common Cartridge includes the Online Instructor Resources. QTI® question banks are available within the Online Instructor Resources for import into your LMS. These prebuilt assessments help you measure student knowledge and track results in your LMS gradebook. Questions and tests can be customized to meet your assessment needs.

Online Instructor Resources (OIR)

Online Instructor Resources provide all the support needed to make preparation and classroom instruction easier than ever. Available in one accessible location, the OIR includes Instructor Resources, Instructor's Presentations for PowerPoint®, and Assessment Software with Question Banks. The OIR is available as a subscription and can be accessed at school, at home, or on the go.

Instructor Resources One resource provides instructors with time-saving preparation tools such as answer keys, chapter outlines, and other teaching aids.

Instructor's Presentations for PowerPoint® These fully customizable, richly illustrated slides help you teach and visually reinforce the key concepts from each chapter.

ExamView® Assessment Suite Administer and manage assessments to meet your classroom needs. The ExamView® Assessment Suite allows you to quickly and easily create, administer, and score paper and online tests. Included in the assessment suite are the ExamView® Test Generator, ExamView® Test Manager, and ExamView® Test Player. G-W test banks are installed simultaneously with the software. Using ExamView simplifies the process of creating, managing, administering, and grading tests. You can have the software generate a test for you with randomly selected questions. You may also choose specific questions from the question banks and, if you wish, add your own questions to create customized tests to meet your classroom needs.

G-W Integrated Learning Solution

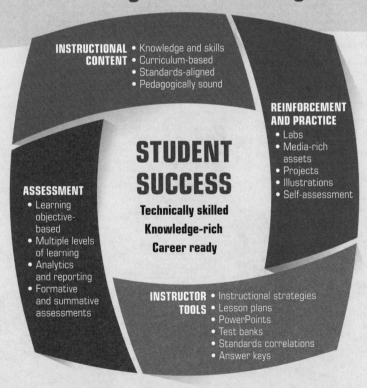

INSTRUCTIONAL CONTENT
- Knowledge and skills
- Curriculum-based
- Standards-aligned
- Pedagogically sound

REINFORCEMENT AND PRACTICE
- Labs
- Media-rich assets
- Projects
- Illustrations
- Self-assessment

STUDENT SUCCESS

Technically skilled

Knowledge-rich

Career ready

ASSESSMENT
- Learning objective-based
- Multiple levels of learning
- Analytics and reporting
- Formative and summative assessments

INSTRUCTOR TOOLS
- Instructional strategies
- Lesson plans
- PowerPoints
- Test banks
- Standards correlations
- Answer keys

The G-W Integrated Learning Solution offers easy-to-use resources that help students and instructors achieve success.

▶ **EXPERT AUTHORS**
▶ **TRUSTED REVIEWERS**
▶ **100 YEARS OF EXPERIENCE**

EMPLOYABILITY SKILLS · TECHNICAL SKILLS · ACADEMIC KNOWLEDGE · INDUSTRY RECOGNIZED STANDARDS

Features of the Textbook

The instructional design of this textbook includes student-focused learning tools to help you succeed. This visual guide highlights these features.

Chapter Opening Materials

Each chapter opener contains a chapter outline, a list of learning objectives, and a list of technical terms. The **Chapter Outline** summarizes the topics that will be covered in the chapter. **Objectives** clearly identify the knowledge and skills to be learned when the chapter is completed. **Technical Terms** list the key words to be learned in the chapter. **Additional Reading** directs the student to more content coverage found in *Modern Refrigeration and Air Conditioning*. **Introduction** provides an overview of the chapter content.

Additional Features

Additional features are used throughout the body of each chapter to further learning and knowledge. **Safety Notes** alert you to potentially dangerous materials and practices. **Pro Tips** provide advice and guidance that is especially applicable for on-the-job situations.

Illustrations

Illustrations have been designed to clearly and simply communicate the specific topic. Illustrations have been completely replaced and updated for this edition. Photographic images have been updated to show the latest equipment.

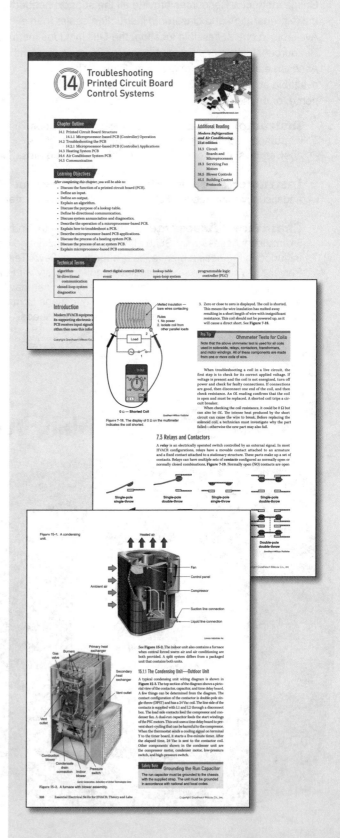

Hands-On Lab Activities

Each chapter includes labs that provide students with the practice to master application and are aligned with the chapter content. Labs include an objective, introduction, equipment list, procedures, and lab questions. The **Objective** informs the student of the goal of the lab and what they will be practicing and learning. **Introduction** provides an overview and key information useful for understanding the lab. **Equipment** lists the materials that a student will need to perform each lab successfully. **Procedures** offer step-by-step instructions to complete the lab. **Lab Questions** direct students in considering the main takeaways from each lab.

End-of-Chapter Content

End-of-chapter material provides an opportunity for review and application of concepts. A concise **Summary** provides an additional review tool and reinforces key learning objectives. This helps you focus on important concepts presented in the text. **Know and Understand** questions enable you to demonstrate knowledge, identification, and comprehension of chapter material. **Critical Thinking** questions develop higher-order thinking and problem solving, personal, and workplace skills.

Brief Contents

Contents

1 Electrical Safety

Chapter Outline

1.1 Standards and Codes
1.2 Personal Hazards
    1.2.1 Working on Electrical Equipment
    1.2.2 Personal Protective Equipment
1.3 Tools and Meters
1.4 Circuit Protection

Additional Reading

Modern Refrigeration and Air Conditioning, **21st edition**

2.1 Safety and the Government
2.2 Hazard Assessment
2.3 Personal Protective Equipment (PPE)

Learning Objectives

After completing this chapter, you will be able to:

- Discuss the need for standards and codes outlined by various industry organizations.
- Define the *National Electric Code* (*NEC*) and the Occupational Safety and Health Administration (OSHA).
- Understand the types of electrical hazards an HVACR technician must consider to ensure their safety.
- Discuss the types of personal protective equipment (PPE).
- Explain the safety requirements for working with electricity.
- Define electrically-insulated tools.

Technical Terms

American Society of Heating, Refrigerating, and Air-Conditioning Engineers (ASHRAE)
dielectric strength
electric shock
ground-fault circuit interrupter (GFCI)

live-dead-live (LDL) method
lockout/tagout (LOTO)
multimeter
National Electric Code (*NEC*)
National Fire Protection Association (NFPA)

National Electrical Manufacturers Association (NEMA)
Occupational Safety and Health Administration (OSHA)
personal protective equipment (PPE)

Introduction

Electrical safety is an important consideration when on the job to protect yourself and others from unintended injury. An HVAC technician must know about electrical hazards in order to prevent harm to themselves, others, and their property and equipment. Understanding the potential dangers and applying the established safety guidelines will provide you with the confidence needed to work safely while not being fearful of electricity. Due to its high level of importance, electrical safety is introduced in this first chapter. The safety content in this chapter will be applied and put into practice in later chapters for you to build a strong safety foundation.

1.1 Standards and Codes

In the 1800s, the emergence of electrical power brought about dangers, such as electrical fires, electrocutions, and property damage. Many of these electrical issues were caused by safety not being a major consideration. Eventually in the late 1890s, standardized electrical installations and methods to promote and ensure safety were instituted to address the crisis.

Standards for electrical work are outlined by several organizations in the electrical field. The *National Fire Protection Association (NFPA)* develops codes and standards to protect the public from hazards. The NFPA 70B standard addresses electrical maintenance while the NFPA 70E standard covers electrical safety in the work place. The NFPA codes provide minimal safety requirements and are superseded by local municipal or state codes. The *National Electric Code (NEC)*, or *NFPA 70*, is published by the NFPA and is a standard that addresses electrical systems and safe installation. The *NEC* is updated every three years to include the latest improvements in safety. With over 100 years of documented experience, the *NEC* is the most widely used electrical code in the United States.

There are many other sources for standards and codes for a technician to understand. These sources include the following:

- *Occupational Safety and Health Administration (OSHA).* Created by Congress in 1970, OSHA mandates workplace safety requirements. OSHA provides training, standard development, and enforcement in the work place.
- *American Society of Heating, Refrigerating and Air-Conditioning Engineers (ASHRAE).* ASHRAE develops standards for the HVACR trade, including electrical applications.
- *National Electrical Manufacturers Association (NEMA).* NEMA develops standards for electrical component specifications.
- *Underwriters Laboratories (UL).* UL tests and certifies equipment/components to ensure design standards are met.

References to these organizations are made throughout this text when applicable.

1.2 Personal Hazards

Electric shock is a discharge of electric current that passes through the human body. In this situation, a person's body becomes a conductor of electric current. Direct physical contact is not required to receive shock.

A person can experience different levels of electric shock. There are many factors that determine the resulting damage of the shock:

- How resistant the skin is at the points of contact with the electrical power source
- The amount of voltage applied
- Whether the electric current is alternating or direct
- The path taken by the current within the body
- The duration of contact
- The individual's current medical or health state

The severity of shock can range from a mild buzzing-type sensation to severe external and internal burns. Current of about 1/10 A through the heart can cause ventricular fibrillation, resulting in loss of consciousness or even death. Shock may also cause muscle contraction where a person cannot let go of the point of contact. This results in a longer exposure to the current and more severe damage from respiratory paralysis or burns. Reactionary accidents are another type of injury that can occur from the electric shock.

1.2.1 Working on Electrical Equipment

Technicians must always be on high alert by surveying their setting and having an action plan to safeguard against injury. In order to prevent electrical incidents like electric shock when working on equipment, ensure that the power is disabled and the circuit is de-energized. This can be checked by using the *live-dead-live (LDL) method*. First, using a meter, measure the voltage on a known live circuit to make sure the meter is working. Next, with the same meter, test the circuit that is turned off. Then, measure again the live circuit to ensure that the meter is still functioning correctly.

Once the circuit is confirmed to be off, *lockout/tagout (LOTO)* can be performed. See **Figure 1-1**. This procedure is used so maintenance or service can be completed in a safe manner. The LOTO process requires a physical lock to be attached to the source disconnect to prevent unintentional activation by another person. Every technician working on the same equipment must secure their own lock to

Ideal Industries, Inc.

Figure 1-1. A lockout/tagout kit for different electrical switch and valve builds.

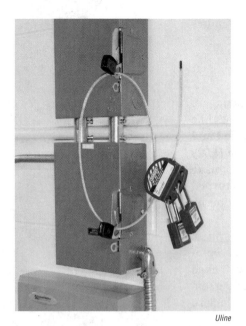
Uline

Figure 1-2. Various forms of electrical lockout equipment are available. All technicians must use a separate lock.

DANGER

DO NOT OPERATE

This lock/tag may only be removed by:

Name : _____

Dept. : _____

Expected Completion : _____

Master Lock.
SAFETY SERIES

Model No. 497A

Uline

Figure 1-3. Include your name and relevant information on a tag.

the electrical disconnect and keep their own key. See **Figure 1-2**. A tag is filled out by the technician and put on the disconnect to inform others that service work is being done. See **Figure 1-3**.

Safety Note

Assisting Another in Contact with a Live Circuit

You should never physically touch another person that is in contact with a live circuit because you too could be shocked. The best way to assist the victim and prevent further injury is to shut off the power.

1.2.2 Personal Protective Equipment

Personal protective equipment (PPE) is worn to protect an individual from a number of safety hazards. Use of PPE and careful planning allows a technician to complete a job more safely than working without PPE. PPE includes gloves, boots, helmets, clothing, face shields, and glasses and are available for the class, or specific type, of electrical work being performed. In many cases, wearing the proper PPE for a specific task is a mandated requirement and not considered optional. Higher-voltage systems can produce explosions and arcs, so the proper PPE is vital for your safety. Jewelry should not be worn when working, as it has high conductivity.

Sometimes a work area can become wet, such as when checking a condensing unit for power in the rain. Situations like these may be unavoidable. A tarp is a type of PPE that can be used to deflect the rain and keep service equipment on a dry surface. Perspiring in a hot attic requires added precaution as skin resistance decreases significantly when coated with perspiration.

Safety Note

Dielectric Ratings

Not all gloves provide adequate electric shock protection. Make sure the dielectric rating on a pair of gloves meets the requirements for the appropriate voltage level needed.

Safety Note

Extinguishing Electrical Fires

Electrical fires require the use of a Class C fire extinguisher. See **Figure 1-4**. Never use water as a substitute.

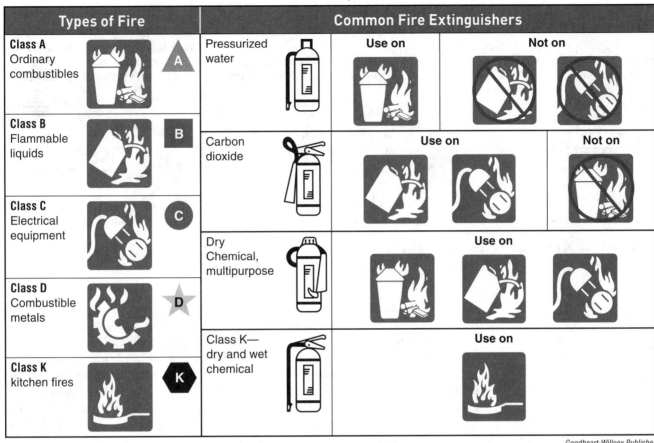

Types of Fire	Common Fire Extinguishers		

Class A Ordinary combustibles

Class B Flammable liquids

Class C Electrical equipment

Class D Combustible metals

Class K kitchen fires

Pressurized water — Use on / Not on

Carbon dioxide — Use on / Not on

Dry Chemical, multipurpose — Use on

Class K— dry and wet chemical — Use on

Goodheart-Willcox Publisher

Figure 1-4. Different fires require different classes of fire extinguisher.

1.3 Tools and Meters

Tools used for electrical work on live circuits must be properly insulated. Electrically insulated tools provide the dielectric strength of the insulating material used. *Dielectric strength* is a measure of a substance's ability to block electric flow. A technician must heed the limits set by the manufacturer.

Although some hand tools are covered in a rubber, vinyl, or plastic material and appear to be insulated, they are not suited for electrical work. Manufacturers often display warnings on these tools, such as "not electrically insulated." These tools are intended for use on nonpowered circuits.

Safety Note

Insulation Ratings

An insulator becomes compromised when it encounters a voltage that exceeds the insulator rating. For example, consider a residential insulation that is typically rated for 600 V. If a voltage beyond 600 V is applied, the insulator may break down and conduct current.

Normal Operation	**Ground Fault with No Equipment Grounding**	**Ground Fault with Equipment Grounding**

Normal Operation

Electricity flows through conductors only.

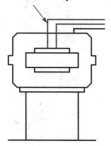

Ground Fault with No Equipment Grounding

Technician touches equipment and becomes a conductor.

Ground fault allows electricity to flow into equipment housing.

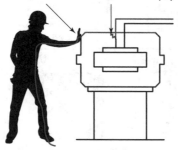

Ground Fault with Equipment Grounding

With equipment grounding conductor attached to housing, fault currents flow through path of least resistance.

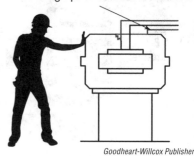

Goodheart-Willcox Publisher

Figure 1-5. An electric motor is grounded to prevent electric shock.

Most plug-in power tools are double insulated and use a two-prong plug. This reduces shock risk to the user if the power shorts to the tool casing handle. Tools like a drill chuck (metal) can still conduct current. Some older tools and larger equipment use a three-prong plug that includes a ground. Inspect the cords for damage before use. The purpose of the ground is to provide a path for current back to the source to prevent it from passing through the user. See **Figure 1-5**. A technician must keep in mind that a receptacle outlet may not contain a ground, or that the ground is faulty. The outlet can be checked with a voltmeter or a convenient outlet tester. The outlet tester displays polarity and ground. See **Figure 1-6**.

The *NEC* requires a ground-fault circuit interrupter as protection for outdoors, basements, garages, and locations near water sources in the home. A ***ground-fault circuit interrupter (GFCI)*** trips, or opens the circuit, when there is a current imbalance between the hot and neutral feeds of a power source. See **Figure 1-7**.

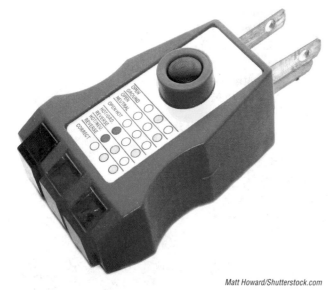

Matt Howard/Shutterstock.com

Figure 1-6. An electrical receptacle tester shows if wires are connected to the correct receptacle terminals and checks for the presence of ground.

Brandon Blinkenberg/Shutterstock.com

Figure 1-7. A GFCI is used as protection during electrical work.

Portable GFCI outlets are used in the field. Always use an extension cord with the proper voltage and current rating to prevent equipment and property damage. See **Figure 1-8**.

Some battery-powered tools deliver voltage of 60 Vdc, which can cause severe electrical hazards. Always follow the manufacturer's instructions for charging, using, and storing these powerful batteries. Lithium batteries can explode if overheated.

Multimeters are instruments used in electrical work, so they must be used safely and properly. Multimeters are rated for maximum safe voltage usage. Read the owner's manual thoroughly to understand the meter functions, capabilities, and safety precautions. Be sure to check leads for damage. Test leads and terminals are also rated for maximum voltage. Ensure that leads are fully connected to the meter and are operational by checking continuity prior to checking live voltage.

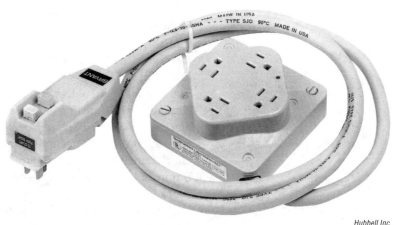

Hubbell Inc.

Figure 1-8. A GFCI circuit breaker and receptacle outlet protects an HVACR worker from an electrical hazard.

1.4 Circuit Protection

Because electric current generates heat, there is a limit to how much current a conductor can carry without generating excessive heat. It is possible that excess heat could ignite nearby materials and lead to fire.

The *NEC* provides ampacity tables listing conductors by gage, material, and insulation type. The table is used to properly size conductors for an application. Some factors determining the proper conductor insulator include ambient conditions and the presence of chemicals. Limitations on the number of conductors in conduits, junction boxes, and panels are also regulated since heat must be dissipated from the conductors.

In the event of excessive currents caused by a short circuit or component malfunction, the current must be interrupted to prevent damage. Circuit breakers and fuses are designed to break the current path when the temperature or magnetic field exceeds the design limits. A circuit breaker is a resettable device. It can be turned back on when circuit conditions are back to normal. A fuse is a one-time device and must be replaced. These devices should never be bypassed or replaced by a product with different specifications. This topic is discussed in Chapter 9, *Power Distribution*.

Summary

- An HVACR technician must be aware electrical hazards to protect themselves, others around them, and their property and equipment.
- There are many organizations that outline electrical codes and standards to be followed.
- The National Fire Protection Association (NFPA) publishes the *National Electric Code* (*NEC*), which develops standards and codes for safe installation of electrical systems.
- The Occupational Safety and Health Administration (OSHA) develops safety standards, training and enforcement of standards.
- Electric shock is a discharge of electric current that passes through the body, causing the body to become a conductor for current.
- Shock injury can range from mild to fatal injury based on several factors.
- Damp and wet locations require extra precaution, including PPE, to avoid shock.
- Use the live-dead-live (LDL) method for ensuring that a circuit is de-energized.
- The lockout/tagout (LOTO) process is used so maintenance and service can be completed in a safe manner. It requires a lock to be attached to a source and a tag placed on the disconnect.
- Personal protective equipment (PPE) is worn to protect an individual from a number of safety hazards. PPE includes gloves, boots, helmets, clothing, face shields, and glasses.
- Insulated tools are rated by dielectric strength.
- A ground-fault circuit interrupter (GFCI) trips a circuit when there is a current imbalance between the hot and neutral feeds of a power source.
- Multimeters have limitations regarding voltage levels. Always check the manufacturer's operating manual for further instruction.
- Circuit breakers and fuses protect circuits from overheating.

Lab 1.1 Digital Multimeter Owner's Manual

This lab will allow you to better locate the instructions and safety precautions set forth in a digital multimeter owner's manual.

Lab Introduction

Before you begin electrical work using a digital multimeter, it is a good practice to review the digital multimeter's owner's manual. This will help you to become a more knowledgeable technician and ensure you are following the proper safety measures associated with the multimeter. Record your findings from the Lab Questions for future use.

Equipment

- Digital multimeter
- Multimeter owner's manual

 Procedure

1. Briefly explore your digital multimeter owner's manual safety notes and warnings.
2. Use the manual to answer the questions outlined below in the Lab Questions.

Lab Questions

1. What is the maximum voltage that should not be exceeded when using your meter? (Check both the manual and on the meter enclosure.)

   _____

2. What is the overload protection when measuring ac voltage?

   _____

3. What is the class rating?

   _____

4. Which organization(s) certify the meter?

   _____

5. What is the recommended calibration frequency? (The meter should be calibrated for stated accuracy per manufacturer by a certified lab.)

   _____

Lab

1.2

Lockout/tagout Performed by Individual

This lab will help you to become familiar with the lockout/tagout (LOTO) safety procedure. You will learn how to lock out a piece of equipment using your own tag.

Lab Introduction

The LOTO procedure is used to ensure the safety of yourself and those around you while electrical equipment is shut down for repair or maintenance. This lab will walk you through the steps in LOTO to prepare you for your career as an HVACR technician.

Equipment

- Electrical disconnect
- Instructions for LOTO components

- Pad lock and key
- Lockout tag

 Procedure

1. Read the LOTO literature for the equipment you are using.
2. Disconnect power by turning handle to off position.
3. Attach lock loop through handle and plate opening and secure lock.
4. Place key in your pocket.
5. Fill out tag with your information.
6. Attach the lock to the disconnect handle so it is visible to yourself and others. See **Figure 1.2-1**.

Figure 1.2-1.   artboySHF/Shutterstock.com

Lab Questions

1. Can the disconnect switch be turned on?

_____

2. Is the tag legible and visible to others?

_____

Lab 1.3 Lockout/tagout Performed by Group

This lab will allow you to become familiar with lockout/tagout (LOTO) safety procedure and how it is performed with a group.

Lab Introduction

When servicing or maintenance is performed on a piece of electrical equipment by a group, there must be LOTO procedures established for the whole group to perform. This lab will walk you through a scenario in which you and your peers can practice LOTO on the same equipment.

Equipment

- Instructions for LOTO components
- Electrical disconnect
- Pad lock and key
- Lockout tag
- Lock hasp

 ## Procedure

1. Read the LOTO literature for the equipment you are using.
2. Disconnect power by turning handle to off position.
3. Attach lock loop through handle and plate opening and secure lock.
4. Place key in your pocket.
5. Each person must use a separate lock and hold their own key.
6. Fill out tag with your information.
7. Attach the lock to the disconnect handle. See **Figure 1.3-1**.
8. Allow each person to practice attaching their own lock to the disconnect. Ensure each log is visible.

Figure 1.3-1. *Red_Shadow/Shutterstock.com*

Lab Questions

1. Can only one person remove their lock and turn the disconnect switch on?

_____

2. Is the tag legible and visible to others?

_____

Know and Understand

_____ 1. Why does an HVACR technician need to thoroughly understand electrical hazards?
   A. To protect themselves from injury.
   B. To protect property from damage.
   C. To protect others from injury.
   D. All the above.

_____ 2. The _____ standard covers electrical installations.
   A. NFPA 70E
   B. NFPA 70
   C. NFPA 70B
   D. NFPA 54

_____ 3. *True or False?* Local codes supersede the *National Electric Code.*

_____ 4. _____ certifies that equipment/component standards are met.
   A. UL
   B. OSHA
   C. ASHRAE
   D. *NEC*

_____ 5. *True or False?* LOTO is only applied to industrial electrical work.

_____ 6. Power should be shut off when working on electrical circuits and equipment. What is the next step?
   A. Perform LOTO.
   B. Call the power company.
   C. Tell someone not to turn power back on.
   D. Send the property owner a text.

_____ 7. During an electric shock, _____.
   A. contact with an electrical source causes a magnetic field that can cause harm
   B. the heat from the electricity contacting passes through the body causing burns
   C. the body becomes a conductor for current flow
   D. electric current is too weak to do harm to a person

_____ 8. *True or False?* The severity of electric shock injury depends on the path taken, duration, amount of current, and the person's health.

_____ 9. *True or False?* The best way to save a person while being shocked is to grab and pull the person away from the electric source.

_____ 10. Never assume that a tool's plastic handle will protect against shock. What determines a material's insulating property?
   A. Thickness of material.
   B. Type of material.
   C. Color of material.
   D. Dielectric strength.

_____ 11. *True or False?* There is a limit to how much voltage a multimeter can safely measure.

_____ 12. Which device(s) disrupts current flow when current flow exceeds the rated amount?
   A. GFCI outlet.
   B. Circuit breaker.
   C. Insulator switch.
   D. Disconnect switch.

_____ 13. *True or False?* A ground-fault circuit interrupter (GFCI) trips a circuit when there is a current imbalance between the hot and neutral feeds of a power source.

_____ 14. Which is the safest way for a technician to use a vacuum pump when the ground is wet?
A. Use the home's outdoor outlet because it should have a GFCI protection.
B. Use a portable GFCI outlet and place the pump on a dry surface.
C. Place the pump on condensing unit and use the home's outlet directly.
D. Use a portable GFCI only. No other precaution is needed.

_____ 15. *True or False?* Fuses and circuit breakers protect a conductor from overheating and potentially causing a fire.

Critical Thinking

1. "It is the current that kills" is a common phrase. What about the voltage? Signs warn "DANGER HIGH VOLTAGE." Is the current phrase a myth, or is there a relationship between the two?

_____

_____

_____

_____

_____

2. Why calibrate the multimeter at the recommended frequency? Is doing so a legal requirement?

_____

_____

_____

_____

_____

Essential Electrical Skills for HVACR: Theory and Labs

2 Electrical Fundamentals

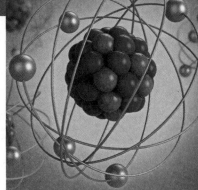

Chapter Outline

Additional Reading

Modern Refrigeration and Air Conditioning, **21st edition**

Learning Objectives

After completing this chapter, you will be able to:

- Define electricity.
- Describe the Law of Charges.
- Discuss how current flows from one atom to another.
- Define the components of electricity.
- Discuss current and its unit of measure.
- Explain voltage and its unit of measure.
- Define resistance and its unit of measure.
- Differentiate between conductors and insulators.

Technical Terms

atom	insulator	voltage
conductor	Law of Charges	
electric current	resistance	

Introduction

This chapter introduces some of the fundamentals of electricity and how electricity is quantified into units of measure. Electricity is a form of energy caused by charged particles. Although energy cannot be created or destroyed, it can be converted into another form of energy. This concept can be applied to the study of basic circuits and applied to troubleshooting electrical systems. An HVACR technician must have a solid understanding of how electricity works, which starts at the microscopic atomic level.

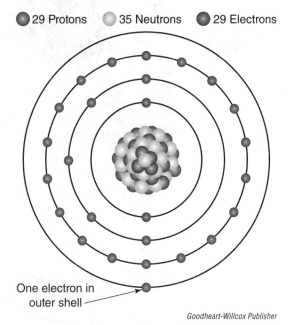

● 29 Protons ◯ 35 Neutrons ● 29 Electrons

One electron in
outer shell

Goodheart-Willcox Publisher

Figure 2-1. A copper atom has 29 electrons and 29 protons. The outer orbital (shell) contains only one electron.

2.1 Atomic Theory

All matter is made up of elements or compounds (a combination of elements). Any natural element is made up of unique atoms. An ***atom*** comprises three subatomic parts: protons, neutrons, and electrons. See **Figure 2-1**. The nucleus, or center of an atom, contains protons, which are positively charged, and neutrons, which are electrically neutral. Negatively charged electrons orbit around the nucleus, much like the planets orbit around the sun. The number of protons, neutrons, and electrons determine the makeup of the atom.

Electrons orbit the nucleus in shells, or orbitals, and each orbital has its own energy level. For conducting electricity, only the outer orbital, called the valence orbital, is of concern. The number of electrons the valence orbital holds determines the electrical properties of a particular element.

Each element varies based on its number of orbitals and electrons. For example, a copper atom's outermost orbital contains only one electron. It also has an equal number of protons and electrons. Like electrons, the number of protons can vary, which also makes each element distinguishable. For example, the smallest hydrogen atom has only one electron and one proton.

2.1.1 Law of Charges

There are two kinds of charges, negative and positive. Based on the ***Law of Charges***, like charges repel one another and unlike charges attract. See **Figure 2-2**.

This law can be best observed by using two magnets. A magnet has a north pole and a south pole, one side that is negatively charged and one that is positively charged. The north pole of one magnet repels the other magnet's north pole. Placing the south pole to the other magnet's north pole causes the magnets to attract.

Because an electron has a negative charge and a proton a positive charge, these two attract since they are unlike charges. The electron is not pulled into the positive nucleus but instead is held in an orbit due to centrifugal force. Centrifugal force acts on an object that is moving in a circular path and propels it outward away from the center. The attraction force is equal in strength to the centrifugal force, but the two forces applied to the electron are in opposite directions. Therefore, the electron stays in orbit and the atom is stable, meaning it has no charge. See **Figure 2-3**.

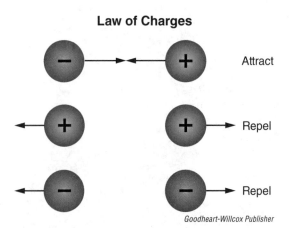

Law of Charges

Attract

Repel

Repel

Goodheart-Willcox Publisher

Figure 2-2. The Law of Charges was established through the study of physics, stating that unlike charges attract and like charges repel.

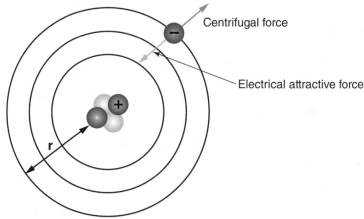

Centrifugal force

Electrical attractive force

r

Goodheart-Willcox Publisher

Figure 2-3. Centrifugal force acts outward on an electron so it does not get pulled to the nucleus.

2.2 Electric Current

An atom wants to be in a neutral state. To be neutral means it must have an equal number of electrons and protons. An atom can pick up or lose electrons, thus affecting its charge. When an electron is displaced from the atom's orbit by an external force, it becomes a free electron. From here, two things occur: the atom now has a net positive charge, since there are more protons than electrons, and the free electron attaches to another atom's outer orbital. The net positive atom then attracts another electron into its orbit to regain neutrality. This whole process creates a chain reaction, or movement of electrons. See **Figure 2-4**. This flow of electrons is called *electric current*.

The atomic structure of an element or material determines how well that material conducts electric current. Generating free electrons is easier with atoms of pure elements. Silver, gold, and copper atoms contain only one electron in the outer orbit, which allows the electron to be easily displaced from the atom's orbital when compared to other elements.

The coulomb, named after 18th century physicist, Charles Coulomb, is the fundamental unit of electrical charge. One coulomb is equal to the charge of 6.24×10^{18} electrons. Electric current—the flow of electricity—is measured in amperes. One ampere is equal to one coulomb of charge passing a point in one second. Abbreviations for the ampere include amps and A.

Current can be measured in amperes using an ammeter, which will be explained more thoroughly in Chapter 10, *How Electric Meters Work*. The capital letter I refers to current in formulas or an abbreviation for current in literature. A lowercase i with an arrow showing direction of current flow may be found in control circuit schematics.

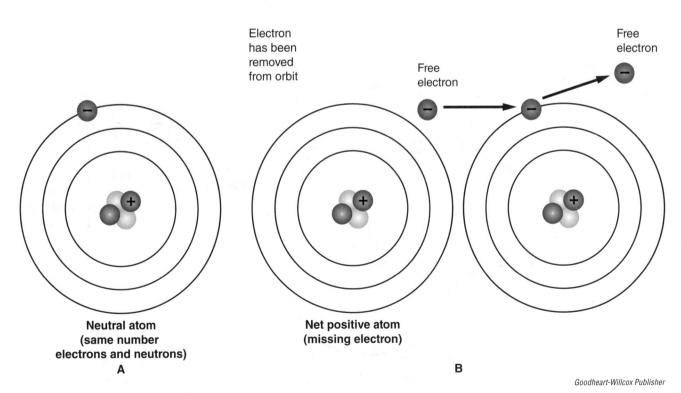

Electron has been removed from orbit

Free electron

Free electron

Neutral atom (same number electrons and neutrons)

A

Net positive atom (missing electron)

B

Goodheart-Willcox Publisher

Figure 2-4. Free electrons bombard electrons in adjoining atoms to create a chain reaction. A—A neutral atom with the same electrons and neutrons. B—When the electron on the outer orbital is removed, the atom has a net positive charge. The free electron moves atom to atom to find stability.

2.3 Voltage

The force behind the movement of electrons is called **voltage**. Also known as *electromotive force (EMF)*, it is a force created by a difference in atomic charge. The unit of measure for voltage is the volt (V). To denote voltage in formulas, *V* or *E* is used.

A difference in "potential" can be explained using a 1.5 V dry cell battery. This battery has a potential difference between the negative terminal and the positive terminal of 1.5 V. That is, the negative side of the battery is 1.5 V more negative than the positive side.

A voltmeter measures the difference in potential between two points. Connecting a light bulb between the opposite terminals causes current to flow between the negative and positive terminals of the battery. As the current flows, the difference in potential decreases as free electrons fill the valence orbital of the net positive atoms.

2.4 Resistance

Resistance is the restriction, or opposition to, current flow. Resistance is required to limit current flow. Without resistance, opposite potentials would generate uncontrolled and very high, destructive current flow would occur.

No materials allow current to flow unrestricted. Consider a small piece of copper wire connected between the positive and negative terminals of the battery, as shown in **Figure 2-5**. The wire offers no appreciable resistance to the current. All the electrons move toward the positive atoms at once, thus creating excessive destructive heat. This can neutralize the battery (or create a dead battery) and result in a short, causing damage. See **Figure 2-6**.

The unit of measure for resistance is called the ohm (Ω). The capital letter *R* represents resistance in formulas. Resistance is measured with an ohmmeter.

Figure 2-5. No resistance results in damaging excess current.

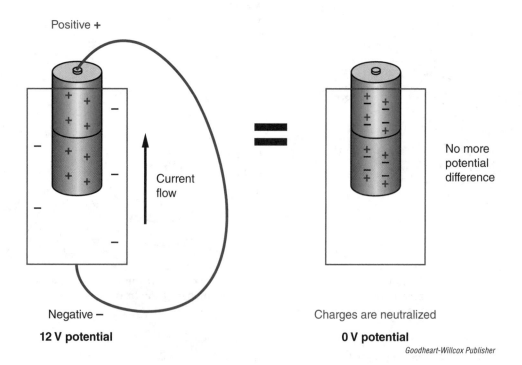

Positive **+**

Current flow

Negative **−**

12 V potential

No more potential difference

Charges are neutralized

0 V potential

Goodheart-Willcox Publisher

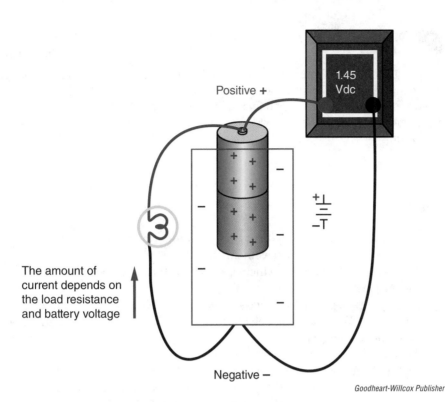

Positive **+**

1.45
Vdc

The amount of
current depends on
the load resistance
and battery voltage

Negative **−**

Goodheart-Willcox Publisher

Figure 2-6. The resistance
offered by the light bulb
safely controls current flow.

2.5 Conductors and Insulators

The conductivity of a material describes its resistive property. ***Conductors*** are materials, typically made of metal, that allow current to flow through them easily. Length, cross sectional area, and temperature are factors that affect a conductor's total resistance to current.

Not all metals conduct at the same rate. Silver, copper, gold, and aluminum are all conductors. Silver is often used for electrical contacts, while copper is used for electrical wire. Gold is often found in low-energy applications, such as low-voltage transducers and printed circuit boards. Aluminum requires special considerations due to its lower corrosion resistance and melting temperature. Each of these conductors have different resistivity and corrosion levels. See **Figure 2-7.**

An ***insulator*** is a material that has an atomic structure that does not readily release electrons. Thus, it restricts the flow of electrons. Insulators are mostly compounds, such as paper, rubber, glass, and porcelain. Pure water is an insulator, but because of minerals and other impurities, it can conduct electricity.

Conductor Characteristics			
	Lowest ⟶		**Highest**
Resistance to Current	Silver	Copper	Gold	Aluminum
Corrosion Resistance	Aluminum	Copper	Silver	Gold
Cost	Aluminum	Copper	Silver	Gold

Figure 2-7. Conductors
characteristics vary, which
make certain conductors
more suitable for a type of
electrical job.

Summary

- Matter is made of pure elements or compounds (a combination of elements).
- Atoms are specific to each element.
- Atoms contain a nucleus, which has both protons and neutrons. Electrons orbit the nucleus in shells.
- Protons have a positive charge, and electrons have a negative charge.
- The flow of free electrons is electric current. Current is measured in amperes and represented by I.
- An ammeter is an instrument used to measure current.
- Voltage is the driving force that moves electrons.
- The unit of measure for voltage (electromotive force) is the volt (V).
- The voltmeter measures the difference in potential as volts.
- Resistance restricts current flow and is measured in ohms (Ω).
- An ohmmeter is used to measure resistance.
- A conductor readily allows current flow.
- A conductor's resistance is determined by type of material, cross sectional area, length, and temperature.
- Good conductors are silver, copper, gold, and aluminum.
- Some good insulators are paper, rubber, glass, and porcelain.

Lab 2.1   Navigating the Digital Multimeter

This lab will walk you through the main functions of a digital multimeter. The goal is to become familiar with the device and start to practice how it is used. The multimeter will be used in chapters moving forward in this textbook. It is also covered in full detail in Chapter 10, *How Electric Meters Work*.

Lab Introduction

A digital multimeter (DMM) is a test tool designed to measure electrical quantities, such as voltage (in volts), current (in amps), and resistance (in ohms). The multimeter is a standard diagnostic device for technicians in the HVACR industry. See **Figure 2.1-1**. There are many manufacturers of DMMs, but all produce similar results. There are a few subtle differences in operating them.

You must be familiar with the operation of a digital multimeter as its use will be required in the field. The following steps will assist a new user in performing basic exploratory functions of the meter.

Equipment

Digital multimeter

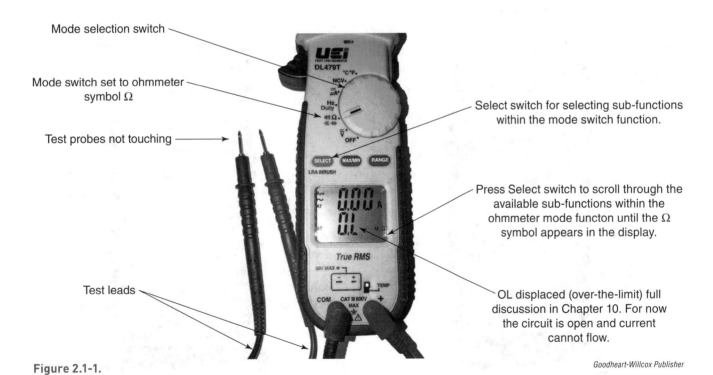

Mode selection switch

Mode switch set to ohmmeter symbol Ω

Test probes not touching

Test leads

Select switch for selecting sub-functions within the mode switch function.

Press Select switch to scroll through the available sub-functions within the ohmmeter mode functon until the Ω symbol appears in the display.

OL displaced (over-the-limit) full discussion in Chapter 10. For now the circuit is open and current cannot flow.

Figure 2.1-1.

Goodheart-Willcox Publisher

Procedure

1. Explore the DMM owner's manual.
2. Ensure an adequate battery charge level.
3. Unpack and straighten out the test leads. Some meters also supply alligator clamps, which may be used in some labs.
4. Thoroughly insert the test leads into DMM jacks (leads and jacks are color coded).
5. Using the Mode selector, select Ohms. See **Figure 2.1-2**. The Ω symbol is commonly used.
6. Do not connect test probes together. The display should show *OL* as shown in Figure 2.1-1.
7. Touch and hold test probes together. The display should show 0, or a value below 1 Ω. The near zero reading indicates that test leads are connected to the DMM and the meter function should be good.
8. Set the Mode selector to Volts. Some meters have Vac (voltage ac) and Vdc (voltage dc). See **Figure 2.1-3** and **Figure 2.1-4**. Other meters use one Volts mode and a selector button to select either ac or dc. You can check the owner's manual to determine which is used.

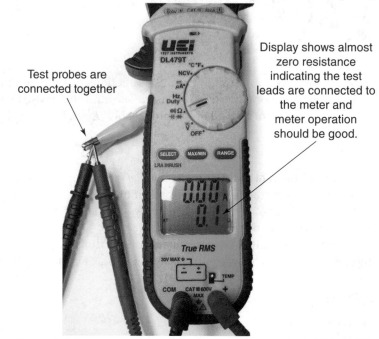

Test probes are connected together

Display shows almost zero resistance indicating the test leads are connected to the meter and meter operation should be good.

Figure 2.1-2. *Goodheart-Willcox Publisher*

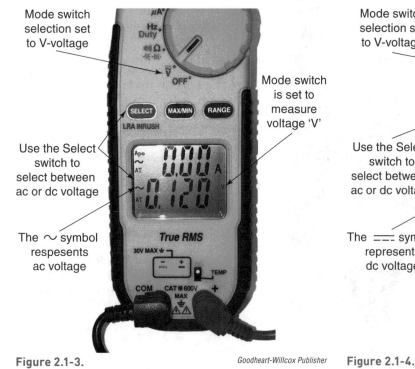

Mode switch selection set to V-voltage

Mode switch is set to measure voltage 'V'

Use the Select switch to select between ac or dc voltage

The ∿ symbol respesents ac voltage

Figure 2.1-3. *Goodheart-Willcox Publisher*

Mode switch selection set to V-voltage

Mode switch is set to measure voltage *V*

Use the Select switch to select between ac or dc voltage

The ⎓ symbol represents dc voltage

Figure 2.1-4. *Goodheart-Willcox Publisher*

Continued▶

Probe Guide

9. Use the ac inductive clamp to measure current. See **Figure 2.1-5**. The currents measured in the labs will be large enough for most meters to display. Some meters have dual displays. As long as the meter is on, the clamp will display the current. Other meters require a Mode selection. See owner's manual for any questions.

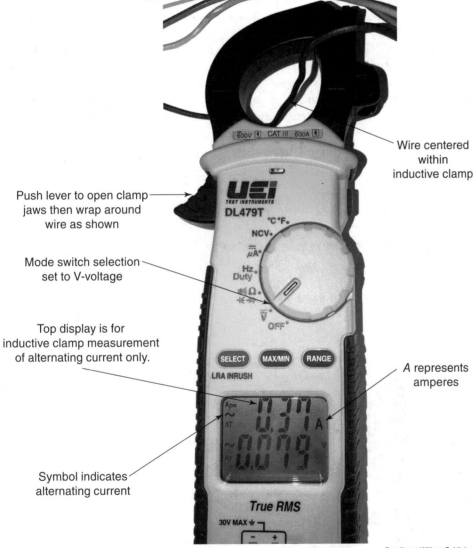

Figure 2.1-5.

Goodheart-Willcox Publisher

Lab 2.2

Distinguishing between Basic Wiring Tools

This lab is designed to help introduce students to the basic wiring tools used in the HVACR field and become familiar with their intended use. Being able to use these tools will be required for most of the labs in this textbook moving forward.

Lab Introduction

Wires are sized according to the American Wire Gage (AWG). This number is found printed on the wire's insulation or for smaller diameter wire, it is written on the packaging label. Wire is either solid core or stranded. Both multipurpose tools and single-purpose tools, such as wire strippers, crimping pliers, long-nose pliers, and screwdrivers, are used to prepare wire for HVACR electrical work.

A multipurpose tool is suited for repair work where only a few wires need to be stripped and terminated. Although this tool has limited use, it is handy and saves space in a technician's toolbox. Single-purpose tools, although more expensive, are best for installations or situations where many wires must be prepared. A spring-loaded stripper has a faster and more comfortable operation for large jobs. Crimping pliers provide more leverage to crimp terminals, and therefore prevent hand fatigue. Long-nose pliers allow greater reach and are suited for many situations.

Use this lab as an opportunity to differentiate between the tools used in the HVACR field.

Equipment

- Multipurpose wiring tool
- Wire strippers
- Crimping pliers
- Long-nose pliers
- Screwdriver

 Procedure

1. Take note of each part of your multipurpose wiring tool using **Figure 2.2-1** as a guide.

Multipurpose Wiring Tool

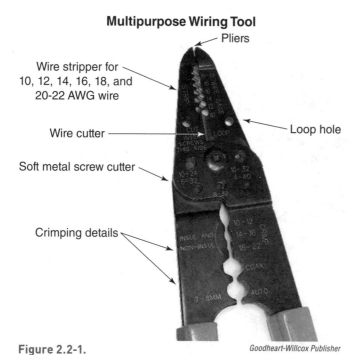

Figure 2.2-1.

Goodheart-Willcox Publisher

Continued ▶

24

Copyright Goodheart-Willcox Co., Inc.

2. Explore the parts of your wire stripper using **Figure 2.2-2** as reference.

Wire Stripper/Cutter

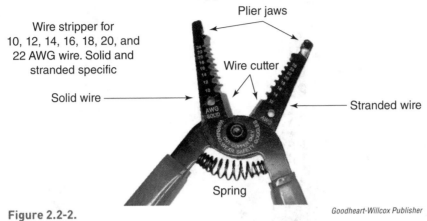

Plier jaws

Wire stripper for 10, 12, 14, 16, 18, 20, and 22 AWG wire. Solid and stranded specific

Wire cutter

Solid wire

Stranded wire

Spring

Figure 2.2-2.

Goodheart-Willcox Publisher

3. Notice the two main parts of your crimping pliers using **Figure 2.2-3** as a guide.

Crimping Pliers

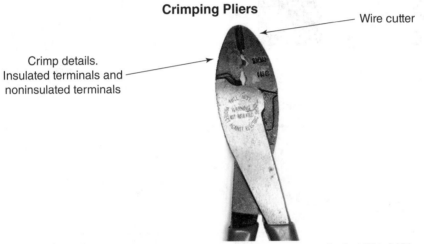

Wire cutter

Crimp details. Insulated terminals and noninsulated terminals

Figure 2.2-3.

Goodheart-Willcox Publisher

4. Review the parts of your long-nose pliers using **Figure 2.2-4** as reference.

Long-nose Plier

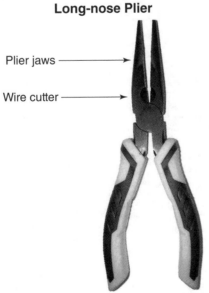

Plier jaws

Wire cutter

Figure 2.2-4.

Goodheart-Willcox Publisher

Continued ▶

5. Compare your screwdriver to the two in **Figure 2.2-5** below. Note any differences.

Screwdriver

Phillips bit   Flat bit

Figure 2.2-5.   *Goodheart-Willcox Publisher*

6. Read the owner's manual and safety notes for each tool.

Lab Questions

1. Which type of wire stripper will be used for the labs?

_____

2. Which type of crimper will be used?

_____

3. List the safety precautions for each tool.

_____

_____

_____

_____

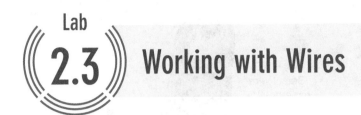

Lab 2.3 Working with Wires

This lab will apply your learning in Lab 2.2. Use specific wiring tools to practice making connections and splices with solid and stranded wire for electrical work.

Lab Introduction

An HVACR technician must be able to prepare wires by making different connections and splices. Connections refer to attaching a wire to a component. Splicing is a method used to join two or more wires together. In this lab, two 14 AWG wires will be spliced together using a wire nut. Wire nuts are rated for the maximum and minimum number of wires that can be used for the same gage wire or combinations of different gages and number of wires. Wire nut manufacturers provide this information on a chart attached to the packaging. Wire nuts are color coded for quick reference. This lab will use a yellow wire nut, which is a standard color for connecting up to three 14 AWG wires.

Spend some time working through this lab to begin to master connecting and splicing wire. As with any skill, practice is necessary. Any connections and splices must be reliable and conform to safety requirements. Neatness is also critical. Be sure to ask your instructor for further guidance if needed.

Equipment

- Wire strippers
- Crimping pliers
- Long-nose pliers
- Optional—multipurpose tool
- 2′ of black or red 14 AWG solid wire
- 1′ of black or red 14 AWG stranded wire

- Light switch – single-pole 120 V 15 A rated (clamp or side connection)
- 14 AWG spade-insulated crimp terminal
- 14 AWG wire nut (yellow) or equivalent for 2–14 AWG wires

 Procedure—Cutting and Stripping Wire

1. Examine the difference between solid and stranded wire using **Figure 2.3-1** for reference.

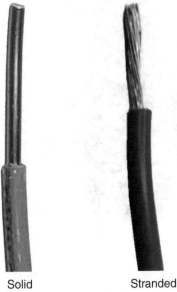

Solid   Stranded

Figure 2.3-1. *Goodheart-Willcox Publisher*

Continued ▶

2. Cut a 6″ piece of 14 AWG black solid wire.
3. Place the piece of wire in the 14 AWG solid hole in your wire strippers. See **Figure 2.3-2**.
4. Position the wire strippers 3/4″ from one end.
5. Squeeze the wire strippers. The sharp edge of the 14 AWG solid hole cuts through the insulation.
6. Release pressure on the handle and rotate wire one quarter turn.
7. Squeeze the wire strippers handle again. This ensures that the insulation is cut around the entire circumference of the wire.
8. Slightly relieve pressure on the handle and slide the tool forward (toward the end of the wire) to remove the insulation.
9. Check the bare wire for cut lines. The wire should be smooth. Using the wrong gage hole will produce a ring (cut line) around the wire. This can lead to wire breakage in the future.

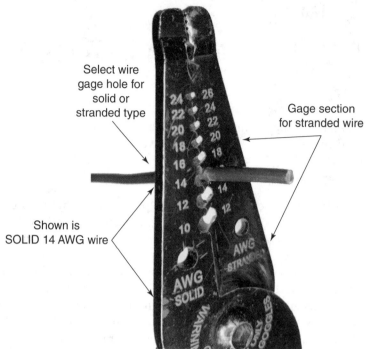

Select wire gage hole for solid or stranded type

Gage section for stranded wire

Shown is SOLID 14 AWG wire

Goodheart-Willcox Publisher

Figure 2.3-2. Stripping wire insulation.

 ## Procedure—Making a Hook Connection

1. Insert the stripped portion of the wire into the hook hole on the wire strippers. See **Figure 2.3-3**.
2. Insert the tip of wire into the hole and about 1/8″ past the tool surface.
3. Turn the tool clockwise until a hook is formed. An alternate method is to use long-nose pliers to form the hook.

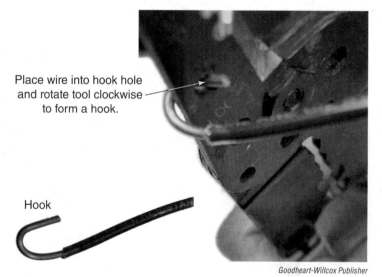

Place wire into hook hole and rotate tool clockwise to form a hook.

Hook

Goodheart-Willcox Publisher

Figure 2.3-3. Forming a hook.

Continued▶

4. Connect the hook end to a switch by using the side connection method, **Figure 2.3-4**. Note all the details in the illustration.

   A. The right side connection is wrong because the open part of the loop will alternate clockwise as the screw is tightened. This will cause a poor connection with not enough surface area making contact.

   B. There must not be bare wire beyond the switch body. The bare wire may contact an unwanted surface when the switch is installed to an electrical box.

   C. Insulation must not be under the screw head surface. This will cause a soft shoulder—a condition where the screw appears to be tight but may loosen over time due to temperature changes acting on the insulation.

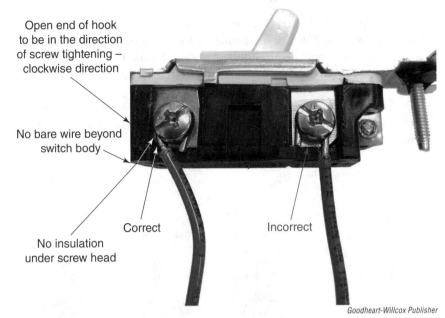

Goodheart-Willcox Publisher

Figure 2.3-4. Making a screw connection. Note the difference between the correct and incorrect method.

5. An alternative connection is the clamp feature shown in **Figure 2.3-5**. There is a strip guide on the switch body showing the amount of insulation to remove for the clamp and side terminal use. Fully tighten the screw when making this type of connection. Torque wrenches can be used to ensure properly tightened connections per manufacturer's specification.

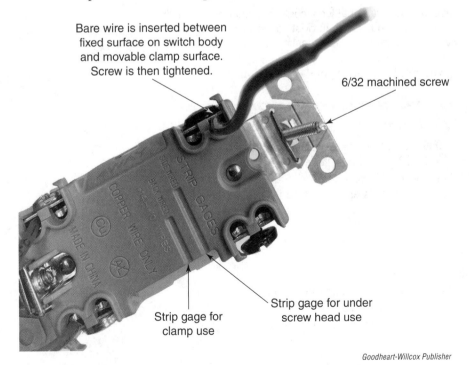

Goodheart-Willcox Publisher

Figure 2.3-5. Making a clamp connection. There are optional connections to the switch available.

Procedure—Making Wire Nut Connections

1. Cut a 6″ piece of 14 AWG black solid wire.
2. Strip between 1/2″ to 5/8″ of insulation from one end.
3. Using the wire that is connected to the switch from the previous procedure, strip between 1/2″ to 5/8″ of insulation from the other end of wire (that is, the end not connected to the switch).
4. Position the two wires next to one another with the bare wire at the ends running next to each other. See **Figure 2.3-6**. The length of bare wire should be closely even.
5. Place a wire nut over the bare ends of the wires and turn the wire nut clockwise until it is fully tightened.

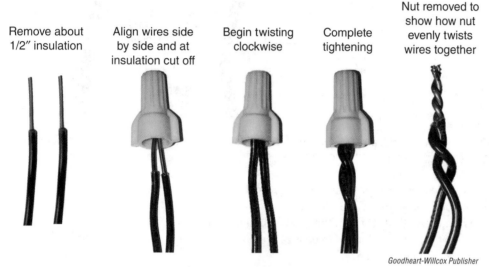

Remove about 1/2″ insulation

Align wires side by side and at insulation cut off

Begin twisting clockwise

Complete tightening

Nut removed to show how nut evenly twists wires together

Goodheart-Willcox Publisher

Figure 2.3-6. Making a nut connection. Be aware that local code may require pre-twisting of wires with pliers before putting on the wire nut and then twisting the wire nut. Wire nut manufacturers may claim pre-twisting is not required. The local code supersedes.

6. Hold the wires and pull on the wire nut to ensure a good connection.

Pro Tip

Wire Nut Connections

Many wire nut manufacturer's instructions state that wires do not need to be twisted together before installing the wire nut. The last illustration in **Figure 2.3-6** shows how well the wire nut twists the wires. However, some local codes require wires to be twisted together before the wire nut is applied. In addition, some local codes require wrapping electrical tape around the wire and wire nut after the connection is made.

---

7. Cut a 6″ piece of 14 AWG black solid wire.
8. Strip between 1/2″ to 5/8″ of insulation from one end of the piece of wire.
9. Cut a 6″ piece of 14 AWG black or red stranded wire.
10. Strip 5/8″ of insulation from one end of the stranded wire. Be sure to use the correct hole for stranded wire.

Continued▶

Pro Tip

Stripping Stranded Wire

Stranded wire requires additional care when stripping insulation to avoid cutting strands of wire. Practice stripping stranded wire to improve your skills. After stripping stranded wire, look inside the removed piece of insulation for broken wire strands. Cut strands will reduce the wire size and reduce current carrying capacity of the conductor.

11. Twist the stranded wire around the solid wire, **Figure 2.3-7**.
12. Position the wire nut on the wires and turn the nut clockwise. Do not force the nut down onto the wires. Instead, allow the threaded insert in the wire nut to capture the wires. Continue turning the nut until fully tightened.
13. Check for proper connection by pulling on the wire nut.

Remove about 3/4″ from stranded wire and twist around solid wire

Slowly twist clockwise while not pushing down. Continue twisting until tight

Finished product should not have strands of wire protruding from nut insert. Check for tightness by holding wires and pulling wire nut.

Goodheart-Willcox Publisher

Figure 2.3-7. Making a wire nut connection from solid wire to stranded wire. Note both solid to stranded or stranded to stranded require pretwisting.

☰ Procedure—Making Spade Terminal Connections

1. Using the solid and stranded wires connected with a wire nut from the previous procedure, strip 1/4″ of insulation from the free end of each wire.
2. Twist the bare end of the stranded wire so that the strands stay together. See **Figure 2.3-8**.
3. Place the barrel end of the spade terminal into the crimping tool (but do not crimp). Crimping tools normally include crimping areas for insulated and uninsulated barrels. Ensure that the steel barrel is located within the crimp detail.
4. Place the bare stranded wire into the barrel end of the terminal. (Some technicians prefer to place the terminal over the wire and then wrap the tool around the terminal barrel. This is a matter of preference and you should try both methods.)

Continued ▶

5. Squeeze the handle of the crimping tool.
6. Relieve pressure on handle, move the terminal, and crimp again. (The barrel length of terminals is larger than the crimp detail in most tools. Thus, multiple crimps may be needed to complete a connection.)
7. Test for proper connection by pulling on the terminal. The terminal should not move relative to the wire.

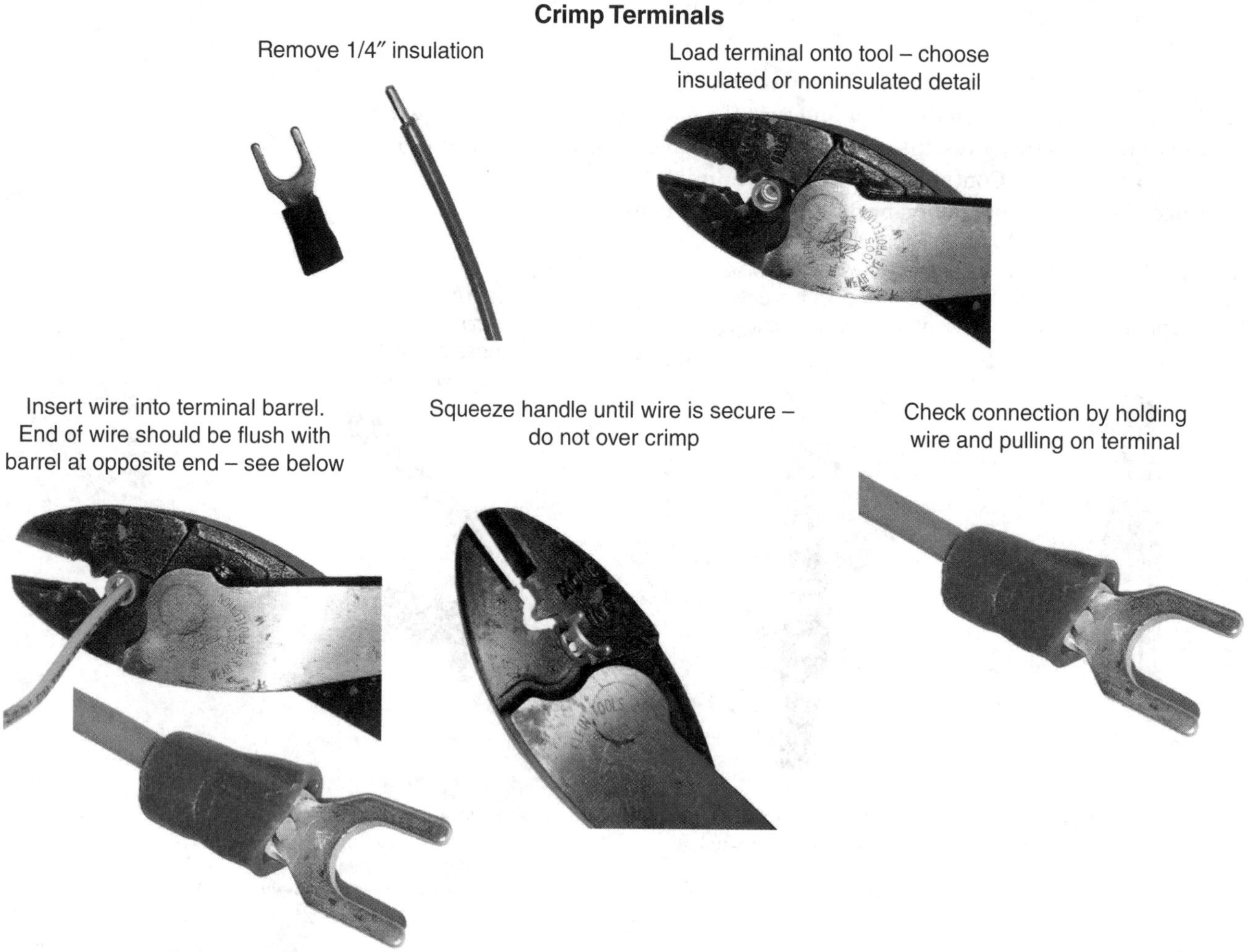

Crimp Terminals
Remove 1/4″ insulation
Load terminal onto tool – choose insulated or noninsulated detail
Insert wire into terminal barrel. End of wire should be flush with barrel at opposite end – see below
Squeeze handle until wire is secure – do not over crimp
Check connection by holding wire and pulling on terminal

Goodheart-Willcox Publisher

Figure 2.3-8.

Pro Tip

Practicing Electrical Connections

Making secure electrical connections is critical for an HVACR technician. Practice the procedures for each type of connection several times until you become proficient. Electrical connections must be safe, reliable, and neat in appearance.

Lab Questions

1. Does your switch connection conform to the illustration details?

   _____

   _____

2. Do your wire nut connections conform to the illustration details?

   _____

   _____

3. Do your space terminal connections conform to the illustration details?

   _____

   _____

Know and Understand

_____ 1. Electricity is a form of _____.

                 A. matter   C. positive charge only

                 B. negative charge only   D. energy

_____ 2. The _____ of an atom has a negative charge.

     A. proton   C. neutron

     B. electron   D. nucleus

_____ 3. Electric current is _____.

     A. the movement of free electrons   C. the flow of voltage

     B. the movement of free protons   D. the flow of neutrons

_____ 4. When an atom's outer orbit electron is displaced, the net charge of the atom is _____.

     A. neutral   C. positive

     B. negative   D. no charge

_____ 5. The unit of measure for current is the _____.

     A. volt   C. ampere

     B. ohm   D. watt

_____ 6. The unit of measure for voltage is the _____.

     A. volt   C. ampere

     B. ohm   D. EMF

_____ 7. *True or False?* Voltage is a force that causes electron flow.

_____ 8. Another term for voltage is _____.

     A. amp   C. EMF

     B. omega   D. resistance

_____ 9. The unit of measure for resistance is the _____.

     A. ohm   C. volt

     B. ampere   D. restricter

_____ 10. *True or False?* High resistance will cause more current to flow.

_____ 11. *True or False?* All metals conduct current equally.

_____ 12. Which of the following is *not* a factor in calculating a conductor's resistance?
A. Voltage rating.
B. Length.
C. Material type.
D. Temperature.

_____ 13. _____ is the material best for low energy system contacts.
A. Copper
B. Lead
C. Stainless steel
D. Gold

_____ 14. *True or False?* An insulator will block current regardless of the voltage level.

_____ 15. *True or False?* Insulators have atomic structures that readily release electrons.

Critical Thinking

1. Is it possible to have more current than voltage? Remember that voltage is force and current is the movement of electrons in a conductor. An example is two amperes and one volt.

_____

_____

_____

_____

2. Explain when an insulator will not protect a person who comes in contact with an insulated wire.

_____

_____

_____

_____

3 The Simple Circuit

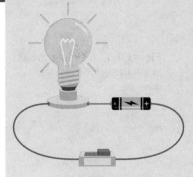

Chapter Outline

3.1 Circuit Requirements
3.2 Open Circuits
3.3 Short Circuits
3.4 Introduction to Reading Schematics

Additional Reading

Modern Refrigeration and Air Conditioning, 21st edition

12.4 Circuit Fundamentals

13.3 Electrical Problems

Learning Objectives

After completing this chapter, you will be able to:

- Define a functional electric circuit.
- Discuss the requirements for a complete circuit.
- Explain the types of circuit failure modes.
- Draw a circuit with its proper components.
- Construct a functioning circuit in lab.
- Use electrical instruments to measure voltage and current throughout a circuit.

Technical Terms

conductor	power source	switch
load	schematic	

Introduction

Chapter 2, *Electrical Fundamentals* discussed the three components of electricity. In this chapter, you will learn how electricity is used in a simple electric circuit to provide useful work. A simple circuit means there is only one component that receives electrical energy to perform this work and only one device to control the operation of the circuit. Understanding electricity in a simple circuit is foundational in being able to interpret more complex schematics, since they are comprised of a multitude of simple circuits. It can also be applied to troubleshooting electrical systems.

3.1 Circuit Requirements

A simple circuit serves as a path for electric current to travel. It must have three main components to operate successfully: a power source, load, and conductors. See **Figure 3-1**. A **power source**, such as a battery or power grid, supplies energy to the circuit. A **load**, or electric device that requires electric energy to operate, uses the power from the source to perform work. A load transforms electrical energy into another form of energy. An example of a load is an incandescent lamp. An incandescent lamp changes electrical energy into heat and light energy. A simple circuit also has **conductors**, such as wire, that complete the path for electron flow. This path connects the source to the load.

Figure 3-1. A simple circuit is the basis for complex devices.

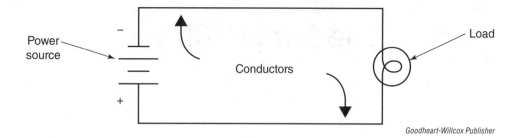

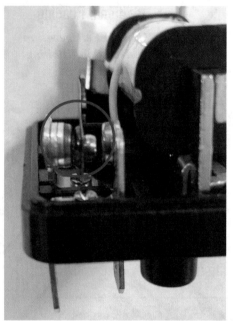

Figure 3-2. Contacts welded together due to excess current. Contacts should be open.

For a circuit to be controlled, a switching device is required. A simple *switch* is a device that connects or disconnects the load to the power source. Circuit control can be more complex than a simple switch, such as dimming a light. The switching action can be either automatic or manual.

Switches, as well as other circuit components, will be covered in later chapters of this text.

3.2 Open Circuits

An operable circuit must have a complete path, which means that the circuit is closed, or complete. An open circuit occurs when current cannot flow due to an interruption in the path. The interruption can be intentional, such as when a switch is in the open position, or unintentional, such as in the case of a broken or loose wire.

When a circuit is open unintentionally, it is considered a type of failure. Another failure can occur if a switch malfunctions and cannot open the circuit because the contacts are fused together. See **Figure 3-2**. Knowing the status of a circuit is critical when troubleshooting. If a technician knows whether a circuit should be open or closed within a given sequence of operation, it is then possible to detect a problem. Troubleshooting and failure modes are discussed in later chapters.

3.3 Short Circuits

A short circuit is a condition where there is not enough resistance offered by a faulty load or the load is bypassed, resulting in excess current flow. In the case of a bypassed load, electrons find a shorter path, or the path of least resistance, to the opposite potential. See **Figure 3-3**. Recall from Chapter 2, *Electrical Fundamentals* that the components of electricity are voltage, current, and resistance. A circuit

Figure 3-3. When a switch is closed, current will flow through the path of least resistance to the opposite potential, and a direct short occurs.

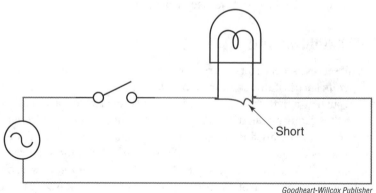

without resistance yields unrestricted, or uncontrolled, current flow, which can destroy the entire electrical system. Loads are designed to provide a specific amount of resistance and to operate within specified electrical characteristics, such as voltage and current. Unwanted changes to the load's internal resistance can cause a change in current flow. Less resistance results in more current, and more resistance yields less current.

Not all shorts are the same. A direct short, also referred to as a dead short, is caused by a direct contact between opposite potentials. This contact can be between the positive and negative polarities of a battery or the line voltage on the hot and neutral feeds on residential power. Other shorts can be between wires in control circuit sensors. Although the term "short" is often misused, this material will provide you with an accurate, thorough understanding of the concept to avoid this confusion.

3.4 Introduction to Reading Schematics

A simple circuit can be represented in print and is shown in a wiring schematic. A wiring *schematic* shows the complete path that makes up a circuit and its related components. It often contains a network of multiple circuits. A technician must be able to interpret the electrical components represented in a schematic to understand a complete electrical system. See **Figure 3-4**.

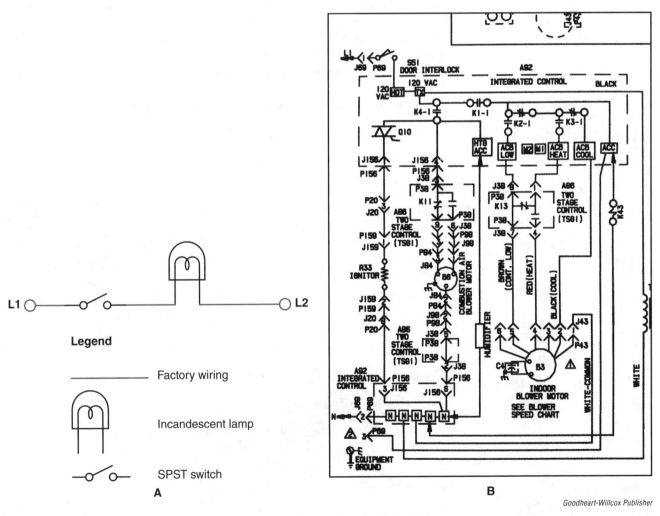

Figure 3-4. A—A simple circuit. B—A multitude of simple circuits comprise a system.

Schematics display all the information needed to determine how a system should be functioning based on the manufacturer's design intent. The lines represent conductors, or the path of electrical flow, that current can travel. Other distinct symbols are used to represent the load and control devices, which are identified in the legend on the schematic. Schematics of specific electrical systems and components will be shown in more detail throughout this text.

Summary

- A circuit requires three elements or components: a power source, load, and conductors.
- A power source supplies electrical energy to the load.
- A load uses the electrical energy to perform useful work.
- A control (switching device) is required to control the operation of an electrical circuit.
- A circuit can be considered either open or closed.
- In a closed circuit, the current is able to flow freely, thus allowing the load to function.
- In an open circuit, the current flow is disrupted and therefore the load does not work.
- A short circuit is a condition where there is not enough resistance offered by a faulty load or the load is bypassed resulting in excess current flow.
- In a direct short, the current flows directly from one potential to the opposite potential.
- A wiring schematic visually displays all of the components that make up one or more circuits through the use of symbols and lines.
- A wiring schematic legend identifies and names the symbols that are used on a schematic.
- Schematics are used for troubleshooting, repairing, and installation.

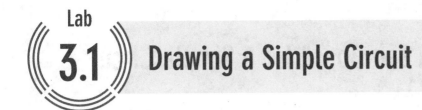

Lab 3.1 Drawing a Simple Circuit

Test your knowledge of the concepts learned in this chapter by drawing a simple, functioning circuit. Ensure the appropriate components and setup for this circuit.

Lab Introduction

A simple circuit has three main components: a power source, load, and conductors. It often includes a switch to control the operation of the circuit. Understanding the requirements for a complete circuit is fundamental for an HVACR technician. Use this opportunity to familiarize yourself with the components of a circuit and how they are set up to function properly. Being able to construct a simple circuit is a key first step in electrical troubleshooting.

Equipment

- 1 sheet of paper
- 1 pen or pencil

Procedure

1. Draw the schematic diagram using the reference symbol legend in **Figure 3.1-1**.

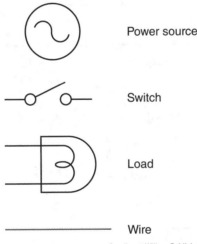

Power source

Switch

Load

Wire

Figure 3.1-1   *Goodheart-Willcox Publisher*

2. Draw lines to represent conductors starting at one end of the power source to the control switch.
3. From the control switch, supply power to the one side of the load.
4. From the load, draw a line back to the other side of the power source to complete the circuit.
5. Consider the switch to be closed. Draw arrows from the power source through the switch and load, then back to the power source. This demonstrates there is a complete path for current flow, starting and ending at the power source.

Lab

3.2 Assembling a Lab Board for a Simple Circuit

This lab will instruct you on how to prepare a lab board to operate a simple circuit. This board setup will be used in additional labs in subsequent chapters. Thus, it is critical for you to become familiar with how it is constructed.

Lab Introduction

Constructing a lab board will begin to provide you with an organized method for performing labs. The board will be reused in future labs and allow for additional components to be added. The lab board uses a $2' \times 2' \times 3/4''$ piece of plywood where electrical components are attached with wood screws.

You will use one of the two layouts provided to begin your lab board construction. **Figure 3.2-1** shows a layout containing fewer parts due to a bulb base that has exposed terminals. Aside from making a connection, these terminals are used to make measurements. **Figure 3.2-2** shows an alternate bulb socket that is mounted on a 4×4 electrical box. The connections are made inside the box and do not allow for a safe way to make measurements. As a result, connections are made to a terminal block first and then to the bulb base.

Some components may need to be rearranged in future labs. All the components are available at a local home improvement store, HVACR supply house, or online.

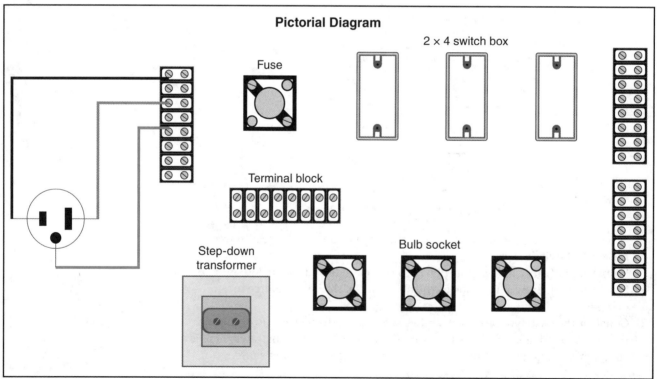

Figure 3.2-1

Goodheart-Willcox Publisher

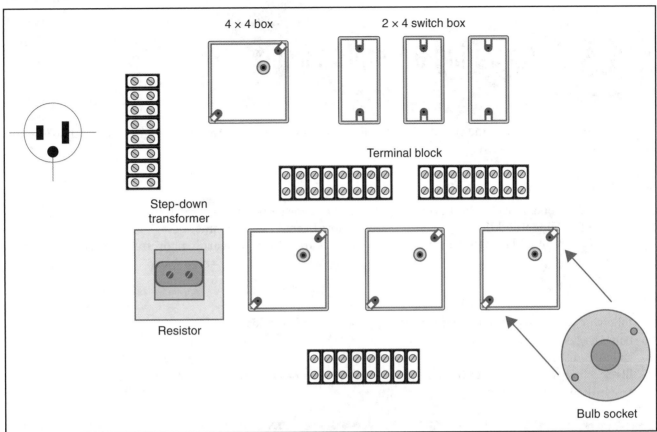

4 × 4 box

2 × 4 switch box

Terminal block

Step-down transformer

Resistor

Bulb socket

Goodheart-Willcox Publisher

Figure 3.2-2

Equipment

- 1—2′ × 2′ × 3/4″ plywood
- 4—terminal blocks (8 position shown)
- 3—2 × 4 switch box (metal)
- 4—4 × 4 electrical box (metal) if using Figure 3.2-2 layout
- 4—bulb base either exposed terminal or hidden terminals

- 1—Edison base 15 A fuse
- 1—14 to 16 AWG 3-wire power cord (length is user defined)
- Wood screws
- 14 AWG solid wire (black, white, and green) single wire
- Spade terminals 14–16 AWG (blue)
- 1—40 W incandescent bulb

Procedure

1. Select either layout from Figure 3.2-1 or Figure 3.2-2. This layout will serve as a guide to assemble the components. The layout is scaled so the parts will fit as shown on a 2′ × 2′ plywood section.
2. Use wood screws to secure components to the plywood. Note the resistors will not be secured to the lab board. Resistors will be introduced and used in later labs.
3. Connect the power plug wires to the upper left terminal block with spade terminals.
4. Insert the 15 A fuse into the bulb socket.
5. The lab board is now ready to be used in the following labs.

Operating a Simple Circuit

Using the lab board created in the previous lab, develop a functioning circuit. Then use a digital multimeter to obtain electrical readings from the circuit.

Lab Introduction

An HVACR technician must be able to construct a circuit and measure electrical readings using a digital multimeter. This is a fundamental skill that you must master before becoming successful in the field. Being able to identify a functioning circuit and recognizing circuit failure modes are integral in learning how to troubleshoot electrical problems.

Equipment

The following tools and equipment are needed to perform these activities:

- Lab board constructed in Lab 3.2
- Digital multimeter
- Black, white, and green 14 AWG solid wire
- 1—14–16 AWG spade terminal

- Option: 16 AWG solid wire
- Single-pole single-throw 120 V 15 A switch
- 1—40 W incandescent bulb

Safety Note

Resistance Check on Multimeter

Before connecting power to the circuit, use an ohmmeter to check circuit resistance. Connect the ohmmeter leads to the ac plug prongs. When the switch is in the open position, the resistance should be over-the-limit, as displayed by the letters *OL* on a digital meter. When the switch is closed, the meter will display the load resistance. A reading of 0 or close to 0 Ω with the switch in either position indicates a short circuit, and therefore it is not safe to plug in to power outlet. The short must be found before proceeding.

Procedure

1. Assemble the lab board from Lab 3.2.
2. Use 16– or 14–gage THNN wire solid wire for components of the board. Use
   Figure 3.3-1 to connect the wires to the components. Refer to Lab 2.3: *Working
   with Wires* for making the connections.

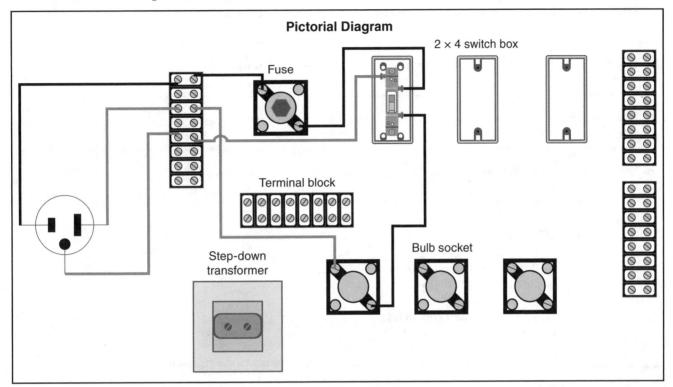

Pictorial Diagram

2 × 4 switch box

Fuse

Terminal block

Step-down
transformer

Bulb socket

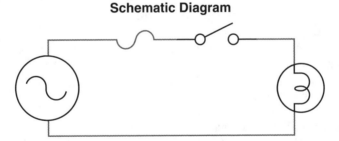

Schematic Diagram

Figure 3.3-1

Goodheart-Willcox Publisher

3. Use Lab 2.3 connection methods for the switch (side connections or clamp style).
4. Use space terminals for exposed terminal-type bulb base.
5. Before connecting the circuit to power, follow the Safety Note requirement.
   Have your instructor approve the circuit before proceeding to the next step.
6. Plug in and visually check for proper operation.
7. Proceed with the electrical measurements outlined in the Lab Questions.

Lab Questions

Use your digital multimeter or other electrical instruments to answer the following questions.

1. What is the voltage across the source?

   _____

2. What is the voltage across the load when the switch is in the open position?

   _____

3. What is the voltage across the load when the switch is in the closed position?

   _____

4. How much current is drawn by the load when the switch is in the open position?

   _____

5. How much current is drawn by the load when the switch is in the closed position?

   _____

Know and Understand

_____ 1. The three parts of an electric circuit are _____.
   A. source, voltage, and switch
   B. control, current, and conductor
   C. source, load, and conductor
   D. control, source, and load

_____ 2. *True or False?* A power source supplies electrical energy to the circuit load(s).

_____ 3. The load _____.
   A. converts electric energy into another form
   B. disrupts the flow of current
   C. allows current to flow faster in a circuit
   D. is a switching device

_____ 4. Conductors in a circuit _____.
   A. block current flow
   B. increase electrical resistance
   C. provide a path for current flow
   D. transform energy to another form

_____ 5. The purpose of a switch in an electric circuit is to _____.
   A. always conduct current   C. never conduct current
   B. limit the flow of current   D. allow the flow of voltage

_____ 6. *True or False?* A switch controls the on and off operation of a load(s).

_____ 7. *True or False?* The electrical status of an electrical circuit must always be closed.

_____ 8. A closed circuit means that _____.
   A. current will flow in the circuit
   B. current will not flow in the circuit
   C. only partial current flow will exist
   D. current flow will be excessive and destroy the load

_____ 9. An open circuit means that ____.
  A. current flows normally in the circuit
  B. it is a special green technology circuit
  C. current will not flow in the circuit
  D. current is at the maximum level

_____ 10. *True or False?* A shorted circuit means that the circuit is closed and is operating normally.

_____ 11. No electrical resistance in a circuit results in ____.
  A. more efficiency
  B. less efficiency
  C. normal operation
  D. a direct short (dead short)

_____ 12. Less resistance than normal requirement in a circuit results in ____.
  A. a more efficient operation
  B. higher than normal current flow
  C. lower than normal current flow
  D. no current flow

_____ 13. When a circuit is described as a short circuit, which is true?
  A. Current will not flow.
  B. Excessive current flow due to a sudden rise in voltage.
  C. The load is bypassed by a near 0 Ω conductor.
  D. More information is needed to determine result.

_____ 14. *True or False?* Schematics are used to troubleshoot, repair, and install HVAC systems.

_____ 15. A schematic legend shows ____.
  A. how components are connected together
  B. how the circuit works
  C. how to troubleshoot the circuit
  D. the meaning of symbols being used

Critical Thinking

1. Why is it good practice to test for resistance in a circuit before applying power?

   _____

   _____

   _____

   _____

   _____

2. When testing for resistance in a circuit, if the resistance is close to 0 Ω and power is then applied, what is likely to occur?

   _____

   _____

   _____

   _____

Essential Electrical Skills for HVACR: Theory and Labs

4 Ohm's Law

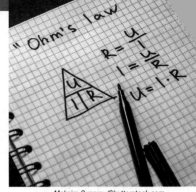

Maksim Gusarov/Shutterstock.com

Chapter Outline

4.1 Ohm's Law
4.2 Relationship between Electrical Variables
4.3 Application of Ohm's Law
4.4 Electrical Power
4.5 Application of the Power Equation

Additional Reading

Modern Refrigeration and Air Conditioning, 21st edition

12.1.3 Ohm's Law
13.1   Electrical Power

Learning Objectives

After completing this chapter, you will be able to:

- Discuss how voltage, current, and resistance are related using Ohm's law.
- Define the directly proportional and inversely proportional relationships in Ohm's law.
- Use Ohm's law to calculate each electrical variable.
- Describe electrical power and how it is calculated.
- Calculate electrical power in resistive circuits.

Technical Terms

directly proportional	inversely proportional	watt
electrical power	Ohm's law	

Introduction

The previous chapters introduced the three components of electricity and their application in a simple circuit. The next step is to conceptualize how these components are related and manipulated to control electricity. Ohm's law, which is at the heart of most electrical and electronic formulas, was developed by German physicist, George S. Ohm in the early 1800s. This law explains the relationship between voltage, current, and resistance through simple math and logic. Understanding and being able to calculate Ohm's law is fundamental for the design and troubleshooting of working electrical systems.

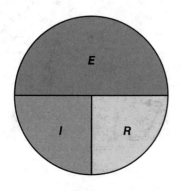

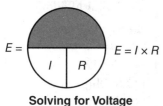

$E =$ $E = I \times R$

Solving for Voltage

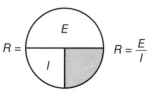

$R =$ $R = \dfrac{E}{I}$

Solving for Resistance

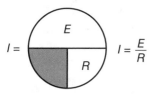

$I =$ $I = \dfrac{E}{R}$

Solving for Current

Goodheart-Willcox Publisher

Figure 4-1. The pie chart is an easy way to use Ohm's law. The horizontal line represents division and the vertical line represents multiplication. Cover the unknown variable and perform the remaining operation.

4.1 Ohm's Law

Ohm's law describes the relationship between the three electrical components of voltage, current, and resistance. This law is expressed as the following equation:

$$E = I \times R$$

where
    E = voltage measured in volts (V)
    I = current measured in amperes (A)
    R = resistance measured in ohms (Ω)

In written word, this equation means voltage equals current multiplied by resistance. By using algebraic application, we can rearrange the three values to derive two other equations:

$$I = \frac{E}{R}$$

$$R = \frac{E}{I}$$

An easier method is to use an Ohm's triangle, or pie chart. See **Figure 4-1**. To use this chart, cover the variable you need to calculate, and you are left with the formula for that variable.

The quantitative definition for the unit of measure for voltage, current, and resistance can be described as follows:

- One volt is the electromotive force required to generate 1 A of current through 1 Ω of resistance.
- One ampere is the amount of current generated by 1 V where the flow is restricted by a 1 Ω resistor.
- One ohm of resistance limits the amount of current generated by 1 V to 1 A.

4.2 Relationship between Electrical Variables

By analyzing each equation, we can find a few relationships between the variables. Voltage and current are ***directly proportional***. This means increasing the voltage when resistance is held constant causes an increase in current. On the contrary, current and resistance are ***inversely proportional***. This means a decrease in resistance while the voltage remains constant causes an increase in current.

A common analogy used to describe this relationship is a town water system. Consider a town that has a city water pressure of 80 psi and another town that has water pressure of 60 psi. Both towns use 1/2″ diameter tubing. This tubing serves as resistance since it limits the amount of water that can flow. In this example, resistance is the same for both water systems. The water pressure represents voltage, and the flow of water represents current. We know the relationship between current and voltage is such that as current increases, so does voltage. Thus, in this water system analogy, a one-gallon pail of water fills faster in the town that has water pressure of 80 psi. The force of 80 psi is greater than the force of 60 psi, and therefore the flow is greater. Likewise, if a one-gallon pail is held constant and the diameter of tubing varies home to home, the homes with larger diameter tubing will fill more quickly. The force is the same, but there is less resistance with larger tubing.

This logical thinking in respect to the relationship between voltage, current, and resistance is necessary for troubleshooting electrical systems.

Solving for Equations

Recall that in a mathematic equation, the equal sign (=) represents equality between the left and right side of the equation. Often times, equations may have an unknown variable on one side of the equation that needs to be solved. Usually, this is called solving equations with the unknown value *x*.

The concept of solving for an unknown variable is key for an HVACR technician and when working with Ohm's law. A multiplication problem where there is an unknown variable on one side can be solved by using division. Consider the following multiplication problem:

$$2 \times x = 4$$

The key to solving for one unknown value is to isolate the unknown on one side of the equation by itself. Whatever operation is performed on one side of the equation must also be performed on the other side. Since division is the inverse of multiplication, dividing both sides by 2 will help us solve for *x*:

$$x = 4 \div 2$$
$$x = 2$$

The next equation is set up as a division problem, but it can be solved by using multiplication.

$$4 \div x = 2$$

The best way to solve this is to multiple both sides by *x*.

$$4 = 2 \times x$$

Then divide both sides by 2.

$$4 \div 2 = x$$
$$x = 2$$

4.3 Application of Ohm's Law

The following examples demonstrate how to use Ohm's law to calculate an unknown value.

Consider a 12 V source is applied to a 6 Ω resistance in a simple circuit with one load. We need to find how much current is flowing in this circuit. To find this, we first must select the correct form of Ohm's law:

$$I = \frac{E}{R}$$

Next, plug in the known values into the equation.

$$I = \frac{12\text{ V}}{6\,\Omega}$$

$$= 2\text{ A}$$

Now, if the voltage doubled and the resistance remained constant, the current still increases.

$$I = \frac{24\text{ V}}{6\,\Omega}$$

$$= 4\text{ A}$$

Consider the scenario where resistance is doubled and voltage is left at 24 V. What happens to current?

$$I = \frac{24\text{ V}}{12\text{ }\Omega}$$

$$= 2\text{ A}$$

In this case, current decreases since it has an indirectly proportional relationship with resistance. For a visual example, see **Figure 4-2**. Doubling the applied voltage to a lamp doubles the current in a circuit. This increase in current causes the lamp to shine brighter.

Now, let's look at solving for a new unknown variable. Consider a resistor with a resistance of 6 Ω and a current of 20 A. Solve for the unknown voltage driving the current:

$$E = I \times R$$

$$E = 20\text{ A} \times 6\text{ }\Omega$$

$$= 120\text{ V}$$

Next consider a standard light emitting diode (LED) used as a pilot lamp in a circuit that has 24 Vdc. The current must not exceed 0.020 A, or the LED becomes damaged. What value resistor is needed to limit the current?

$$R = \frac{E}{I}$$

$$R = \frac{24\text{ V}}{0.020\text{ A}}$$

$$= 1200\text{ }\Omega$$

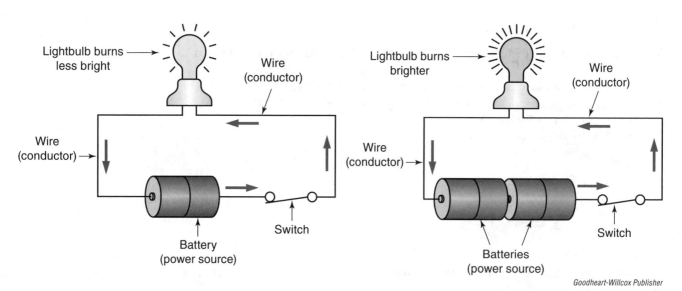

Goodheart-Willcox Publisher

Figure 4-2. Doubling the voltage doubles the current if resistance remains constant. The increase in current increases bulb brightness.

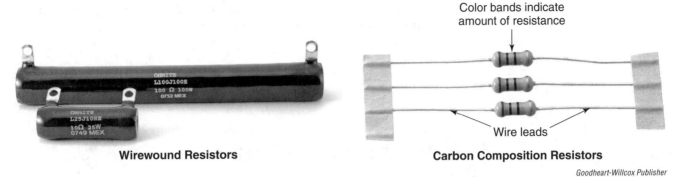

Wirewound Resistors

Color bands indicate
amount of resistance

Wire leads

Carbon Composition Resistors

Goodheart-Willcox Publisher

Figure 4-3. Two of the many types of resistors used in electrical circuits in HVACR equipment.

4.4 Electrical Power

Electrical power exists when current flows due to a potential difference (voltage). The amount of electrical power consumed in a circuit is a variable that can be calculated. Electrical power is measured by a unit of power called the ***watt*** and is calculated using the power equation. The power equation is defined as follows:

$$P = I \times E$$

where
P = power measured in watts (W)
I = current measured in amperes (A)
E = voltage measured in volts (V)

To put this equation into application, consider an electric heating element that has a pure resistance of 12 Ω and requires 240 V of supply voltage and 20 A of current to produce its rated Btu of heat energy. How much electrical power is being consumed by the element?

$$P = I \times E$$
$$P = 20\,A \times 240\,V$$
$$= 4800\,W$$

This amount, 4800 W, of electrical energy can be transformed to 16,382 Btu of heat energy by the load (electric heating element).

Note the similarities between the power equation and Ohm's law. Ohm's law uses $E = I \times R$, while the power equation uses $P = I \times E$. Like Ohm's law, a pie chart can be used as a tool for solving the variables within the power equation. See **Figure 4-4**.

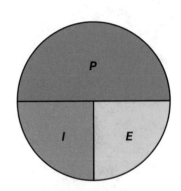

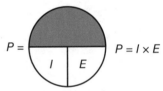

$P =$ $P = I \times E$

Solving for Power

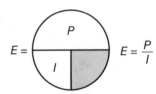

$E =$ $E = \dfrac{P}{I}$

Solving for Voltage

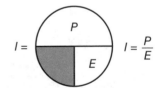

$I =$ $I = \dfrac{P}{E}$

Solving for Current

Goodheart-Willcox Publisher

Figure 4-4. A pie chart can also be used to simplify the power equation in the same way as the Ohm's law pie chart.

The variables of the power equation can be algebraically rearranged so each can be solved. The other forms of the power equation are:

$$I = \frac{P}{E}$$

$$E = \frac{P}{I}$$

4.5 Application of the Power Equation

Consider an electric heating element that consumes 4800 W and is operating with 240 V. Calculate the current drawn by this heating element.

First, we must find the correct equation that is used to solve for current.

$$I = \frac{P}{E}$$

Then we can plug in the known values to solve for I.

$$I = \frac{4800 \text{ W}}{240 \text{ V}}$$

$$= 20 \text{ A}$$

Note the power equation here applies to direct current power sources and alternating current power sources where only pure resistance is used. There are three types of resistance with ac power sources that will result in modification of the power equations. These types of resistance will be covered in more detail later in this textbook.

Summary

- The relationship between voltage, current, and resistance must be clearly understood by HVACR technicians to design and troubleshoot HVACR electrical systems.
- Ohm's law describes the relationship between the voltage, current, and resistance.
- The equations in Ohm's law allow us to calculate any of the unknown components of electricity if the other two values are known.
- Voltage and current are directly proportional. If voltage increases then current increases, providing that resistance is held constant.
- Current and resistance are indirectly proportional. If resistance increases then current decreases, providing that voltage is held constant.
- The unit of measure for electrical power is the watt (W).
- Electrical power is calculated by multiplying current by voltage: $P = I \times E$.
- The power equation can be rearranged to solve for power, current, or voltage.

Lab 4.1

Verifying Ohm's Law through Simple Circuits

This multi-part lab will use hands-on application to build upon the lessons learned in this chapter. Its purpose is to give you practice in verifying Ohm's law calculations by using a power source, loads, and a multimeter.

Lab Introduction

There will be three parts to this lab. This first part will use a battery and two different value resistors to compare calculated values to measured values. Since most HVACR meters only measure dc microamperes, large value resistors must be used with a 9 V battery.

The second part includes observing and evaluating measured values. The third part of this lab uses 24 Vac and power resistors to compare calculated values to measured values. The 24 Vac allows for the use of an ac inductive clamp ammeter. Both the second and third lab will use a step-down transformer to supply approximately 24 Vac, which is called the secondary voltage. This secondary voltage is a reduced voltage that is produced when the step-down transformer plugs into a 120 V.

Use the lab diagrams to construct the circuits. The equipment below contains the materials for all three parts of the lab.

Equipment

- 1—Digital multimeter with clamp-on ammeter
- Lab board. See Lab 3.2 for complete instructions.
- 1—100,000 Ω 1/4 W resistor (no less than 5% tolerance)
- 1—200,000 Ω 1/4 W resistor (no less than 5% tolerance)
- 1—9 V dry cell battery
- 2—60 W incandescent bulb

- 16 or 14 AWG solid wire
- 14 AWG wire nuts
- 14 AWG spade terminals
- 25 Ω 25 W power resistor
- 50 Ω 25 W power resistor
- 2—jumper wires with alligator clips

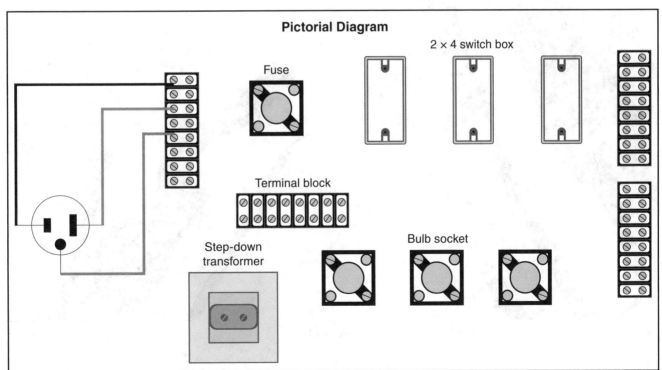

Figure 4.1-1.

Goodheart-Willcox Publisher

Resistors

Resistors are rated by ohmic value, tolerance, and wattage. A 10 Ω, 1/4 W resistor burns if 6 V are applied because the current 0.6 A (6 V/10 Ω) multiplied by the voltage of 6 V results in 3.6 W. A 1/4 W (0.250 W) resistor is too small for the application. The correct wattage resistor must be used by first calculating the resistor power consumption.

Procedure—Part One

1. Review **Figure 4.1-2**.
2. Connect one jumper end to the positive terminal of the battery and the other end to one end of the 100,000 Ω resistor.
3. Connect a second jumper to the negative end of the battery and the end to the opposite end of the 100,000 Ω resistor.
4. Measure and record the battery voltage (after resistor is connected).
5. Connect the 200,000 Ω resistor to the battery.
6. Measure and record the battery voltage (after resistor is connected).
7. Rewire circuit to include ammeter in circuit. See **Figure 4.1-3**.
8. Connect jumper to positive battery terminal and meter positive probe.
9. Connect jumper to negative meter probe and any one side of resistor.
10. Connect jumper to opposite side of resistor and to the negative battery terminal.
11. Complete Lab Questions for the 100,000 Ω resistor.
12. Replace the resistor with the 200,000 Ω resistor.

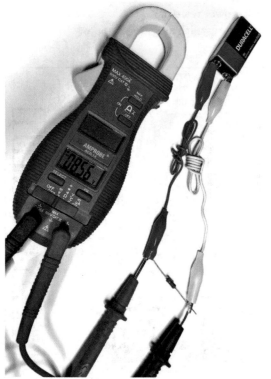

Figure 4.1-2.   *Goodheart-Willcox Publisher*

Figure 4.1-3.   *Goodheart-Willcox Publisher*

Continued▶

Resistance Check on Multimeter

Lab Questions

Use your digital multimeter or other electrical instruments to answer the following questions.

1. Use the correct equation to calculate the current through R_1 when battery voltage is applied. What is the calculated current?

   _____

2. Use the multimeter to measure the current in the circuit. Use the dc microammeter scale and inline method. HVACR meters have this range to measure flame rectification outputs. Read the owner's manual for instructions. Is the calculated value close to the measured value? (Note that values will not match 100% due to sources of error not yet discussed.)

   _____

3. Use the correct equation to calculate the current through R_2 when battery voltage is applied. What is the calculated current?

   _____

4. Use the multimeter to measure the current in the circuit. Are the calculated value and measured value the same?

   _____

Procedure—Part Two

1. Use the simple circuit already constructed in the lab board.
2. Connect the transformer 120 V primary to one 60 Watt bulb.
3. Connect another 60 W bulb to the 24 V secondary of the transformer. See **Figure 4.1-4**.
4. Use the terminal strip to make splices. Do not attach two wires to the same bulb base terminal. The single pole switch will control the on/off operation of both bulbs.
5. Before powering up the circuit, get approval from your instructor.

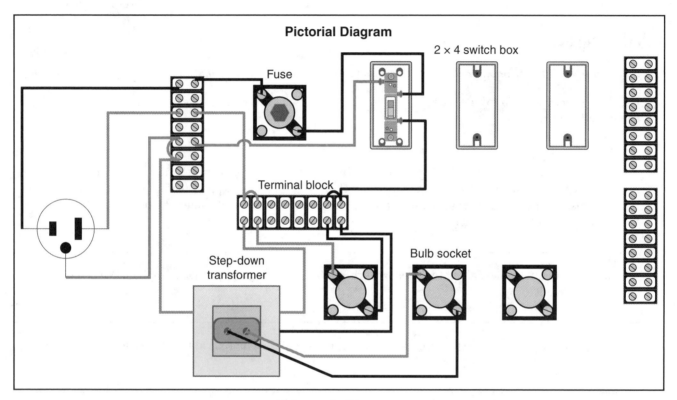

Pictorial Diagram

Schematic Diagram

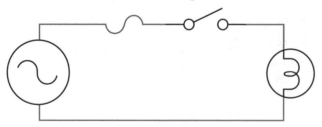

Figure 4.1-4.

Goodheart-Willcox Publisher

Lab Questions

Use your digital multimeter or other electrical instruments to answer the following questions.

1. Plug in the lightbulbs. Notice one bulb is brighter. Why?

   _____

   _____

   _____

   _____

2. Measure the current through each bulb and the applied voltage. The bulbs are both rated for 60 W. Why is one brighter?

   _____

   _____

   _____

   _____

 Procedure—Part Three

1. Remove the 24 V bulb connections to the transformer.
2. Use jumper wires to connect a 25 Ω 25 W resistor.
3. Answer the Lab Questions.
3. Disconnect the 25 Ω resistor, and then use the same jumper wires to connect the 50 Ω resistor. See **Figure 4.1-5**.

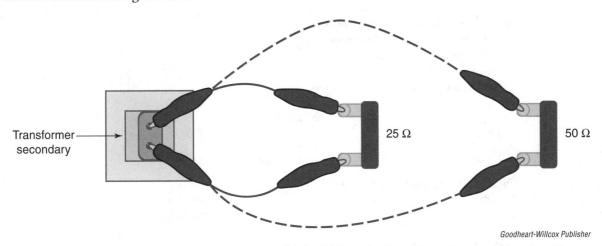

Transformer secondary

25 Ω

50 Ω

Goodheart-Willcox Publisher

Figure 4.1-5. The dotted lines represent first using the 25 Ω resistor then disconnecting to use the 50 Ω. Note resistors are not polarity sensitive. Connect resistors to jumper wires and transformer as shown.

Lab Questions

Use your digital multimeter or other electrical instruments to answer the following questions.

1. What is the secondary voltage?

_____

_____

2. Using the actual secondary voltage and the 25 Ω resistor, calculate the current.

_____

_____

3. With only the 25 Ω resistor connected to the secondary, plug in the transformer and use the inductance clamp ammeter to measure the actual current draw of the resistor. Record value.

_____

_____

4. Are the calculated and measured values of current approximately the same?

_____

_____

5. Repeat the above steps using the 50 Ω resistor. Do the measured values support Ohm's law?

_____

Know and Understand

_____ 1. Ohm's law states a relationship between _____.
A. voltage, resistance, and current
B. volts, resistance, and current
C amperes, volts, and ohms
D. resistance, voltage, and amperes

_____ 2. The _____ is the unit of measure for voltage.
A. EMF   C. ampere
B. ohm   D. volt

_____ 3. The _____ is the unit of measure for current.
A. ohm   C. volt
B. ampere   D. intensity

_____ 4. The _____ is the unit of measure for resistance.
A. ohm   C. volt
B. ampere   D. watt

_____ 5. *True or False?* Resistance limits the current flow in a circuit.

_____ 6. Voltage is directly proportional to _____.
A. resistance   C. current
B. power   D. amperes

_____ 7. *True or False?* Voltage is always greater than current.

_____ 8. Current is indirectly proportional to ____.
A. voltage
B. resistance
C. ohms
D. amperes

_____ 9. How much voltage is required to generate 2 A through a 300 Ω resistance?
A. 120 V
B. 240 V
C. 600 V
D. 1000 V

_____ 10. How much current flows through a 20 Ω load when 240 V are applied?
A. 10 A
B. 24 A
C. 12 A
D. 6 A

_____ 11. How much resistance is required to limit a current to 1/2 A if 120 V is applied?
A. 60 Ω
B. 100 Ω
C. 120 Ω
D. 240 Ω

_____ 12. *True or False?* An easy way to remember and use Ohm's law is by using a pie chart.

_____ 13. *True or False?* One volt is required to generate one ampere of current through a 2 Ω resistance.

_____ 14. *True or False?* One ohm of resistance will limit current flow to 2 A if the applied voltage is 2 V.

_____ 15. How much power is used by a load with 120 V applied and drawing 12 A?
A. 1440 Ω
B. 1440 W
C. 10 W
D. 720 W

Critical Thinking

1. What happens to the current when resistance is 0 Ω and voltage is 120 V?

_____

_____

2. Find current when voltage and power are the only known values.

_____

_____

_____

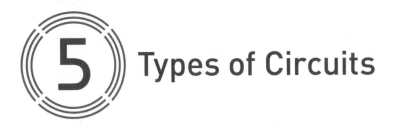

5 Types of Circuits

Chapter Outline

Additional Reading

Modern Refrigeration and Air Conditioning, **21st edition**

12.4 Circuit Fundamentals

Learning Objectives

After completing this chapter, you will be able to:

- Define a series circuit.
- Explain what occurs when a series circuit fails.
- Apply the series circuit rules to calculate voltage, current, and resistance.
- Define a parallel circuit.
- Explain what occurs when a parallel circuit fails.
- Apply the parallel circuit rules to calculate voltage, current, and resistance.
- Define a complex circuit.
- Explain how to calculate voltage, current, and resistance in a complex circuit.
- Define the American Wire Gage and its wiring standards.
- Discuss *NEC* ampacity tables and wiring.

Technical Terms

American Wire Gage (AWG)	Kirchhoff's current law (KCL)	solid wire
branch	Kirchhoff's voltage law (KVL)	stranded wire
complex circuit	parallel circuit	
cross-sectional area	series circuit	

Introduction

As discussed in the previous chapter, a simple circuit contains only one load. This chapter will apply what you have learned with simple circuits to circuits that have more than one load connected to a power source. These circuits can be arranged in series, parallel, or complex configurations based on the electrical system design.

With each configuration, there are unique rules that apply for finding current, voltage, and resistance. An HVACR technician must be able to differentiate between the types of circuits and understand how voltage, current, and resistance are affected and calculated.

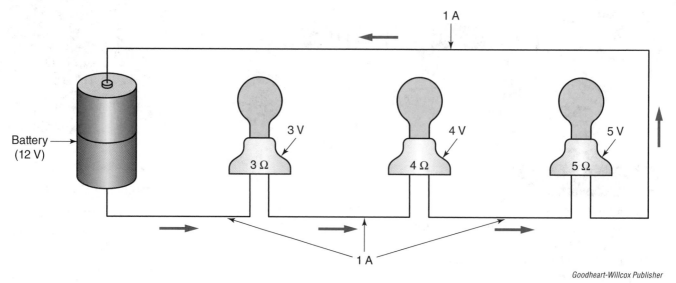

1 A

Battery
(12 V)

3 V
3 Ω

4 V
4 Ω

5 V
5 Ω

1 A

Figure 5-1. A series circuit has only one path for current to take.

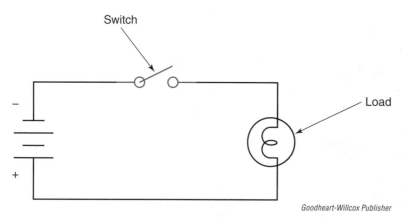

Switch

−

+

Load

Figure 5-2. The switch is wired in series with the load. When the switch is open, there is no current flow. Current flows when the switch is closed.

5.1 Series Circuits

A *series circuit* has only one path for current flow. The current travels from one load to the next. A simple circuit is a series circuit with only one load. See **Figure 5-1**.

Because there is only one path for current in a series circuit, any break along the circuit causes a disruption in current flow through the load. For that reason, a switch that controls the operation of the load is wired in series with the load. See **Figure 5-2**. There are limited applications for series circuit loads. Two examples are motor windings and electric heating elements.

Because there is only one path or direction for current to travel, the amount of current in the circuit is the same throughout. This is one of the rules of a series circuit, which can be represented mathematically by the following equation:

$$I_t = I_1 = I_2 = I_3 = I_n$$

where

I_t = the total current of the circuit
$I_{1,2,3}$ = the current flowing through each load
I_n = the current flowing through any load in the circuit

There is also a general rule to calculating the resistance in a series circuit. Recall that a load has resistance. The resistance of each individual load adds up to create the total circuit resistance. This can be expressed as the following:

$$R_t = R_1 + R_2 + R_3 \ldots + R_n$$

where

R_t = the total resistance of the circuit
$R_{1,2,3}$ = the resistance of each load
R_n = the resistance of additional loads in the circuit

Like total current, the total resistance (R_t) is the value used when calculating current or voltage in Ohm's law.

Because voltage is the force that drives current, it decreases in force the further it goes through the circuit. This means voltage drops across each resistance in the circuit. The series circuit rule for total voltage is as follows:

$$E_t = E_1 + E_2 + E_3 \ldots + E_n$$

where

E_t = the total voltage of the circuit
$E_{1, 2, 3}$ = the voltage drop across each load
E_n = the voltage drop across any load in the circuit

In other words, the sum of the voltage drops in a series circuit equals the source voltage (E_t). This is also known as **_Kirchhoff's voltage law (KVL)_** of conservation. This law is useful when troubleshooting. Consider a motor is performing poorly. The voltage measured at the motor terminals is 110 Vac, while the voltage measured at the circuit breaker panel is 120 Vac. We know there is a problem since the voltage is not the same. The voltage that is dropped within the circuit must be due to resistance of contacts, lose or corroded connections, or wrong gage or faulty wires. An HVACR technician must be able to identify and correct this problem.

5.1.1 Using Ohm's Law in a Series Circuit

Before being able to calculate voltage, current, or resistance, the totals of each value must be determined in a circuit. Consider **Figure 5-3**. In this circuit, the resistance for each load and source voltage (E_t) are given.

To solve for the unknown values, first calculate the total resistance, R_t.

$$R_t = R_1 + R_2 + R_3$$
$$R_t = 6\,\Omega + 2\,\Omega + 4\,\Omega$$
$$= 12\,\Omega$$

Now that total resistance is found, current can be determined. This is found by using Ohm's law:

$$I_t = \frac{E_t}{R_t}$$
$$I_t = \frac{12\text{ V}}{12\,\Omega}$$
$$= 1\text{ A}$$

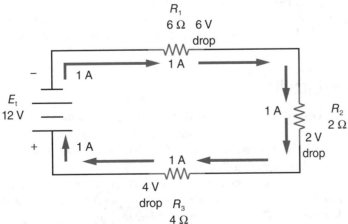

Figure 5-3. An example of a series circuit. Given resistance and source voltage values, the remaining values are found by using series circuit equations and Ohm's law.

Goodheart-Willcox Publisher

This is the total current for the circuit. We also know this since it is a series circuit, and the total current is the same through each load regardless of individual resistance value.

The voltage drop across each load in a series circuit is based on the load's resistance and the current in the circuit. Ohm's law can be used to calculate each voltage drop:

$$E_n = I_t \times R_n$$

$$E_1 = 1\,A \times 6\,\Omega$$
$$= 6\,V$$

$$E_2 = 1\,A \times 2\,\Omega$$
$$= 2\,V$$

$$E_3 = 1\,A \times 4\,\Omega$$
$$= 4\,V$$

Therefore, the voltage drops are 6 V for R_1, 2 V for R_2, and 4 V for R_3. The sum of the voltage drops equals the source voltage, which means Kirchhoff's law is satisfied. This proves that the calculated results are correct. An important observation about a series circuit is that the largest resistance drops the most voltage. More force is required to move current through the 6 Ω resistor than the 2 Ω resistor. Accordingly, the smallest resistor drops the least voltage.

An unintentional break in a circuit is one type of failure that an HVACR technician must troubleshoot. A switch contains contacts that in time deteriorate and act as a type of resistance. This resistance from the switch is in series with the load. The rules for series circuits will apply, thus causing the switch to drop voltage. This voltage drop is small due to a small value of resistance. However, a further increase in switch contact resistance can result in a significant voltage drop that negatively affects the load and can eventually lead to system failure. See **Figure 5-4**.

This voltage drop can occur in all types of connections where two surfaces meet due to environmental factors, overcurrent, wear, and poor workmanship. This is called contact resistance. These connnections include relay and contactor contacts, terminals, wire nuts, a circuit breaker, fuses, and switch contacts.

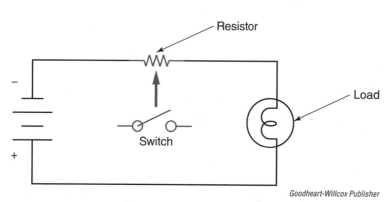
Goodheart-Willcox Publisher

Figure 5-4. Deteriorating switch contacts become an added resistance that is in series with the load. Voltage is dropped by the contact resistance resulting in less available voltage for the load.

5.2 Parallel Circuits

Most loads are wired in parallel where different rules apply for finding current, voltage, and resistance. A ***parallel circuit*** has more than one path for current flow. Each individual path that makes up the circuit are called ***branches***. A branch may contain one or more loads. See **Figure 5-5**.

Since there is more than one path in a parallel circuit, the current through each individual branch adds up to the total current supplied and returned to the power source. This is known as ***Kirchhoff's current law (KCL)***.

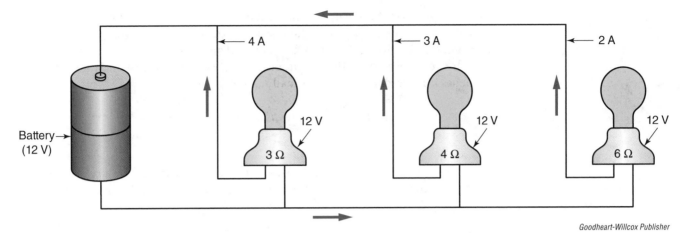

Figure 5-5. A parallel circuit has multiple paths for current to take.

The equation for this rule is as follows:

$$I_t = I_1 + I_2 + I_3 \ldots + I_n$$

where

I_t = the total current of the circuit
$I_{1,2,3}$ = the current through each branch
I_n = the number current through additional branches in the circuit

Each branch is independent from the other branches in the sense that each branch has its own path for current flow. Because we know voltage is the force driving current down a path, the voltage is the same across each branch. See **Figure 5-6**.

This rule is reflected in the mathematical representation for voltage in parallel circuits:

$$E_t = E_1 = E_2 = E_3 = E_n$$

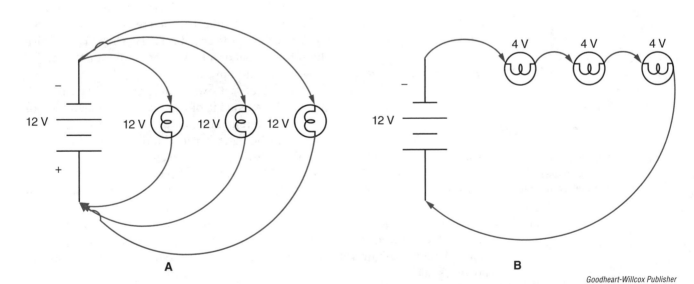

Figure 5-6. A—Lamps are wired in parallel. Voltage is the same across each parallel branch. The illustration shows how all the branch loads are connected to the power source. Thus, voltage is the same for each branch. The loads are independent of each other. Parallel loads are connected by two common points. B—In contrast, voltage drops across each load in a circuit. Series loads are connected by one common point.

Because current adds up in a parallel circuit, resistance must decrease as more branch resistances are added. Total resistance thus is less than the smallest branch resistance in the circuit—even if by a fractional amount.

This reciprocal, or indirect relationship, between resistance and the number of branches is represented mathematically as such:

$$R_t = \frac{1}{\dfrac{1}{R_1} + \dfrac{1}{R_2} + \dfrac{1}{R_3} \ldots + \dfrac{1}{R_n}}$$

There is a different formula for calculating total resistance for a parallel circuit with only two branches, or for calculating two branches of a multi-branch circuit:

$$R_t = \frac{R_1 \times R_2}{R_1 + R_2}$$

If a circuit has the same resistance for all the branches, the following equation can be used:

$$R_t = \frac{\text{branch resistance}}{\text{quantity of branches}}$$

5.2.1 Using Ohm's Law in a Parallel Circuit

Consider a parallel circuit that has three branches, and each branch has a resistance of 6 Ω. In order to solve for the total resistance in the circuit, we first select the correct equation and then solve:

$$R_t = \frac{\text{branch resistance}}{\text{quantity of branches}}$$

$$R_t = \frac{6\,\Omega}{3}$$

$$R_t = 2\,\Omega$$

Note that the total resistance in the circuit (2 Ω) is less than the smallest branch resistance of 3 Ω. This total resistance then can be used in Ohm's law to solve for total current or total voltage, assuming that one of these values is known.

Most loads are wired in parallel. Each branch load is independent of others, which means it can draw a different amount of current, and it is not affected when another load fails or is shut off. Consider the example of a parallel circuit shown in **Figure 5-7**. There are three branches with different resistance and current amounts through each branch.

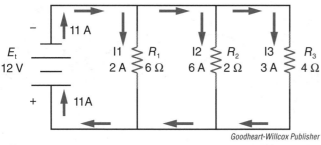

Figure 5-7. A sample of a parallel circuit. Given resistance and source voltage values, the remaining values are found by using parallel circuit equations and Ohm's law.

Goodheart-Willcox Publisher

Assume we want to solve for current through each branch. We know voltage is constant at 12 V through the circuit, and there are varying resistances through each branch. We can use Ohm's law to solve for each individual current by performing the following steps:

$$I_n = \frac{E_t}{R_n}$$

$$I_1 = \frac{12\,\text{V}}{6\,\Omega}$$

$$= 2\,\text{A}$$

$$I_2 = \frac{12\,\text{V}}{2\,\Omega}$$

$$= 6\,\text{A}$$

$$I_3 = \frac{12\,\text{V}}{4\,\Omega}$$

$$= 3\,\text{A}$$

Therefore, we know that branch 1 draws 2 A of current, branch 2 draws 6 A, and branch 3 draws 3 A. We quickly can determine the total current (I_t) by adding these three branches together:

$$I_t = I_1 + I_2 + I_3$$
$$I_t = 2\,\text{A} + 6\,\text{A} + 3\,\text{A}$$
$$= 11\,\text{A}$$

Next, assume we want to solve for the total resistance through each branch. This can be accomplished now that our current through each branch is known.

$$R_n = \frac{E}{I_n}$$

$$R_1 = \frac{12\,\text{V}}{2\,\text{A}}$$

$$= 6\,\Omega$$

$$R_2 = \frac{12\,\text{V}}{6\,\text{A}}$$

$$= 2\,\Omega$$

$$R_3 = \frac{12\,\text{V}}{3\,\text{A}}$$

$$= 4\,\Omega$$

Now we must calculate total resistance. Because there are more than two branches and the resistance is not the same through each branch, the following equation is used:

$$R_t = \frac{1}{\dfrac{1}{R_1} + \dfrac{1}{R_2} + \dfrac{1}{R_3}}$$

$$R_t = \frac{1}{\dfrac{1}{6}\,\Omega + \dfrac{1}{2}\,\Omega + \dfrac{1}{4}\,\Omega}$$

$$= \frac{1}{0.167 + 0.500 + 0.250}$$

$$= \frac{1}{0.917}$$

$$= 1.090\,\Omega$$

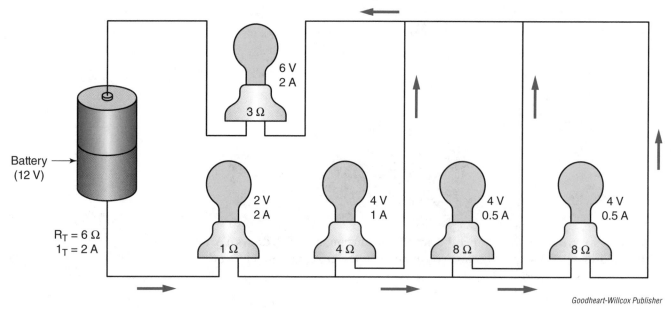

Goodheart-Willcox Publisher

Figure 5-8. A complex circuit is a combination of series and parallel circuits.

5.3 Complex Circuits

A *complex circuit* is a combination of series and parallel circuits. Other terms for the complex circuit are *combination circuits* and *series-parallel circuits*. See **Figure 5-8**.

Recall that switch contacts can become a resistance. Consider two parallel loads controlled by one simple on/off switch. Initially, the resistance from the switch contact is negligible, so when the switch is closed, the circuit is solely parallel. At some point, the switch contacts develop significant resistance, and now a complex circuit is created. The switch contact resistance is now in series with two parallel branches, **Figure 5-9**. This affects the voltage across the load and the current draw of the load. Voltage drops across the switch, and the remainder of the voltage is applied to the parallel branches. The voltage drop can cause load malfunctions.

To solve for unknown values in a complex circuit, the rules for series and parallel circuits are used in addition to logical procedure. Use **Figure 5-10** as an example.

Figure 5-9. Deteriorating switch contacts become an added resistance that is in series with the two loads. Voltage is dropped by the contact resistance resulting in less available voltage for the loads.

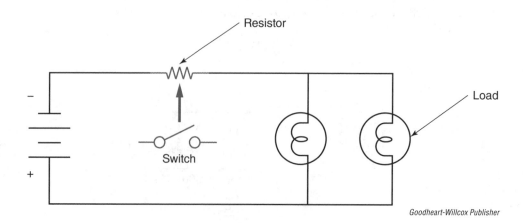

Goodheart-Willcox Publisher

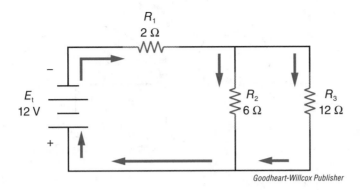

Figure 5-10. An example of a complex circuit. The resistance and source voltage values are provided. Thus the remaining values can be found by using parallel and series circuit equations and Ohm's law.

Goodheart-Willcox Publisher

First, solve for resistance. To do this, the circuit is simplified by using the parallel resistance equation to combine the two resistors.

$$R_t = \frac{R_1 \times R_2}{R_1 + R_2}$$

$$R_t = \frac{6\,\Omega \times 12\,\Omega}{6\,\Omega + 12\,\Omega}$$

$$= \frac{72\,\Omega^2}{18\,\Omega}$$

$$= 4\,\Omega$$

Next, the new equivalent resistor, E_{eq}, (4 Ω) can be used to calculate total resistance. See **Figure 5-11**. The circuit is now redrawn in series with the switch contact resistor. The series circuit rules can now be used to produce the total circuit resistance.

$$R_t = R_1 + R_2$$
$$R_t = 2\,\Omega + 4\,\Omega$$
$$= 6\,\Omega$$

Now the total current can be solved by using Ohm's law.

$$I_t = \frac{12\,V}{6\,\Omega}$$

$$= 2\,A$$

Applying this current to the series switch resistor and parallel equivalent resistor (E_{eq}) allows the voltage drops to be calculated. See **Figure 5-12**.

$$E_1 = 2\,A \times 2\,\Omega$$
$$= 4\,V$$

$$E_{eq} = 2\,A \times 4\,\Omega$$
$$= 8\,V$$

Notice that the sum of the voltage drops equals the source voltage.

Redraw the original circuit, labeling the voltage drops for the series portion of the circuit R_1 and the parallel branches containing R_2 and R_3. See **Figure 5-13**. The total circuit current flows through the switch and divides into the two branch loads.

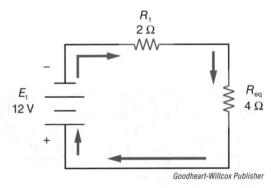

Goodheart-Willcox Publisher

Figure 5-11. Use the series circuit portion of the complex circuit to calculate the total resistance.

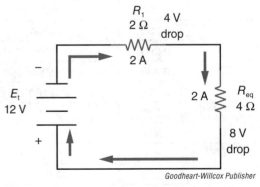

Goodheart-Willcox Publisher

Figure 5-12. The voltage drops can be found across the series circuit. Recall the sum of the voltage drops equal the source voltage.

Figure 5-13. Once the voltage drops are calculated, redraw the original circuit, and label the correct quantities. Kirchhoff's current and voltage laws can be confirmed to ensure the calculations are correct.

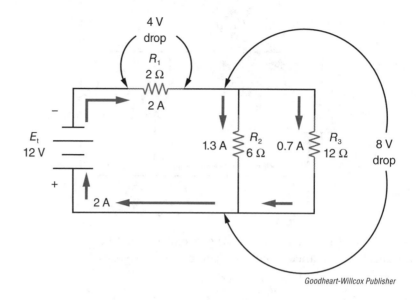

Goodheart-Willcox Publisher

Since the individual load resistances and the voltage across the parallel branches are known, the current through each is found using Ohm's law.

$$I_t = \frac{4\,V}{2\,\Omega}$$

$$= 2\,A$$

We now see that the sum of the parallel branch currents equals the total circuit current.

5.4 Residential and Commercial Conductors

In addition to contact resistance in a circuit, there is the added resistance of a conductor itself. That means voltage drops in a circuit because of a conductor's resistance. The resistance depends upon the type of material, size (diameter or cross section), length, and temperature. Putting these facts into a formula yields the resistance for a particular conductor length.

5.4.1 American Wire Gage

A wire resistance table can be used as an easier method. These tables are typically provided by conductor manufacturers. See **Figure 5-14**. The **American Wire Gage (AWG)** is the standard for wire size in the United States. It is based on the number of extrusion details used to extrude, or stretch the wire to its final diameter.

Based on AWG wire sizing, the bigger a gage number, the smaller the diameter of the wire. For example, a 14 AWG wire has a larger diameter than an 18 AWG wire. As the diameter increases, so does the ability to conduct more current. This is reinforced through our understanding of the relationship between current and resistance. The diameter of the wire is also given.

Conductor Sizes and Resistance Properties		
Size (AWG)	Diameter	Direct Current Resistance at 75°C (167°F) [Ω/1,000 feet]
18	0.040″	8.08 Ω
16	0.051″	5.08 Ω
14	0.064″	3.19 Ω
12	0.081″	2.01 Ω
10	0.102″	1.26 Ω
8	0.128″	0.786 Ω
6	0.184″	0.510 Ω
4	0.232″	0.321 Ω
3	0.260″	0.254 Ω
2	0.292″	0.201 Ω
1	0.332″	0.160 Ω
1/0	0.372″	0.127 Ω
2/0	0.418″	0.101 Ω
3/0	0.470″	0.0797 Ω
4/0	0.528″	0.0626 Ω

Goodheart-Willcox Publisher

Figure 5-14. A wire resistance table can be used to determine the best wire for an electrical job.

Some charts show the **cross-sectional area**. This is the area of a flat slice of the wire, or the area of the circle you see when you look straight into a wire. Manufacturers in other countries use the metric system and measure diameter in millimeters. The resistance is given for 1000′ of wire based on 167°F. Since ambient temperatures do not rise above 167°F in most HVACR applications, the chart is adequate for most use. Temperatures beyond 167°F require the use of formulas.

Consider the scenario where a 50′ 18 AWG extension cord is connected to a 120 V outlet. A load that draws 5 A is connected to the cord. The 18 AWG wire's resistance is 8.08 Ω per 1000′. The 50′ cord has two conductors, which means the total length is 100′.

Since 100 is 1/10 of 1000, we divide 8.08 Ω by 10 to get 0.808 Ω for 100′ of wire. By using Ohm's law, the voltage drop caused by wire resistance can be found:

$$E_{drop} = 5\,A \times 0.808\,\Omega$$
$$= 4.04\,V$$

If we subtract this voltage from 120 V, we see that the load is instead getting 116 V instead of 120 V due to wire resistance. If the extension cord was 100′, then the voltage drop would double to 8 V, and the load gets only 112 V, which is a problem.

5.4.2 *NEC* Ampacity Tables

The *National Electrical Code (NEC)* has established codes for safe wire usage reflected in ampacity tables. These *NEC* ampacity tables list a conductor's safe current carrying capacity. Factors used in the tables are gage, insulation types, wire material, and temperature. *NEC* tables list the various types of insulating materials and conduits to carry the wires. Typical residential wires use an insulator that is rated for 167°F and is moisture resistant. Refer to the *NFPA 70, Chapter 9* for more information on the NFPA website.

Conductors are made of solid or stranded wires. See **Figure 5-15**. **Stranded wires** provide flexibility and are used in applications where there is vibration or movement. HVACR equipment uses stranded wires as there are many motor applications. The flexibility allows wire management within the equipment. Multiple wires have to be routed in tight spaces. Wires larger than 10 AWG are typically stranded. HVACR thermostat wires are mostly **solid wires**, as these are small diameter 18 AWG wires. Wiring in homes and light commercial buildings use solid wire. The rigid wires are better for routing in structures. Solid wire is also less expensive than stranded wire.

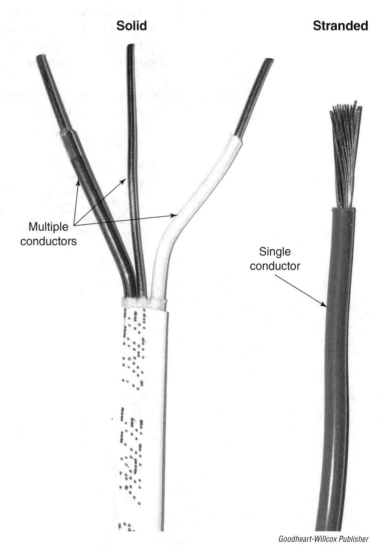

Solid   **Stranded**

Multiple conductors

Single conductor

Goodheart-Willcox Publisher

Figure 5-15. Solid and stranded wire are shown. Solid wire is typical for house wiring.

Summary

- A series circuit has only one path for current flow.
- Series loads are dependent on each other. If one load fails to conduct, the others will fail as well.
- Switches are wired in series with loads.
- In a series circuit, individual load resistances add up to equal total circuit resistance.
- Voltage drops across each load add up to equal the source voltage in a series circuit. This is known as Kirchhoff's voltage law.
- A switch can add additional resistance to the circuit and becomes in series with the load. This negatively affects the load as voltage drops across the switch.
- Most loads are connected in parallel. A parallel circuit contains more than one path for current flow.
- The parallel current paths are called branches.
- Branch currents add up to equal the total current in the circuit. This is called Kirchhoff's current law.
- Voltage is the same across each parallel branch and source.
- Total resistance decreases as more parallel branch resistances are added.
- Total resistance is less than the smallest branch resistance in the circuit.
- Parallel loads are independent from each other. If one fails, the other loads can continue to work.
- A complex circuit is made up of series and parallel portions.
- Wire resistance tables are used to calculate voltage drop due to the wire's resistance.
- The standard for wire size in the United States is the American Wire Gage (AWG).
- The larger the AWG number, the smaller the wire diameter.
- Conductors are made of solid or stranded wires.

Lab 5.1 Series, Parallel, and Complex Circuit Evaluation

This lab will allow you to further investigate the rules of a series, parallel, and complex circuit by measuring current and voltage drops.

Lab Introduction

A series circuit always has one path for current to travel, while a parallel circuit will have more than one path. Complex circuits are a combination of series and parallel. An HVACR technician must recognize and understand the type of circuit that is being analyzed. One type of circuit can change into another type because of a fault.

 This multi-part lab will show how loads are affected by a specific circuit type. Light bulbs are used for the loads, so students can visually observe how the loads are affected. Current, voltage, and resistance must be quantified. Measurements will be taken to prove the theories learned and evaluate the circuit.

 Use the circuit shown below in the lab. Visually observe how the loads in series are affected. They then can be compared to the parallel and combination results in the second and third part of the lab. Note: Do not measure bulb resistance at this time.

Equipment

- Lab board. See Lab 3.2 for complete instructions.
- Digital multimeter
- 1—60, 40, and 15 W bulbs
- 2—additional lamp sockets

Safety Note

Resistance Check on Multimeter

Before connecting power to the circuit, use an ohmmeter to check circuit resistance. Connect the ohmmeter leads to the ac plug prongs. When the switch is in the open position, the resistance should be over-the-limit, as displayed by the letters *OL* on a digital meter. When the switch is closed, the meter will display the load resistance. A reading of 0 or close to 0 Ω with the switch in either position indicates a short circuit, and it is therefore not safe to plug in to a power outlet. The short must be found before proceeding.

Procedure—Part One

1. Add two lamp sockets to the lab board. Use **Figure 5.1-1** for reference. Note that the circuit should be wired in series.
2. Use 60 W, 40 W, and 15 W incandescent bulbs.

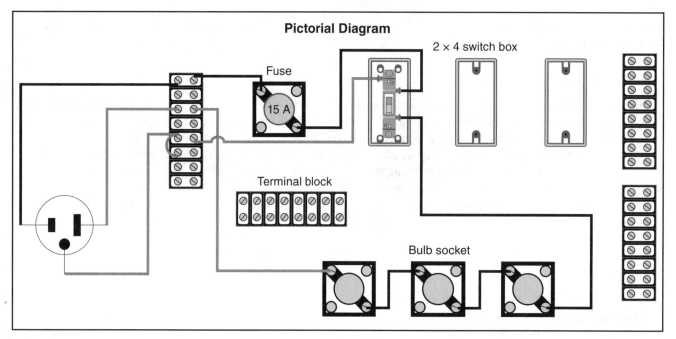

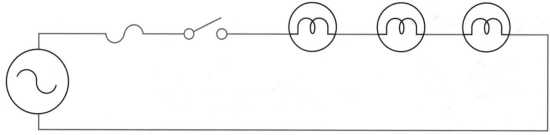

Figure 5.1-1. Series circuit.

Goodheart-Willcox Publisher

Lab Questions

Use your digital multimeter or other electrical instruments to answer the following questions.

1. Are the three bulbs equally illuminated?

2. Explain why the bulbs are or are not equally illuminated.

3. Measure and record the source voltage.

4. Measure and record the voltage across each bulb.

_____

5. Measure the circuit current with an inductive clamp meter at different locations throughout the circuit.

_____

6. Do the measured values support the series circuit rules for voltage?

_____

7. Is Kirchhoff's voltage law supported by the voltage measurements?

_____

8. Does the current measurement support the series circuit rules for current?

_____

Procedure—Part Two

1. Rewire the circuit as shown in **Figure 5.1-2**. Note that the same loads are used as in Lab 5.1 but are now reconfigured in parallel.

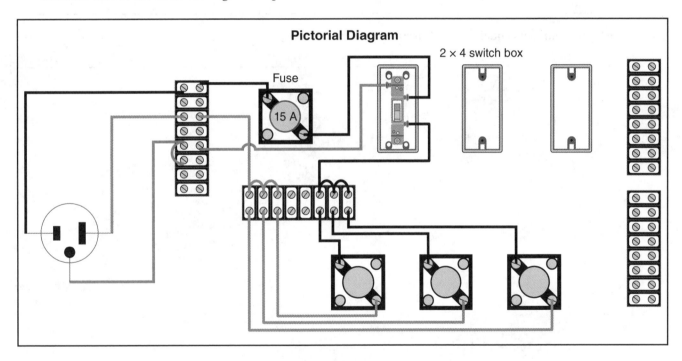

Pictorial Diagram

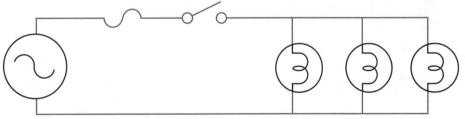

Schematic Diagram

Figure 5.1-2. Parallel circuit.

Goodheart-Willcox Publisher

Lab Questions

Use your digital multimeter or other electrical instruments to answer the following questions.

1. Are the three bulbs equally illuminated?

_____

2. Measure and record the source voltage.

_____

3. Measure and record the voltage across each bulb.

_____

4. Measure the circuit current with an inductive clamp meter for each load and record the values.

_____

5. Measure the total circuit current with an inductive clamp before the first bulb.

_____

6. Add up the two branch currents and record the values.

_____

7. Do the branch currents equal the total circuit current?

_____

Procedure—Part Three

1. Rewire the circuit as shown in **Figure 5.1-3**. Note that the same loads are used but are now reconfigured as a complex circuit.

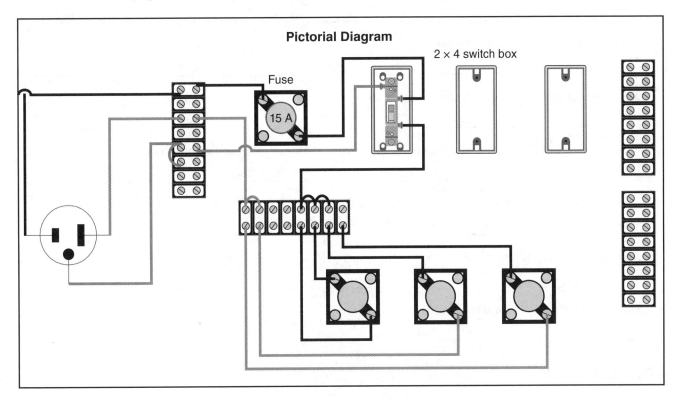

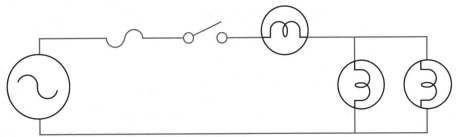

Figure 5.1-3. Complex circuit.

Goodheart-Willcox Publisher

Lab Questions

Use your digital multimeter or other electrical instruments to answer the following questions.

1. Are the three bulbs equally illuminated?

   _____

2. Measure and record the source voltage.

   _____

3. Measure and record the voltage across each bulb.

   _____

4. Measure the circuit current with an inductive clamp meter for each load and record the values.

   _____

5. Measure the total circuit current with an inductive clamp before the first bulb.

   _____

6. Add up the two branch currents and record the values.

   _____

7. Do the branch currents equal the total circuit current?

   _____

Know and Understand

_____ 1. *True or False?* The series circuit is the most commonly used circuit in HVACR.

_____ 2. How does voltage behave in a series circuit?
   A. It has the same value across each load.
   B. It drops across each load.
   C. It is the same throughout the circuit.
   D. It is not a factor in series circuits.

_____ 3. When more resistances are added in a series circuit, the current will _____.
   A. increase
   B. not be affected
   C. be calculated by finding the average value of the resistances
   D. decrease

_____ 4. *True or False?* An interruption of current anywhere in a series circuit will disrupt current flow throughout the entire circuit.

_____ 5. *True or False?* Switch contacts are in series with a load to control its operation.

_____ 6. If three loads are wired in series and one load fails to conduct current, what will happen?
   A. The other two loads will operate normally.
   B. None of loads will operate.
   C. The load next to the failed one will not work but other one will operate.
   D. Only the load closest to the source will operate.

_____ 7. Which statement is *not* true about voltage in a parallel circuit?
   A. Voltage drops across each branch.
   B. Voltage is the same across all branches.
   C. Source and branch voltages are the same.
   D. The sum of branch currents equals the total circuit current.

_____ 8. Which is true about resistance in parallel?
   A. It adds up to a total value.
   B. It is averaged and used to find current.
   C. The total value decreases as more are added in paralleled branches.
   D. Only the largest value is used to find current.

_____ 9. Which statement is *not* true about current in a parallel circuit?
   A. The current is the same in all parts of the circuit.
   B. A branch current depends on the resistance in the branch.
   C. Branch currents add up to equal the total circuit current.
   D. The branch with the largest resistance will have the least amount of current flow.

_____ 10. If three loads are wired in parallel and one fails to conduct current, what will happen?
   A. The other two loads will operate normally.
   B. None of the load will operate.
   C. The load next to the failed load will not operate but the other one will.
   D. The other two loads will operate poorly due to the increased current flow.

_____ 11. Which statement is true about a complex circuit?
   A. It contains more than 10 parallel branches.
   B. The current is the same throughout all parts of the circuit.
   C. It contains more than 10 series loads.
   D. It has parallel and series circuit portions.

_____ 12. The _____ is the standard for wire sizing in the United States.
   A. NEC
   B. NFPA
   C. AWG
   D. ASHRAE

_____ 13. *True or False?* The diameter of a 20 AWG wire is larger than a 14 AWG wire.

_____ 14. The wrong gage wire (diameter too small) was used to supply power to a piece of equipment. Which statement is *not* correct?
   A. Overall wire resistance is too great.
   B. Too much voltage will be dropped.
   C. There will be less voltage supplied to the equipment.
   D. The equipment will operate better.

_____ 15. *True or False?* Solid conductors are preferred when making connections to motors.

Critical Thinking

1. What are the benefits of wiring loads in parallel?

   _____

   _____

   _____

   _____

   _____

2. Describe what happens if two 100 W incandescent bulbs are wired in series? Describe how their illumination, voltage, and current would be affected.

   _____

   _____

   _____

   _____

6 Alternating and Direct Current

Goodheart-Willcox Publisher

Chapter Outline

6.1 Alternating Current
    6.1.1 AC Generators and Alternators
6.2 Direct Current
    6.2.1 Sources of Direct Current
    6.2.2 DC Generators
    6.2.3 Batteries

Additional Reading

Modern Refrigeration and Air Conditioning, 21st edition

12.1 Types of Electricity
12.5 Magnetism
12.6 Electrical Generato

Learning Objectives

After completing this chapter, you will be able to:

- Define alternating current.
- Discuss how current is generated from magnetism.
- Explain the basic operation of an ac generator.
- Explain the basic operation of an alternator.
- List the characteristics of alternating current.
- Define direct current.
- Discuss how direct current is produced.
- List the characteristics of direct current.
- Define some uses for direct current.
- Explain the basic operation of a DC generator.
- Define different types of batteries.

Technical Terms

ac generator	direct current	peak emf
alternating current	frequency	peak value
alternator	generator	root mean square (rms)
battery	Hertz (Hz)	sine wave
current electricity	magnetic flux	static electricity
dc generator	magnetic induction	

Introduction

Current is the flow of negatively charged electrons moving toward positively charged protons in an atom's nucleus. Thus, the direction of current flow is always negative to positive. Current only flows when there is a difference in potential, such as the negative and positive polarities of a battery.

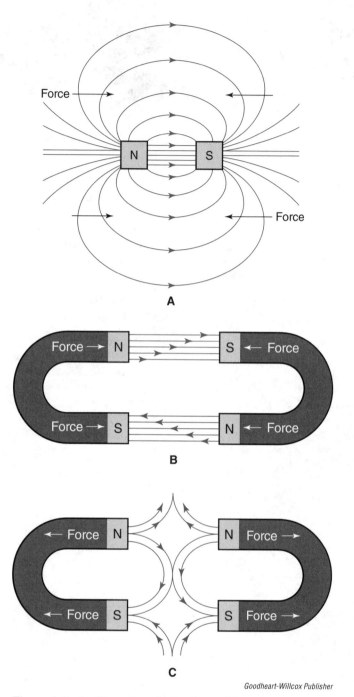

A

B

C

Goodheart-Willcox Publisher

Figure 6-1. A—The magnetic flux lines of force between two unlike magnetic poles. B—Attracting forces since the poles are opposite polarities. C—Repelling forces since the poles are the same polarities.

Both generators and alternators are methods used to generate current. The method is based on whether alternating current or direct current is used. *Alternating current (ac)* changes direction at a periodic rate, while *direct current (dc)* always flows in the same direction. A generator internally produces alternating current, which can be supplied to loads as ac or dc by modifying the output connection configuration. This chapter will introduce both ac and dc as well as the methods used to generate them.

6.1 Alternating Current

Alternating current can be best understood by first examining magnetism. All magnets have a north pole and a south pole. Based on the law of charges, opposite poles of a magnet attract while the same poles repel. This attraction and repelling is caused by lines of force called *magnetic flux*, which connects the north and south poles of a magnet. See **Figure 6-1**. A magnetic field is the volume that the magnetic flux takes up. Similar to a balloon filled with air. When a conductor cuts across magnetic flux lines, the magnetic force frees electrons in the conductor. This results in current flow through the conductor, and electromagnetic force (emf) is generated. Both physical and magnetic forces are required for this process, which is called *magnetic induction*. Magnetic induction is the principle used to generate electricity in many devices, such as a generator.

A *generator* is a device with a wire loop arrangement. The loop rotates within the magnetic poles. See **Figure 6-2**. When the wire loop is in a vertical position, there is no induced current. One revolution of the wire loop from the vertical position completes one full cycle. As the wire loop rotates clockwise toward the south pole, emf increases. When the loop reaches the horizontal position, the maximum, or *peak emf* is generated, **Figure 6-2B**. As the loop rotates further, less emf is generated. As the loop re-approaches the vertical position, it has now moved half a turn, or 180°, and there is zero emf. See **Figure 6-2C**. This means the bottom of the original loop is on top. Continued loop rotation causes the current to flow in the opposite direction compared to the first half turn, **Figure 6-2D**. Peak emf is reached at 270°. As the loop rotates past 270° and onward to 360°, the emf continues to decrease, **Figure 6-2E**.

Start of First Half Cycle

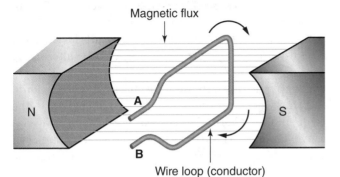

Rotation Angle 0°

First Half Cycle EMF Peak

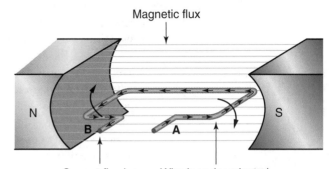

Rotation Angle 90°

Start of Second Half Cycle

Current begins to change direction

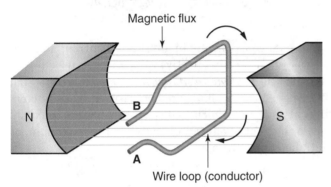

Rotation Angle 180°

Second Half Cycle EMF Peak

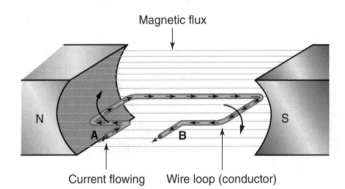

Rotation Angle 270°

End of Cycle

Full cycle complete at 360° and ready to start new cycle at 0°

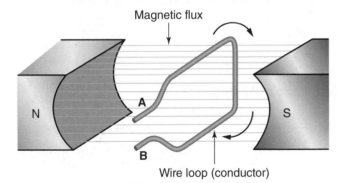

Rotation Angle 360°

Figure 6-2. A—Start of cycle 0° angle and no emf generated. Point A of the loop is on top. B—1/4 revolution 90° and maximum emf and current produced. Note direction of current shown by the arrows. C—1/2 revolution 180° and no emf generated. Point B of the loop is now on top. D—3/4 revolution 270° and maximum emf and current produced. Note current is flowing in the opposite direction. E—Full revolution 360° and no emf produced cycle is complete and a new cycle will begin started at 0°.

One complete turn of the wire loop is considered one cycle of alternating current. This cycle can be graphed to form a *sine wave*. See **Figure 6-3**. An oscilloscope is an electronic test instrument that is used to view and measure waveforms, such as the sine wave. A sine wave shows two characteristics of ac. First, emf rises and falls for each half cycle. Second, the current reverses direction each half cycle. The symbol for alternating current is the sine wave (~).

The number of cycles that occur per section is the alternating current's *frequency*. Frequency is measured in *Hertz (Hz)* where 1 Hz equals one cycle per second. In the United States, the standard frequency for alternating current is 60 Hz, or 60 cycles per second. Other countries use 50 Hz or 55 Hz as their standard. The importance of frequency will be further addressed in Chapter 13, *Electric Motors* and other chapters.

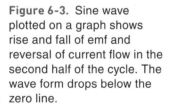

Pro Tip

Power Frequencies

Lower ac power frequencies are used for special applications, such as electric trains and radar equipment. Frequencies lower than 40 Hz can be detected by the human eye, so the power companies maintain a very precise frequency for power distribution. For example, a lamp powered by a frequency less than 40 Hz flickers.

Because voltage and current change frequently in alternating current as it follows the up and down values in a cycle, a steady, effective value must be used to calculate the values. This value is called the *root mean square (rms)*, and it is the actual value of electricity performing work. Root mean square is measured by voltmeters when working with alternating current, so technicians do not often have to calculate it. To calculate the rms voltage or current of alternating current, the maximum, or *peak value*, in the cycle is multiplied by 0.707.

$$rms = Peak \times 0.707$$

If rms is known, the peak can be found by multiplying rms by 1.414.

$$Peak = rms \times 1.414$$

A typical household power receptacle supplies 120 V. The 120 quantity is the rms value of the voltage, which is also the value measured by a voltmeter. The peak value can be measured with an oscilloscope of special function meter.

Using the same equation, the peak value of this receptacle can be found:

$$Peak = 120\,V_{rms} \times 1.414$$
$$= 169.680\,V_{peak}$$

Figure 6-3. Sine wave plotted on a graph shows rise and fall of emf and reversal of current flow in the second half of the cycle. The wave form drops below the zero line.

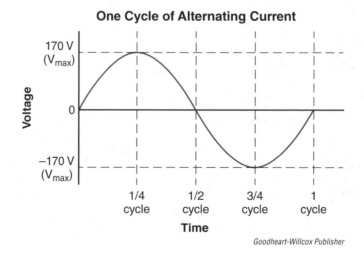

6.1.1 AC Generators and Alternators

Both generators and alternators convert mechanical energy to electrical energy. The two vary, however, on the method in which electrical energy is produced. A generator can generate ac or dc depending on the connection configuration. For *ac generators*, slip rings and brushes are electrically conductive materials that are used to keep the end loop wires from twisting together as it turns within the magnetic flux. Slip rings rotate along with the wire loop. Brushes are spring loaded and pushed up against the slip rings to make proper contact. The brushes are connected to the circuit being powered, and current flows to the circuit. This is shown in **Figure 6-4**. For an ac generator, each wire of the loop is connected to a separate slip ring, causing current to flow. See **Figure 6-5**.

An *alternator* produces the same sine wave as an ac generator. See **Figure 6-6**. The major difference is that its wire loops are fixed and make what is called a stator. The magnet rotates instead. This makes an alternator less expensive as there no slip rings and brushes that wear over time and need to be replaced. The alternator produces alternating current only as opposed to a generator that can be configured to supply ac or dc.

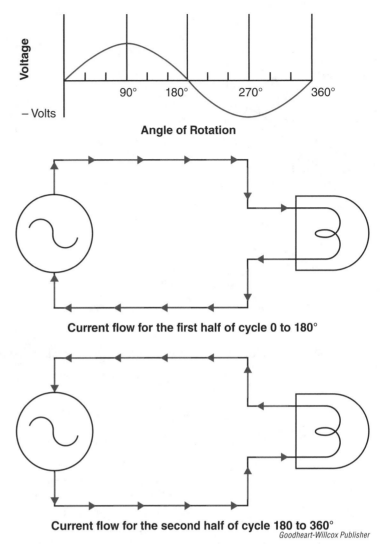

Current flow for the first half of cycle 0 to 180°

Current flow for the second half of cycle 180 to 360°

Goodheart-Willcox Publisher

Figure 6-4. Alternating current flowing in a circuit. Current changes direction in the second part of the cycle as shown in the bottom illustration.

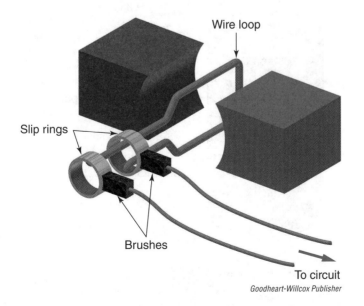

Goodheart-Willcox Publisher

Figure 6-5. A generator. Slip rings and brushes connect alternating current to a circuit.

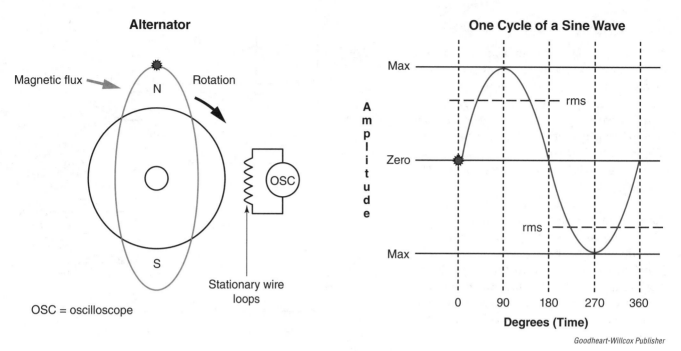

Alternator

Magnetic flux →   Rotation

N

S

OSC = oscilloscope

Stationary wire
loops

One Cycle of a Sine Wave

Max

rms

Zero

rms

Max

0   90   180   270   360

Degrees (Time)

Amplitude

Goodheart-Willcox Publisher

Figure 6-6. An alternator generates ac but uses a round rotating magnet and stationary wire loops called a stator.

6.2 Direct Current

Direct current (dc) is the flow of electrons in one direction, from negative to positive. It is generated by chemical reactions between different materials in a power source, such as a battery. See **Figure 6-7.** In a battery, the electrolyte has a negative charge, and the carbon electrode has a positive charge. Since the polarities of a dc power source stay in the same position once it is in a circuit, direct current flows in the one set direction.

Figure 6-7. Direct current is produced by chemical reaction that cause opposite potentials between different materials.

Positive terminal

+

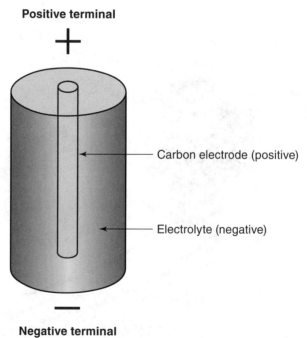

Carbon electrode (positive)

Electrolyte (negative)

—

Negative terminal

Goodheart-Willcox Publisher

Because dc, unlike ac, does not change periodically, the voltage, current, and power remain at a constant level. This can be expressed as a flat line on a graph. See **Figure 6-8**. These values, however, can decrease as a battery discharges. Consider the example of a flashlight that begins to dim as batteries are worn down.

6.2.1 Sources of Direct Current

Two sources for dc are current electricity and static electricity. *Current electricity* provides power to circuit, so current can flow through it. Current electricity can be either ac or dc. Sources of direct current (from current electricity) and their common applications include the following:

- Batteries (dry and wet cells) found in portable tools and test equipment and backup power
- Solar panels found in supplemental power to the power grid
- Photo cells found in sense light intensity
- Rectified alternating current found in circuit boards in electronics
- DC power supplies found in dc motors and controls
- DC generator found in welding and other special equipment

Another source of direct current is *static electricity*, which is the buildup of electrical charge on an object or surface. Static electricity is only dc. An object with a static charge is either negative or positive and attracts the opposite charge from another object. Most people have experienced static electricity. If you walk across a carpet in the winter when relative humidity is low, your body can absorb electrons released by the carpet. The negative charge is stored until you come into contact with a more positive surface. Then electrons are released to the positive surface. The rapid release of electrons to the positive surface causes a mild, or sometimes intense, shock sensation. This sensation can provide a visible or audible spark. The opposing surfaces can be two people if they contain opposite charges.

Static electricity can destroy sensitive electronic components on circuit boards. The troubleshooting sections ahead will cover preventive methods for the HVACR technician to follow.

6.2.3 DC Generator

A *dc generator* uses a special slip ring that is split in half. The gaps between the two halves are insulated. See **Figure 6-9**. It is called a commutator. This allows the current to flow in one direction only.

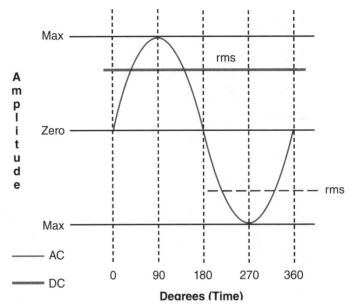

Goodheart-Willcox Publisher

Figure 6-8. Direct current plots as a straight line since emf and current are constant and current flows in one direct as opposed to alternating current. Emf and current amplitude vary throughout the cycle and current reverses direction.

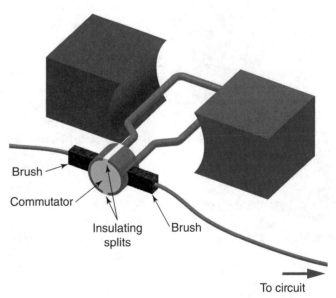

Goodheart-Willcox Publisher

Figure 6-9. The commutator, a split slip ring insulated in the splits, ensures current flow in one direction into a circuit and back into the generator on the opposite wire to complete the circuit.

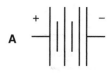

A

One cell

B

or   9 V

Goodheart-Willcox Publisher

Figure 6-10. A—Battery symbol shown with positive and negative signs. The positive and negative signs are sometimes omitted. B—One pair of long and short lines represent one cell. C—9V battery shown using multiple cell convention and the more practical method of listing battery voltage.

6.2.4 Batteries

A **battery** is a type of dc power source that consists of one or more cells. It is either a wet cell, that is rechargeable and contains liquid, or a dry cell. A battery has both a positive and negative terminal. The positive terminal of a battery is labeled with a plus sign (+), and the negative with a minus sign (–). The symbol for a battery is shown in **Figure 6-10A**.

When represented on a schematic, the plus and minus labels are sometimes omitted. Notice that the plus sign is on the long line side, and the minus sign is on the short line. The long and short line pair represent a single battery cell, **Figure 6-10B**.

The standard voltage for a dry cell is 1.5 V and 1.2 V for a rechargeable cell. A nonrechargeable D-cell battery is only one cell and provides 1.5 V. A nonrechargeable 9 V battery is made up of six 1.5 V cells. **Figure 6-10C** shows how it is more practical to write the battery voltage value than to use six cell symbols. This is the common practice. A wet cell provides 2 V and is used in automotive, marine, and other equipment. A car battery has six cells.

A variant of a wet cell is the gel cell, which is also 2 V. The gel cell battery is ideal for back-up power in various types of equipment. Unlike wet cells, gel cell batteries do not leak if tipped over.

Summary

- Alternating current is produced by magnetic induction. It reverses direction every half cycle.
- Voltage and current rise and fall during the ac cycle.
- The sine wave graphically represents the generation of ac during one cycle.
- Root mean square (rms) is the actual value of electricity performing work. A less technical expression for rms is effective voltage or current.
- The number of cycles that occur per section is the alternating current's frequency.
- Frequency is measured in Hertz (Hz). The standard frequency in the United States is 60 Hz.
- One cycle of a sine wave per second equals 1 Hz.
- An ac generator produces alternating current through the use of slip rings and brushes.
- Direct current is produced by chemical reactions.
- Direct current flows in only one direction from negative to positive.
- A dc generator produces direct current through the use of one special slip ring and brushes called a commutator.
- Static electricity is caused by stored charges of opposite polarities on two different surfaces that come in contact with each other. Electrons rapidly move to the positively charged surface.
- Batteries are a main power source of direct current for portable equipment.

Lab 6.1 Evaluate the Effects of AC on a Load

This lab is broken down into three parts and will allow you to observe alternating and direct current more closely. Assemble the circuit provided in each lab part below. Then, answer the guiding Lab Questions to master your understanding.

Lab Introduction

An HVACR technician must be knowledgeable about both direct and alternating current. System controls are increasingly using digital controls. Motor technology is moving away from traditional pure ac motors to high efficiency motors that require knowledge of rectified ac into dc.

The purpose of the lab is to demonstrate how ac and dc differ as each supplies power to a simple incandescent 60 W light bulb. For safety, the lab will use a 24 Vac power source. The 24 Vac will be rectified by a full-wave bridge rectifier. In the first lab, ac will be applied to a 60 W incandescent bulb to take voltage and current measurement and observe the illumination level. **Figure 6.1-1** shows an oscilloscope view of the voltage applied to the bulb. The bulb will then be powered by dc to take voltage and current measurement sand observe the illumination level. **Figure 6.1-2** shows an oscilloscope view of unfiltered dc applied to the load. **Figure 6.1-3** shows an oscilloscope view of the voltage applied to the bulb and includes a capacitor to improve the power factor.

Although, rectifiers and capacitors theory and labs will be covered in future chapters, it is necessary to use these devices now to evaluate ac and dc theory.

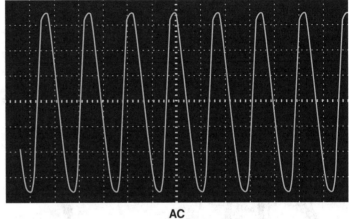

AC

Figure 6.1-1.   Goodheart-Willcox Publisher

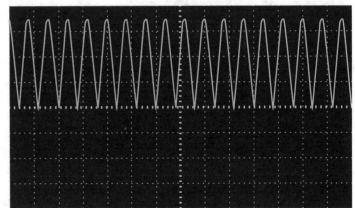

DC without filter capacitor

Figure 6.1-2.   Goodheart-Willcox Publisher

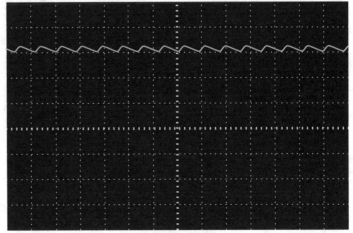

DC with filter capacitor

Figure 6.1-3.   Goodheart-Willcox Publisher

Safety Note

Danger of DC

Direct current is falsely considered not harmful by many people. This is because dc generally uses low voltage batteries—1.5 V through 12 V. However, higher dc voltages can be more dangerous than ac. Always follow safety protocol regardless of types or levels of electricity.

Equipment

- Lab board. See Lab 3.2 for complete instructions.
- 1—24 V step-down transformer
- 1—60 W incandescent bulb
- 1—Full-wave bridge IC. See **Figure 6.1-4**.
- 470 μF, 50 V electrolytic capacitor. See **Figure 6.1-4**.

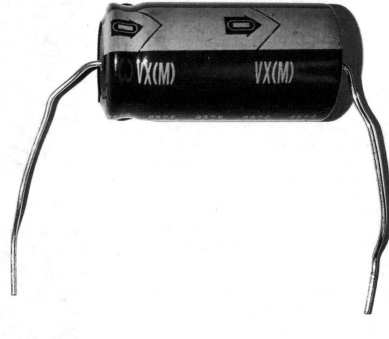

Figure 6.1-4.

Goodheart-Willcox Publisher

Procedure—Part One

1. Assemble the circuit per **Figure 6.1-5**.
2. Have instructor verify circuit before plugging in.
3. Take measurements based on the Lab Questions provided below.

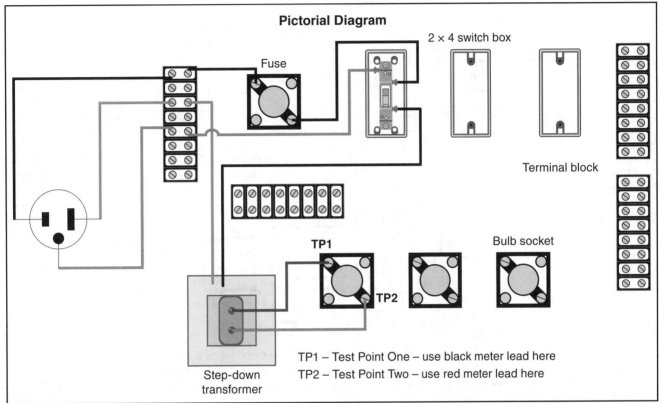

Figure 6.1-5.

Lab Questions

Use your digital multimeter or other electrical instruments to answer the following questions.

1. Measure and record the ac voltage across the bulb.

   _____

2. Measure and record the current with an inductive clamp.

   _____

3. Observe bulb illumination.

   _____

4. Calculate power $P = I \times E$. Record answer.

   _____

Procedure—Part Two

1. Rewire the circuit as shown in **Figure 6.1-6**.
2. Use the Lab Questions below to guide your evaluation.

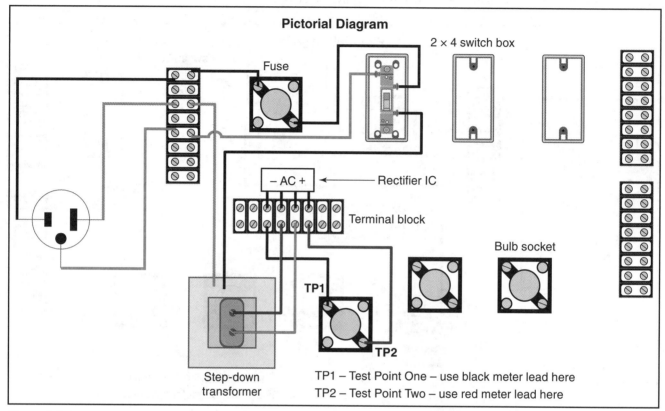

Pictorial Diagram

Fuse

2 × 4 switch box

– AC + → Rectifier IC

Terminal block

Bulb socket

TP1

TP2

Step-down transformer

TP1 – Test Point One – use black meter lead here
TP2 – Test Point Two – use red meter lead here

Figure 6.1-6.

Lab Questions

Use your digital multimeter or other electrical instruments to answer the following questions.

1. Measure and record the dc voltage across the bulb.

2. Measure and record the dc current with an inductive clamp. (Note: Meter must have dc milliamp capability. If meter is not available, do only Steps 1 and 3.)

3. Observe bulb illumination and oscilloscope view. See **Figure 6.1-2**.

4. Calculate power $P = I \times E$. Record answer.

Procedure—Part Three

Procedure—Part Three

1. Rewire the circuit as shown in **Figure 6.1-7.**
2. Use the Lab Questions below to guide your evaluation.

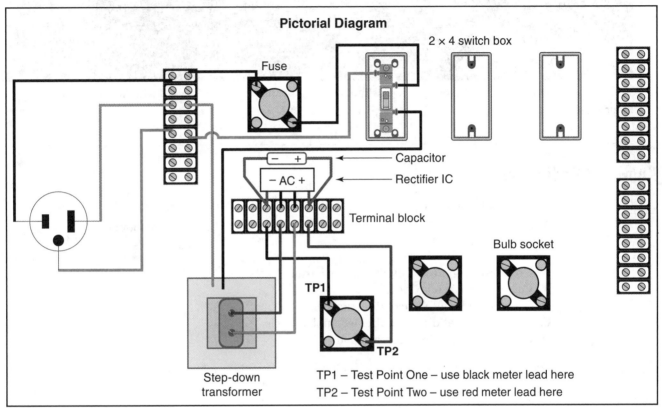

Pictorial Diagram

2 × 4 switch box

Fuse

Capacitor

Rectifier IC

Terminal block

Bulb socket

TP1

TP2

Step-down
transformer

TP1 – Test Point One – use black meter lead here
TP2 – Test Point Two – use red meter lead here

Figure 6.1-7.

Goodheart-Willcox Publisher

Lab Questions

Use your digital multimeter or other electrical instruments to answer the following questions.

1. Measure and record the dc voltage across the bulb.

   _____

2. Measure and record the dc current with an inductive clamp. (Note: Meter must have dc milliamp capability.
   If meter is not available, do only Steps 1 and 3.)

   _____

3. Observe bulb illumination and oscilloscope view. See **Figure 6.1-3.**

   _____

4. Calculate power $P = I \times E$. Record answer.

   _____

5. Which power source results in more power to the load? Use visual observation, physical measurements, and **Figure 6.1-8** to evaluate the question.

Comparison of Waveforms

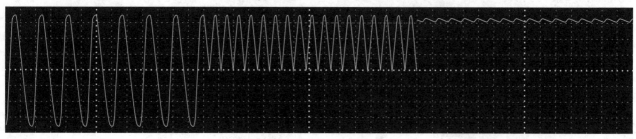

| AC | DC without filter capacitor | DC with filter capacitor |

Figure 6.1-8.

Know and Understand

_____ 1. Alternating current is generated by _____.
  A. chemical reaction
  B. batteries
  C. magnetic induction
  D. solar panels

_____ 2. Which answer best describes alternating current?
  A. Current flows in one direction.
  B. Current randomly changes direction.
  C. Current changes direction after a complete cycle.
  D. Current changes direction each half cycle.

_____ 3. *True or False?* All generators produce alternating current.

_____ 4. Which is *not* true of a sine wave?
  A. A graphical representation of one cycle of generated alternating current.
  B. It represents direct current.
  C. Is used as a symbol for alternating current.
  D. It starts at 0° and ends at 360°.

_____ 5. _____ is the standard ac frequency in the US.
  A. 50 Hz
  B. 60 Hz
  C. 40 Hz
  D. 55 Hz

_____ 6. Since the generated emf rises and falls during the ac cycle, which value describes the actual work done?
  A. Rms.
  B. Peak emf.
  C. Maximum emf.
  D. All of the above.

_____ 7. If peak voltage is known, what is the equation for effective voltage?
  A. Rms = 1 / frequency
  B. Rms = Peak × 1.414
  C. Effective Voltage = Peak × 1.414
  D. Rms = Peak × 0.707

_____ 8. What is the approximate peak voltage of a 120 V$_{rms}$ power source?
  A. 170 V
  B. 90 V
  C. 120 V
  D. 240 V

_____ 9. *True or False?* The peak voltage is consistent throughout the entire cycle of a sine wave.

_____ 10. *True or False?* Alternators produce alternating current by turning a round magnet around a fixed coil of wires.

_____ 11. Which statement is *not* true about direct current?
  A. Current oscillates back and forth.
  B. Current flows in only one direction.
  C. Direct current is generation by chemical reaction.
  D. The amplitude of direct current is constant.

_____ 12. Which of the following is *not* considered a source of current electricity?
  A. Battery.
  B. Rectifier.
  C. Solar Panel.
  D. Rubbing two different types of surfaces together.

_____ 13. *True or False?* Alternators are used to produce direct current without the use of rectifiers.

_____ 14. *True or False?* Alternating current can be rectified to direct current.

_____ 15. *True or False?* Direct current is generated by magnetic induction.

_____ 16. Which of the following best describes the graphical representation of direct current?
  A. Sine wave.
  B. Half of a sine wave.
  C. Straight line.
  D. Cannot be represented graphically.

_____ 17. What is the standard frequency of direct current?
  A. 60 Hertz
  B. 55 Hertz
  C. 50 Hertz
  D. No frequency. It is constant.

_____ 18. A dc generator delivers power to a circuit by ____.
  A. two distinct slip rings and brushes
  B. one continuous slip ring and two brushes
  C. a commutator
  D. any of the above will work

_____ 19. Which would not produce static electricity?
  A. Walking across a carpet when humidity is low.
  B. Rubbing a latex balloon on cloth.
  C. A solar panel.
  D. Accumulated opposite charges on two different surfaces.

_____ 20. What is a nonrechargeable battery dry cell voltage?
  A. 1.2 V
  B. 2.0 V
  C. 1.0 V
  D. 1.5 V

Critical Thinking

1. Generators and alternators require mechanical energy to turn the rotor that is made of loops of wires called winding or a round magnet. What are the sources of mechanical energy?

   _____

   _____

   _____

2. Describe what would happen as the brushes of a generator wear down?

   _____

   _____

   _____

7 Electromagnetic Devices

Mehmet Cetin/Shutterstock.com

Learning Objectives

After completing this chapter, you will be able to:

- Define electromagnetism and electromagnets.
- Discuss the effects of direct current applied to a coil.
- Evaluate the effects of alternating current applied to a coil.
- Explain the operation of a solenoid, relay, and contactor.
- Define contact configuration.
- Define the difference between relays and contactors.
- Discuss how relays and contactors are used to control loads.
- Define symbols used to represent electromagnetic devices.
- Explain the operation of a transformer.
- Define transformer VA rating.

Technical Terms

armature	inductive load	solenoid
back emf	inductive reactance (X_L)	step-down transformer
coil	iron core	step-up transformer
contactor	primary coil winding	transformer
contacts	relay	turns-ratio
electromagnet	resistive load	volt-ampere (VA) rating
electromagnetic induction	secondary winding coil	winding
electromagnetism	self-induction	

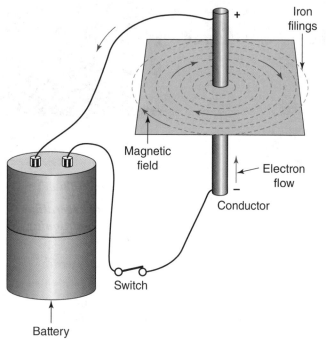

Goodheart-Willcox Publisher

Figure 7-1. Current flows from negative at the bottom end of the conductor to the positive top end. The magnetic flux, shown by the iron fillings, radiates outward from the wire and rotates in a clockwise direction around the wire.

Introduction

In the previous chapter, we studied magnetism and how it can be used to generate electric current. This chapter introduces *electromagnetism*, which is a type of magnetism generated by electric current. This current produces a magnetic field around a conductor. Wrapping the conductor around an iron core causes the core to become an electromagnet as current flows through the conductor. An *electromagnet* is a temporary magnet, controlled by turning power on and off. This feature makes electromagnets ideal for controlling the operation of electrical and mechanical loads.

Electromagnets are used to make electromagnetic devices in all trade industries. Common devices that use electromagnets in the HVACR field are solenoids, relays, contactors, transformers, sensors, and motors. An HVACR technician must understand both the electrical and mechanical aspects of electromechanical devices, which will be covered in this chapter. These concepts are critical in troubleshooting systems, especially intermittent type failures.

7.1 Electromagnetism

Current flowing through a conductor, such as a wire, induces a magnetic field around it. A magnetic field is made of radiating flux lines, which radiate outward from the conductor. The strength of a magnetic field is determined by the amount of electric current—more current induces a stronger magnetic field. See **Figure 7-1**. Although a magnetic field is invisible, iron fillings can be used to show the rotational pattern of the field.

> **Pro Tip**
>
> ## Electromagnetic Interference
>
> A radiated magnetic field can be heard. Plug in a toaster oven or any other high-current-draw appliance to a wall outlet. Tune a battery-operated radio to a weak AM station. While the oven is on, place the radio near the power cord or even along the wall. There will be a distinct whining and buzzing sound. The AM tuner antenna picks up ac frequency and in turn produces the sound. The noise produced is called electromagnetic interference. This interference can be disruptive in communication systems used in some HVACR systems. This theory will be used in later chapters to discuss sensors, the multimeter inductive clamp, and signal interference.

For a magnetic field to be used in practical applications, the field must be strengthened. Twisting or turning the wire to form a loop increases the field onefold. Likewise, ten loops increases the field ten times. These current-carrying wires designed to produce an effective magnetic field are called *coils*. Wire coils may contain hundreds of loops. To further strengthen the field, the wire can be wrapped around an *iron core*, which concentrates the magnetic flux around the coil and produces the electromagnet. Laminated cores and soft iron produce the strongest fields. See **Figure 7-2**. The loops around the core must be tightly wound, so that the field is concentrated with a given area for greater effectiveness.

Solid core

Thin iron sheet

Multiple thin
iron sheets

Laminated core
sheets secured
together

Goodheart-Willcox Publisher

Special polymers are used to insulate the wires. The wires are known as magnet wire and have great insulating properties. See **Figure 7-3.** Only a thin layer is required, which allows the wires to be close together. It is also clear in color, which causes the wire to appear uninsulated. Some manufacturers use colored dyes on their magnet wire.

7.1.1 Electromagnetic Devices

Like permanent magnets, electromagnets have north and south poles. A magnetic north pole is produced at the positive electrical polarity, and a south pole is produced at the negative electrical polarity. Current flows in the direction from negative to positive. See **Figure 7-4.** By changing the current flow direction, the magnetic poles also change.

Gabriel Pacce/Shutterstock.com

Figure 7-3. Clear insulating material on magnetic wire.

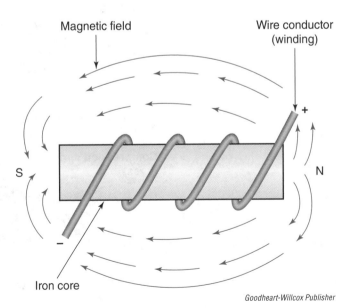

Goodheart-Willcox Publisher

Figure 7-4. Magnetic north and south poles form due to current flow direction.

The magnet flux also changes direction. This understanding is extremely valuable when studying motors in future chapters.

These magnetic poles attract ferrous (iron) metals in the same way permanent magnets attract them. This attraction causes a metal object to move toward the electromagnet's pole. Electromagnets are controlled by turning power on or off. The coil of an electromagnet is considered energized, or turned on, when current is flowing through it. The coil is de-energized, or turned off, when the current is stopped.

Electromagnets vary based on their configuration. **Figure 7-5** illustrates an electromagnet arrangement where the coil is wrapped around a hollow iron core. Inside the hollow core is a plunger that can move freely within the core. The left side of the plunger rests on a spring. The opposite end of the spring is attached to an unmovable surface. In **Figure 7-5A,** the coil is de-energized. Notice that the plunger is in the fully extended position. **Figure 7-5B** shows the coil now energized. The magnetic field generated by the coil current attracts the plunger toward the spring. The magnetic field force applied to the plunger then compresses the spring while retracting the plunger. A device connected to the open end of the plunger is displaced, or moved, a certain distance. **Figure 7-5C** shows the coil de-energized. When the current is stopped, the compression spring releases its stored energy and pushes the plunger to the extended position. The spring provides a counterforce. If a counterforce did not exist, then the plunger would remain in the retracted position when the coil is de-energized.

Figure 7-5. A—Coil is not energized and plunger is fully extended. B—Coil is energized and plunger is retracted and the compression spring is compressed. C—Coil is de-energized and the compression spring returns the plunger to its extended position.

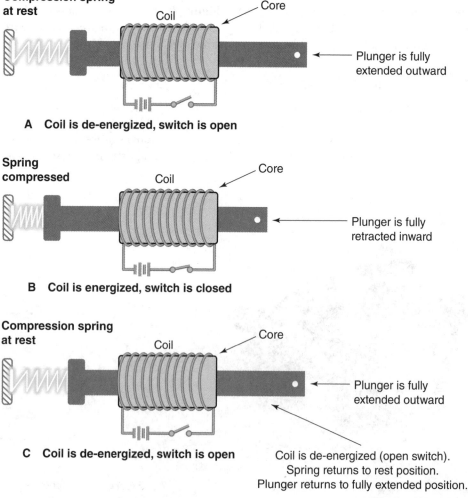

Compression spring at rest

Coil

Core

Plunger is fully extended outward

A Coil is de-energized, switch is open

Spring compressed

Coil

Core

Plunger is fully retracted inward

B Coil is energized, switch is closed

Compression spring at rest

Coil

Core

Plunger is fully extended outward

C Coil is de-energized, switch is open

Coil is de-energized (open switch). Spring returns to rest position. Plunger returns to fully extended position.

Goodheart-Willcox Publisher

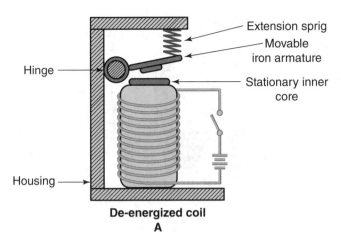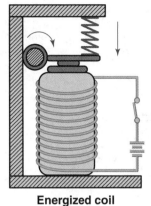

Extension sprig

Movable
iron armature

Hinge

Stationary inner
core

Housing

De-energized coil
A

Energized coil
B

Goodheart-Willcox Publisher

Figure 7-6. A—Coil is de-energized and swinging armature is held up by an extension spring. B—Coil is energized and armature is attracted to the magnetized inner core. The extension spring is stretched.

Any device connected to the plunger then returns to its original position. This configuration is typically used in solenoid valves.

A second arrangement uses a movable plate called an ***armature***. An armature is hinged on one side, which allows it to rotate. **Figure 7-6** shows this type of configuration. An extension spring holds the armature away from the core when the coil is de-energized. When the coil is energized, the core becomes magnetized and pulls in the armature. This causes the spring to stretch. When the coil is de-energized, the spring retracts and pulls the armature away from the core. This arrangement is used for relays.

In the third arrangement, the armature slides along a fixed track, **Figure 7-7**. When the coil is de-energized, the extension spring holds the armature away from the core. When the coil is energized, the magnetic core pulls in the armature, stretching the spring. Once the coil is de-energized again, the spring pulls the armature away from the core. This arrangement is used for contactors and some relays.

A modified version of the third electromagnet configuration is shown in **Figure 7-8**. Here the spring counterforce is replaced by gravity. The position of the device is critical for the correct operation. When the coil is energized, the armature is pulled up by the magnetic force. Then when de-energized, the force of gravity pulls the armature back down. This arrangement is used for current magnetic relays used to start small hermetic compressor motors. Note that the other arrangements are not gravity sensitive.

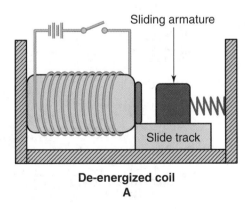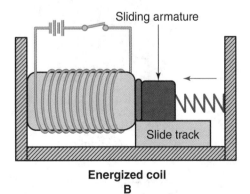

Sliding armature

Sliding armature

Slide track

Slide track

De-energized coil
A

Energized coil
B

Goodheart-Willcox Publisher

Figure 7-7. A—Coil is de-energized and the extension spring pulls the armature away from the core. B—Coil is energized and the armature is attracted to the core. The extension spring is stretched.

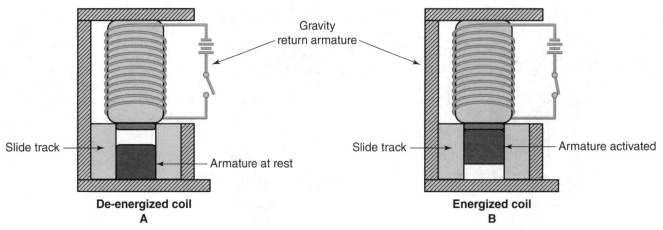

Figure 7-8. A—Coil is de-energized and gravity pulls the armature away from the core. B—Coil is energized and the magnetic core attracts the armature, thus overcoming the force of gravity.

7.1.2 Effects of DC Applied to Coils

Direct current supplies current in one direction at a constant level. As a result, providing dc to a conductor causes the magnetic field to stay at a constant level. The magnetic poles do not alternate location.

When direct current is applied to coils and then stopped, the collapsing magnetic field induces a reverse current flow into the coil circuit. See **Figure 7-9** as an example. A bulb is wired in parallel to the coil with a closed switch, providing a normal level of illumination. This means the coil is energized, thus the magnetic field is at its maximum strength, **Figure 7-9A**. When the switch is opened, current stops, and the magnetic field collapses. Any change in a magnetic field's strength induces current in the conductor. The collapsing field generates a **back emf** that induces a reverse flowing current, **Figure 7-9B**. The process of back emf, also known as *counter emf,* is called **self-induction**. The reverse current continues to flow through the bulb until the magnetic energy is dissipated. Since energy cannot be destroyed, the stored magnetic energy in the coil is converted to electrical energy and finally to heat and light energies.

Figure 7-9. A—Direct current causes a constant induced magnetic field around the coil. B—When the coil is de-energized, the magnetic energy is released back into coil as back emf causing current to flow in the opposite direction through to the light bulb.

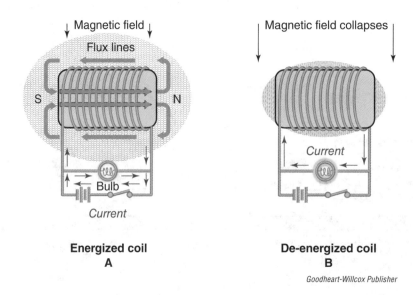

Essential Electrical Skills for HVACR: Theory and Labs

Energy stored in a dc-powered coil can be dangerous even with relatively low voltage if a large coil is used. The coil stores a magnetic field that, when collapsed, generates a back emf, which can be greater than the supplied emf. **Figure 7-10A** shows back emf forcing current across the open contacts. Current leaping through the air gap between the contacts is called an arc. The amplitude of the back emf can be hundreds or thousands of volts. To prevent arcing, a resistor can be added in parallel to the coil, **Figure 7-10B**. The reverse current then flows through the resistor and is converted to heat energy. This is a common method for dc relay coils found in circuit boards and other applications. Coils are specifically made to supply high voltage spark to ignite gas-fired furnaces and boilers. This principle is applied to internal combustion engine spark plugs.

7.1.3 Effects of AC Applied to Coils

Alternating current changes in amplitude and direction every half cycle. A magnetic field strengthens and weakens along with the rise and fall of current. As current level rises to its peak at 90°, so does the associated magnetic field. Between 90° and 180°, the current and magnetic field fall toward 0. At this point, a back emf is generated and induces a reverse flowing current. This current opposes the supplied forward-moving current, **Figure 7-11**. The reverse current serves as resistance to the supplied current. The same effect occurs in the second half of the sine wave.

This opposing resistance to current changing flow is called *inductive reactance* (X_L). It only occurs in ac circuits with coils because dc has no cycle while power is on. Self-induction in dc circuits occurs only when the power is shut off. X_L is called hidden resistance because it cannot be measured with a standard ohmmeter. Detection of X_L is used in some control circuits covered in later chapters.

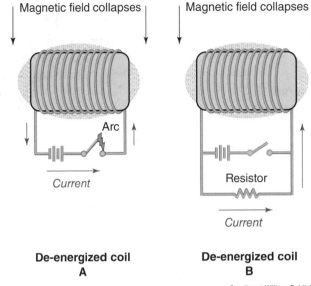

De-energized coil
A

De-energized coil
B

Goodheart-Willcox Publisher

Figure 7-10. A—A coil can produce a high voltage back emf and produce a dangerous arc across switch contacts. B—The solution is to connect a resistor in parallel with the coil. The back emf energy can now be dissipated by the resistor and prevent the arc across the contacts. The resistor is the path of least resistance for the back emf current produced.

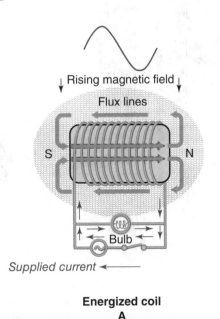

Energized coil
A

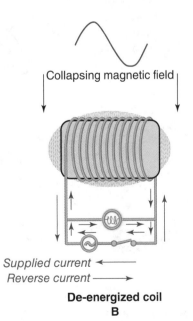

De-energized coil
B

Goodheart-Willcox Publisher

Figure 7-11. A—During the first quarter of a sine wave, the coil is building up a magnetic field and reaches a maximum at the peak voltage. B—During the second quarter of the sine wave, the magnetic field collapses and generates a back emf.

Figure 7-12. Coil symbols.

7.2 Solenoids

A **solenoid** is an electromagnetic device that is used in many trade industries and varies based on its intended purpose. In HVACR, a solenoid is mainly used to open or close a valve to control fluid flow. It can control the opening and closing of dampers in duct systems. Common coil symbols are shown in **Figure 7-12**.

Solenoid valve complexity depends on system pressures and temperatures. A valve assembly is connected to the open end of the plunger to control fluid flow. Solenoids are either normally closed or normally open. A normally closed solenoid must be energized to open the valve and allow fluid flow. A normally open solenoid allows fluid when the coil is not energized.

There are many different types of solenoids. These include two-way, three-way, four-way, water line, and duct damper solenoids. See **Figure 7-13**. Solenoids are designed specifically for ac or dc voltages, typically ranging from 24 V to 240 V.

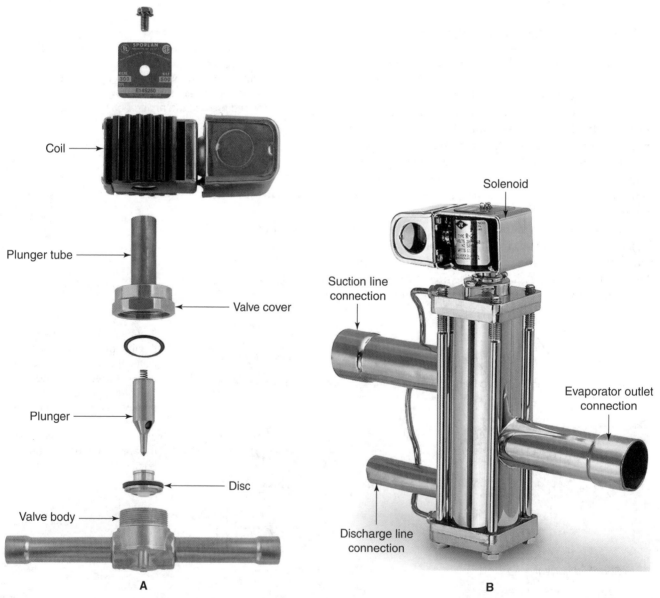

Figure 7-13. A—Exploded view of a solenoid valve used in HVACR. B—A three-way solenoid. Allows flow between two ports when de-energized and when energized allows flow between one original port and a third port. *(Continued)*

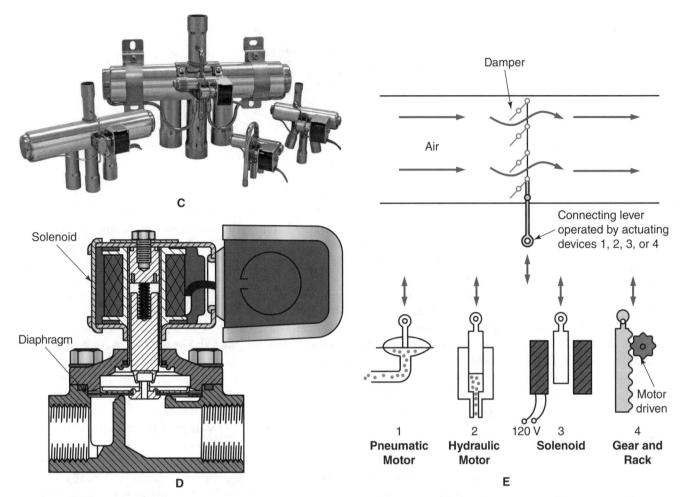

Figure 7-13 (Continued). C—A four-way solenoid to reverse the flow or refrigerant in heat pumps. D— Solenoid used for water lines. E— Solenoid used to control a duct damper.

Solenoid coils are also rated by wattage and temperature. The wattage depends on the minimum and maximum operating pressure differential across the valve. The temperature rating indicates the maximum ambient temperature. The coil must be able to dissipate the heat it produces.

The coil assembly is usually separate from the valve body assembly. This allows the technician to replace only the coil instead of the entire assembly. An ac coil should not be energized until it is attached to the valve assembly. The coil would only draw inrush current, and it would short out and burn up.

Pro Tip

Component Specifications

Use the exact part number, or equivalent, when changing components. The component specifications must match. For example, if a 12 W coil is being replaced, do not use a 10 W coil.

7.2.1 Solenoid Failure

There are several problems that can occur with solenoids, especially those that involve excess current draw. It is critical to first understand proper operation before these problems can be identified and repaired.

Figure 7-14. Inrush current exists when the solenoid coil is energized and the plunger is extended.

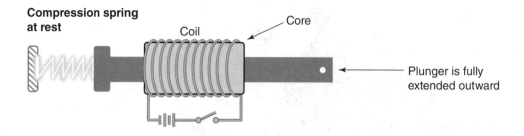

Compression spring at rest

Coil

Core

Plunger is fully extended outward

For ac-driven solenoid coils, current is described as inrush and holding. Inrush current is drawn by the coil when it is first energized and the plunger is extended. When this occurs, the magnetic field around the hollow core has minimum strength. Thus, inductive reactance is at its lowest, and current flow is high. See **Figure 7-14**. As the plunger moves to its retracted position, the magnetic field inside the core increases, and X_L increases to its maximum value. The current through the coil, now considered holding current, decreases. See **Figure 7-15**. The plunger movement is not instantaneous, so the inrush current lasts a fraction of a second.

There are various mechanical problems that can lead to the electrical failure of the coil. These include residue buildup in the valve seat or restriction that limits plunger movement. If the plunger cannot fully retract into the core, then excess current is drawn because X_L does not reach maximum resistance. The current generates excess heat that then melts the magnetic wire insulation. The damaged insulation causes additional current draw and heat until the coil is destroyed.

Consider an example where a coil is formed with 100′ of magnetic wire. When excess current is drawn and the heat melts the insulation of the coil, the actual wire length is reduced. Bare wires then make contact. This causes decreased wire resistance and increased current. Another mechanical problem is when the plunger does not fully retract due to higher than normal system pressures that are beyond the valve's pressure rating. For example, a high-condensing pressure caused by a dirty condenser results in high-liquid-refrigerant pressure pushing against the valve assembly. This restricts plunger retraction into the core. Since the plunger is not fully retracted, a higher-than-normal current is drawn.

> **Pro Tip**
>
> ### Component Specifications
>
> Operating outside of specification can cause component and system damage. This can also void warranty. Be sure to read a manufacturer's installation instructions and component specifications.

Another cause of coil failure is the applied power. Voltage above the maximum voltage rating of a solenoid coil produces excess current and overheats the coil.

Figure 7-15. Holding current occurs when the plunger of a solenoid is fully extended.

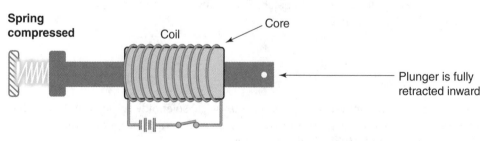

Spring compressed

Coil

Core

Plunger is fully retracted inward

Typically, the coil can operate normally with 10% above voltage supply. As we know from Ohm's law, high voltage increases current flow, and low voltage reduces current and magnetic field strength. The weakened magnetic field does not hold the plunger in place. The low voltage may not be able to overcome inertia and pull in the armature at all. The applied voltage should not be lower than about 80% of the rated voltage. This minimum operating voltage is also called holding voltage. Operating at the extreme ends of the tolerance for long periods causes higher operating temperature that shortens coil life. Operating at the low end of tolerance compounded by plunger restriction leads to early failure. Always refer to a manufacturer's specifications for power requirements.

7.2.2 Checking a Solenoid Coil

There are many factors in determining a wire's resistance. A coil's resistance is based on magnetic strength and operating voltage. A solenoid coil is made with one continuous thin wire that can be over 100′ in length and therefore has a certain amount of resistance. Since a coil's resistance is not easily found, the technician may not know the specific value. A good coil must have some measurable resistance because the wire is continuous. A broken wire shows infinite resistance or *OL* on digital meters.

There are three possible results when checking a coil with an ohmmeter. The following three ohmmeter readings are used to check the coil condition:

1. A measurable resistance is displayed (for example: between 10 Ω to 60 Ω). This indicates that the coil should be good but requires powering up to check operation and current draw. See **Figure 7-16**.
2. The display shows *OL* or infinite resistance. This means the coil is open, the wire is broken, or there is no electrical continuity. See **Figure 7-17**.

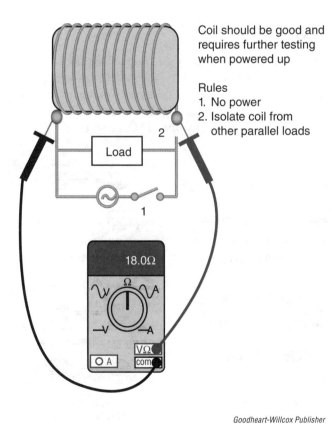

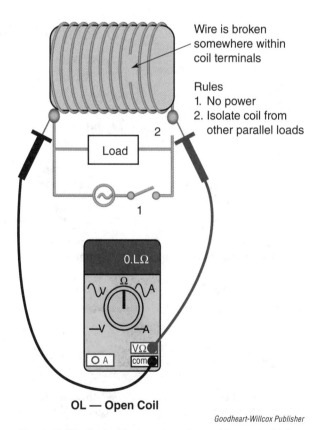

Figure 7-16. The coil resistance is between 10 Ω and 60 Ω, which indicates a good and functioning coil.

Figure 7-17. *OL* is displayed on the multimeter, which means the wire is broken.

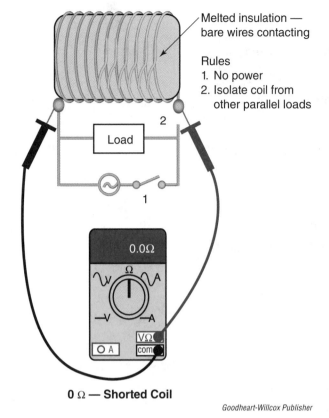

Melted insulation — bare wires contacting

Rules
1. No power
2. Isolate coil from other parallel loads

0.0Ω

0 Ω — Shorted Coil

Goodheart-Willcox Publisher

Figure 7-18. The display of 0 Ω on the multimeter indicates the coil shorted.

3. Zero or close to zero is displayed. The coil is shorted. This means the wire insulation has melted away resulting in a short length of wire with insignificant resistance. This coil should not be powered up, as it will cause a direct short. See **Figure 7-18**.

> **Pro Tip**
>
> ## Ohmmeter Tests for Coils
>
> Note that the above ohmmeter test is used for all coils used in solenoids, relays, contactors, transformers, and motor windings. All of these components are made from one or more coils of wire.

When troubleshooting a coil in a live circuit, the first step is to check for its correct applied voltage. If voltage is present and the coil is not energized, turn off power and check for faulty connections. If connections are good, then disconnect one end of the coil, and then check resistance. An *OL* reading confirms that the coil is open and must be replaced. A shorted coil trips a circuit breaker.

When checking the coil resistance, it could be 0 Ω but can also be *OL*. The intense heat produced by the short circuit can cause the wire to break. Before replacing the solenoid coil, a technician must investigate why the part failed—otherwise the new part may also fail.

7.3 Relays and Contactors

A *relay* is an electrically operated switch controlled by an external signal. In most HVACR configurations, relays have a movable contact attached to an armature and a fixed contact attached to a stationary structure. These parts make up a set of contacts. Relays can have multiple sets of *contacts* configured as normally open or normally closed combinations, **Figure 7-19**. Normally open (NO) contacts are open

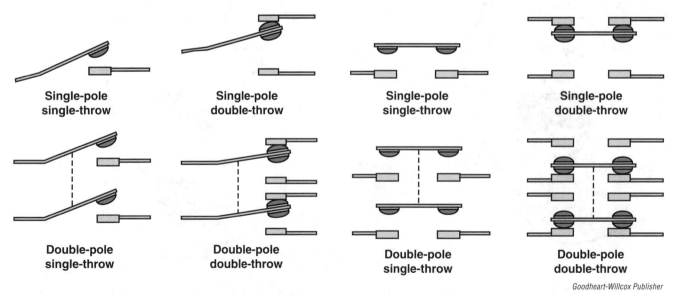

Single-pole single-throw

Single-pole double-throw

Single-pole single-throw

Single-pole double-throw

Double-pole single-throw

Double-pole double-throw

Double-pole single-throw

Double-pole double-throw

Goodheart-Willcox Publisher

Figure 7-19. Contact configurations.

in their de-energized state. The contacts are in their normal state (disconnected). Normally closed (NC) contacts are closed in their de-energized state and open when the coil is energized. When a coil becomes energized, the armature is pulled into the magnetic core, thus closing the contacts. See **Figure 7-20.**

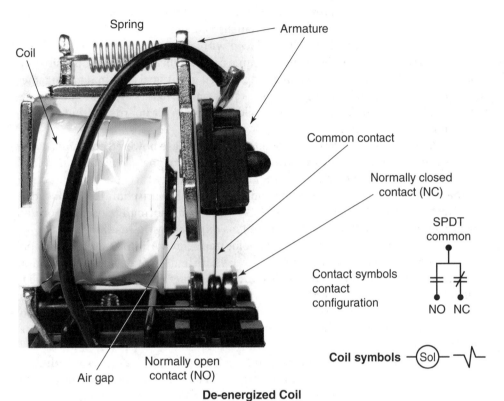

De-energized Coil

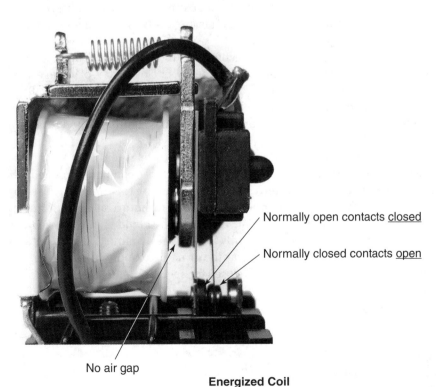

Energized Coil

Figure 7-20. A—De-energized relay coil. The armature is held away from the core – air gap is visible. The common contact is making contact to the closed terminal. So, the normally closed contacts are closed and the normally open contacts are open. Note the contact and coil symbols. B—Coil is energized and the armature is pulled into the core. The normally closed contacts open and the normally open contacts close. The extension spring is stretched. When the coil is de-energized the spring will retract and pull the armature away from the core.

Goodheart-Willcox Publisher

Relays are rated by coil voltage, dc or ac, and contact current capacity. Relays are used for light duty, such as fractional horsepower motors and currents under 30 A. The contacts are rated for inductive loads and resistive loads. *Inductive loads* are those made of coils of wire and powered by alternating current. These include solenoid coils, relay coils, contactor coils, transformers, and electric motors. AC *resistive loads* are heating elements for refrigeration evaporator defrost, hermetic compressor crankcase, and resistors. Contact current ratings are lower for inductive loads because inductive loads suffer from inrush currents.

When a relay fails, it should be replaced with the appropriate specifications of the former relay. This includes using the same part number. If this is not possible, coil voltage (ac or dc) or contact current ratings, and the physical terminal arrangement must be compatible. Relays plug into a socket assembly. Relays are not to be repaired and must be changed. Some relays are soldered on circuit boards. In that case, the entire circuit board is changed.

Contactors are used for larger loads and three-phase powered equipment. Contactors use the sliding armature configuration. Their magnetic coils are rated for ac or dc voltages of 24 V and above. Contact ratings typically start at 30 A. Contacts are normally open and have between one and four sets. Auxiliary contacts can be added to many contactors. The auxiliary contacts can be either normally closed or normally open. They are usually rated for less current as required by the load.

Contactors operate the same as solenoids and relays. All three use a coil to produce a magnetic field. A main difference is relays are light-duty devices, while contactors are heavy duty. Relays also can have various contact configurations, are ideal for control circuits, and supply power to small loads. Contactors supply power to larger loads. Contactor cost increases in proportion to current ratings, thus more expensive contactors can be rebuilt. The stationary and movable contacts, springs, and coil can all be replaced.

Contacts for relays and contactors are made by coating a base metal with silver. Recall that silver is an excellent conductor of electricity and resists corrosion. Cadmium is added to the silver to reduce arching and contacts sticking together. If oxidation or dirt accumulates on the contacts, do not use a file because the file removes the silver and cadmium. Contact burnishing tools are available. These tools contain a mild abrasive and do not remove metal.

Contacts eventually break down and become a source of voltage drop. A voltmeter can be used to easily check this. There should be 0 V, or close to 0 V, of voltage drop across contacts. Literature states that a voltage drop up to 3% of the supply voltage is acceptable per contact. This must be considered with caution since there are other sources of voltage drop in a circuit. The total voltage drop is significant and results in reduced voltage to the load.

Failures that occur with solenoid coils apply for both relay and contactor coils. They both have inrush current for ac coils, but it is less intense compared to solenoids since there is a smaller air gap. Chattering is a distinct failure that may be caused by low supply voltage to the relay and contactor coils. The armature as a result makes a loud buzzing or ringing sound. An HVACR technician should check coil voltage. There is not enough magnetic energy to fully pull in the armature. Dirty contacts may display the same chattering symptom, as the armature cannot fully engage.

The following relays and contactors are used for specific applications:

- Potential relay
- Current magnetic relay
- Current sensing relay
- Latching relay
- Motor starter (line starter)
- Solid-state relay

These components will be covered in later chapters in this textbook.

7.4 Transformers

Transformers use coils of wire to produce magnetic fields and transfer ac from one coil to the other. They do not contain armatures or any other moving parts. A basic transformer is made from two coils wound around a laminated metal core. The coils are called *windings* when used in transformers. The first coil of wire is called the *primary coil winding*. The second coil of wire is called the *secondary coil winding*. Transformers only work with ac since ac has the rising and collapsing of a magnetic field. Recall that current is induced only when a magnetic field collapses. When the two coils are placed in close proximity and one coil becomes energized, it induces an emf and current into the second coil. This process is called *electromagnetic induction*, or often referred to as *mutual inductance*.

The amount of secondary voltage produced in a transformer is determined by the number of turns used to form the coils. This can be explained by the *turns-ratio* between the primary and secondary coils. For example, a 2:1 ratio produces half the primary voltage in the secondary. See **Figure 7-21**. When the primary coil contains more turns than the secondary, this is a *step-down transformer*. Note that the two coils are physically connected. A *step-up transformer* produces a higher secondary voltage than the primary voltage, **Figure 7-22**.

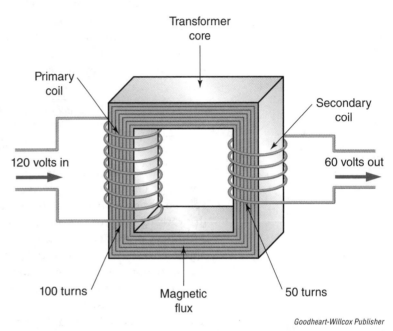

Goodheart-Willcox Publisher

Figure 7-21. Step-down transformer has more turns in the primary and less in the secondary. This is a 2:1 turns-ratio. 120 V applied to the primary will induce 60 V in the secondary.

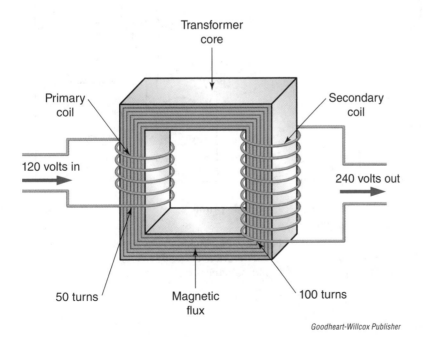

Figure 7-22. Step-up transformer has less turns in the primary and more in the secondary. This is a 1:2 turns-ratio. 120 V applied to the primary will induce 240 V in the secondary.

Goodheart-Willcox Publisher

Multi-primary transformers allow a technician to use different primary voltages as required by the equipment. These are ideal for service parts, **Figure 7-23**. The unused primary wires must be individually insulated from each other. If the unused wires come in contact, the winding becomes damaged. **Figure 7-24** shows a multi-secondary winding transformer.

Transformers are rated by primary and secondary voltages and a volt-ampere (VA) rating. The ***volt-ampere rating*** is the apparent power rating of the transformer. Apparent power will be covered in Chapter 9, *Power Distribution*. The VA is used to find the maximum safe current a secondary winding can supply without causing destruction to the winding.

Consider a transformer with 120 V primary, 24 V secondary, and 40 VA, which is a popular transformer for HVACR applications. To find the limiting secondary current, divide the VA value by the secondary voltage.

$$\frac{VA}{secondary\ voltage} = secondary\ current$$

$$\frac{40\ VA}{24\ V} = 1.667\ A$$

Therefore, the secondary circuit load(s) should not draw current above 1.667 A. Higher currents produce excess heat that affect coil wire insulation. Secondary voltage varies with respect to primary voltage as well as secondary load current draw. Using the previous 40 VA example, when the load draws less than 1.667 A, the secondary is greater than 24 V. As the load current approaches 1.667 A, the voltage approaches 24 V. The opposite is true when the secondary current is above 1.667 A. The secondary voltage drops below 24 V and potentially under supplies the load that in turn causes current increase.

A step-up transformer does not produce free energy. A 120 V primary produces 240 V in the secondary of a step-up transformer. The secondary voltage is double, but the secondary current is half of the primary current. Because energy cannot be created or destroyed, power in must equal power out.

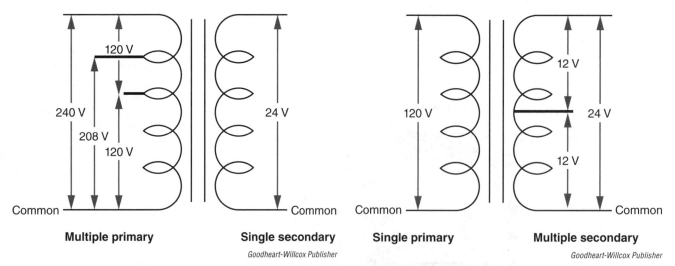

Multiple primary — Single secondary

Goodheart-Willcox Publisher

Single primary — Multiple secondary

Goodheart-Willcox Publisher

Figure 7-23. Multi-tap primary transformer allows for the choice of three different primary voltages. This is an ideal service part because it can be used with different supply voltages.

Figure 7-24. Multi-tap secondary transformer supplies two different secondary voltages as shown. Other multi-tap transformers can supply more than two voltages.

Use the example of a 100 VA step-up transformer with 120 V primary and 240 V secondary. The product of primary voltage and current equals 100 VA, and the product of secondary voltage and current equals 100 VA. The primary and secondary currents can be found through the following:

$$120 \text{ V} \times \text{A primary} = 100 \text{ VA}$$

Thus

$$\text{Primary A} = 100 \text{ VA} \div 120 \text{ V}$$
$$= 0.833 \text{ A}$$

$$240 \text{ V} \times \text{A secondary} = 100 \text{ VA}$$

Thus

$$\text{Secondary A} = 100 \text{ VA} \div 240 \text{ V}$$
$$= 0.416 \text{ A}$$

In conclusion

$$120 \text{ V} \times 0.833 \text{ A} = 240 \text{ V} \times 0.416 \text{ A}$$

Not all of the primary power is transformed into the secondary since some energy is lost in the form of heat energy. A step-down transformer transforms high voltage and low current from the primary to a lower voltage and higher current in the secondary. The step-up transformer transforms low voltage and high current to a higher voltage and lower current.

When checking a transformer in a live circuit, the first step is to check the secondary voltage with a voltmeter. Then, check the secondary current. The current should be below the maximum safe limit. To check an un-powered transformer, follow the same testing procedure for solenoid, relay, and contactor coils. More troubleshooting methods will be covered in later chapters.

Summary

- Electric current flowing through a conductor induces a magnetic field around the conductor.
- The magnetic field is increased by forming the wire into a coil wrapped around an iron core.
- When this coil is energized, it produces a magnetic field that will attract an iron armature.
- Alternating current applied to a coil produces self-induction that produces a back emf into the coil. This results in inductive reactance.
- The coil and armature make up an electromagnetic device.
- Common electromagnetic devices in HVACR are solenoids, relays, contactors, and transformers.
- The armature of a solenoid is connected to a valve plunger assembly to control the flow of fluids.
- Solenoids are rated by voltage (ac or dc), wattage, and maximum fluid pressure.
- Inrush current is drawn by a solenoid coil when initially energized.
- When the plunger is fully pulled into the core, the current is reduced to a lower amount called the holding current.
- Solenoid mechanical problems can lead to electric coil failure.
- Solenoid coil voltage tolerance is typically ±10% of rated voltage.
- The armature of relays and contactors are connected to electrical contacts to supply power to control circuits and loads.
- Relays and contactors both use a magnetic coil that when energized pulls in an armature to close or open sets of contacts.

- Contacts are either normally open or normally closed.
- Normally open contacts close when the magnetic coil is energized.
- Normally closed contacts open when the magnetic coil is energized.
- Contacts are made of silver and cadmium.
- Contacts should not be cleaned with a file as the cadmium and silver will be removed. A burnishing tool can be used for oxidation and dirt.
- Contacts should be replaced when they cause a significant voltage drop.
- Relays are not repairable and are used for light loads.
- Relays should be replaced if their contacts cause a significant voltage drop.
- Contactors are used for larger loads and can be repaired by replacing internal components such as contacts and coil.
- Relay and contactors are rated by coil voltage (ac or dc) and contact current capacity.
- Relay and contactor coil voltage tolerance is typically ±10% of rated voltage.
- Transformers use a primary coil to induce a voltage into a secondary coil.
- Transformer secondary voltage depends on the turns-ratio between the primary and secondary windings.
- Transformers are rated by primary voltage, secondary voltage, and volt-ampere (VA) rating.
- The VA rating is used to calculate the maximum safe operating secondary current.
- A step-down transformer transforms high voltage and low current from the primary to a lower voltage and higher current in the secondary.
- A step-up transformer transforms low voltage and high current to a higher voltage and lower current.
- The coils of wire used to make solenoids, relays, contactors, and transformers can be tested with an ohmmeter to determine coil condition.
- The first step in checking all the above devices in a live circuit is to check for the correct applied voltage.

Lab 7.1 Inductive Reactance and Transformer Evaluation

The resistive effect to current flow will be demonstrated in this lab by Ohm's law calculations and live circuit measurements with an inductive clamp digital multimeter. The transformer required in the lab will also be evaluated through multimeter measurements.

Lab Introduction

An ac-powered relay coil is used to analyze the existence of the hidden resistance called inductive reactance. Both the actual coil resistance and applied voltage are measured and used to calculate current. The calculated current is then compared to the actual current.

The lab also shows a method to measure currents that are too small for most inductive clamp meters to measure. The transformer will then be tested with power off and with power on using a multimeter.

Equipment

- Lab board. See Lab 3.2 for complete instructions.
- Inductive clamp multimeter
- Step-down transformer 24 V secondary 20 or 40 VA (wired or terminals–low voltage)
- Relay, general-purpose 24 Vac coil SPDT or DPDT (using SPDT only), 1/4 in spade terminals
- Insulated 1/4 in box crimp terminals for relay terminals
- Connecting wire, line voltage 14 AWG solid and 18 AWG for low voltage (thermostat wire)
- 2—single-pole 15 A light switch (solid wire per code and proper connection, or stranded can be used for lab.)

1. Assemble the circuit on lab board. Rest relay on lab board and secure (tie strap through drilled holes in board). Use **Figure 7.1-1** for reference.

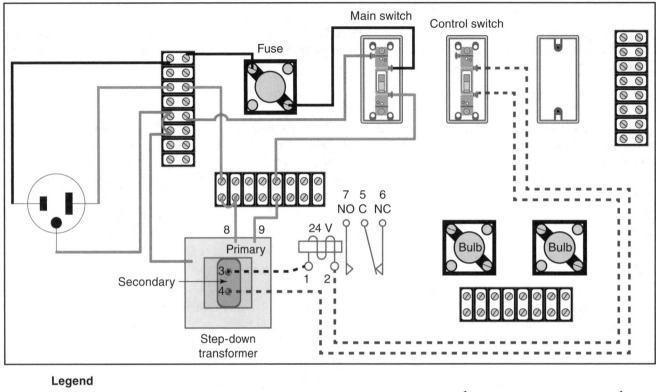

Legend

— Line voltage 120 V

- - - - - Line voltage 24 V control circuit

Numbers denote test points

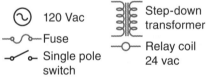 120 Vac

Fuse

Single pole switch

Step-down transformer

Relay coil 24 vac

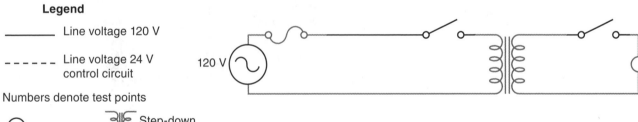

Figure 7.1-1.

2. Complete resistance measurements before plugging in the lab board and turning on the power. When completed, have your instructor check your work before proceeding.
3. Check the multimeter and test lead operation. Refer to Chapter 3, *The Simple Circuit,* if necessary.
4. Check the transformer primary resistance and record. There should be a measurable resistance. Refer to 7.2.2 *Checking a Solenoid Coil.*
5. Check the transformer secondary resistance and record. If the primary and secondary windings show a measurable resistance, proceed with lab.
6. With power off and the relay coil physically isolated from any parallel load, measure the coil resistance.

Continued ▶

Pro Tip

DC Resistance

The resistance measured in Step 6 is called the dc resistance. Note that if the control switch is ON, the coil is in parallel with the secondary transformer winding. Recall that total resistance in parallel is less than the smallest branch resistance. If the coil was open, the meter would read the winding value. That would lead to a wrong diagnosis. Measure the coil resistance—test points 1 and 2 and record value.

7. Ensure both switches are off. Plug in the board, and turn on main switch.
8. Measure the secondary voltage at test points 3 and 4 (secondary voltage output). Record voltage.
9. Turn on the control switch. Check and record secondary voltage test points 3 and 4 again and record value.
10. Use the secondary voltage from Step 8 and the relay coil resistance from Step 6 to calculate coil current draw. Use the equation $I = E/R$. Record the value.
11. Measure the relay coil current with an inductive clamp multimeter. Use either the red or black control circuit wires for the clamp. Record value. The coil current may not be large enough for the inductive clamp to measure. This is due to meter capability and current draw of relay coil used. The solution is to wrap ten turns of wire around the clamp jaw. Divide the displayed value by 10. This will be very close to the actual current. This requires breaking the circuit to place the ten-turn loop in series with the coil, **Figure 7.1-2**. The measured current will be significantly less than the calculated value.

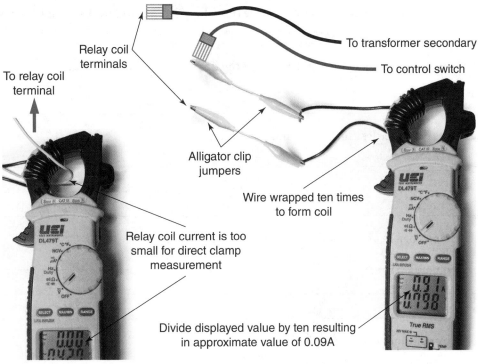

Figure 7.1-2.

Goodheart-Willcox Publisher

Lab Questions

Use your digital multimeter, or other electrical instruments, to answer the following questions.

1. What is the relay coil resistance (dc resistance)?

   _____

2. What is the secondary voltage before energizing the coil?

   _____

3. What is the secondary voltage after energizing the coil?

   _____

4. What is the calculated current based on the above voltage and resistance?

   _____

5. What is the measured inductive clamp amperage?

   _____

6. Is the measured current less than the calculated current? Why?

   _____

   _____

   _____

7. Calculate the inductive reactance resistance (X_L) of the coil. Divide the voltage from Step 2 by the actual current.

   _____

8. Calculate the resistance due to inductive reactance (X_L). Subtract the dc resistance from the total resistance from Step 7.

   _____

9. Why is the secondary winding resistance less than the primary winding?

   _____

   _____

   _____

10. Did the secondary voltage change before and after the relay coil was energized? Why?

    _____

    _____

    _____

11. What is the VA rating of the transformer used in this lab?

    _____

12. Calculate the safe secondary current. Is the relay coil current under this value?

    _____

    _____

    _____

Lab 7.2 Relays

This lab will examine the operating characteristics of a relay by comparing the values measured by the student to the relay specifications.

Lab Introduction

Use **Figure 7.2-1** as a guide to perform the current measurements required in this lab. The measurements shown on the DMM were taken using a relay with the following manufacturer's specifications and will be explained in the procedure section below. Note only the pertinent specifications for this lab are shown. Other specifications will be introduced and examined later in this textbook.

Nominal rated voltage . . . . 24 Vac
Coil, Frequency . . . . 50/60 Hz
Coil, Termination . . . . 1/4″ quick connect
Coil, Operate . . . . 85% of nominal coil voltage;
                                                    110% maximum safe operation
Coil, Duty cycle . . . . Continuous
Coil, DC resistance . . . . 90 Ω
Coil, Nominal current MA 125
Nominal VA sealed 3
Inrush VA 4
Contact ratings voltage . . . . 125/250 Vac
Contact ratings current . . . . Inductive; 8 A continuous, 25 A inrush
                                                    Resistive; 16 A continuous
Contact configuration . . . . SPDT

Note that the lab pictorial diagrams show numbered test points. These numbers will not coincide with the relay you are using. The numbers are there as guidance throughout the labs.

Contact switching action is checked by taking resistance measurement when power is off. When contacts close, continuity exists between them, allowing current to flow. The meter should display approximately 0 Ω. Switching action is observed by illuminating light bulbs and current draw.

Equipment

- Inductive clamp multimeter
- 2—incandescent light bulbs (preferable low wattage)
- 14 AWG black and white wires
- Relay specifications sheet for relay being used (that shows the above information). At a minimum, the student must know the terminal identification and coil voltage that is found on most relay covers.

Procedure

1. Use the same lab board set up from Lab 7.1. Perform initial preparation:
   A. Energize the relay coil.
   B. Check resistance between the normally closed contacts 5 and 6 and normally open contacts 5 and 7.
   C. Record these values.
   D. Locate contact terminals for the specific relay used.
2. Turn off the control switch to de-energize the relay coil. Check resistance between the normally closed contacts 5 and 6. Record value.
3. Measure the coil current with the inductive clamp directly or use the ten wire wrap method. This current is called the holding current. Record value.
4. Measure the voltage across the coil terminals 1 and 2. Record.
5. De-energize the coil.
6. Set up the inductive clamp to measure inrush current. Many meters have a button you hold until an icon or letters LRA or Inrush is displayed, **Figure 7.2-1**. Some meters have MAX hold only and may be used for this purpose.
   A. Open the clamp and wrap around either of the low voltage wires feeding the coil.
   B. Energize the coil. The inrush current will be displayed until the inrush function is turned off. Recall the inrush current occurs in a fraction of a second. Without the inrush feature, you may see a flicker and then the coil holding current will be seen.
   C. Record the inrush current.
   D. Turn off the coil.

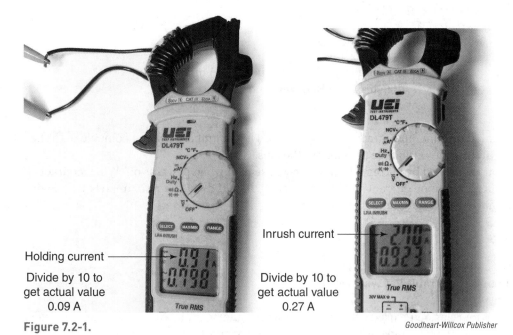

Holding current
Divide by 10 to get actual value 0.09 A

Inrush current
Divide by 10 to get actual value 0.27 A

Figure 7.2-1.

Goodheart-Willcox Publisher

Pro Tip

Drop-Out Voltage

The voltage applied to the coil must not drop below a given voltage. If it does, it can cause the armature to drop-out since the magnetic field also decreases. This low voltage level is called the drop out voltage. Sometimes it is given as a percentage of rating voltage as with the specifications shown in the 7.2 Lab Introduction. The rated voltage is 24 Vac so 85% of 24 equals 20.4 V. Applied voltages under 20 V will cause the armature to drop out (open the NO contacts and close the NC contacts).

Continued ▶

7. Connect a 100 Ω 1/2 W (any tolerance) resistor in series with the relay coil. See
Figure 7.2-2. Observe the sound difference. The armature begins to chatter.
Check voltage across the coil and the current. Record the values.

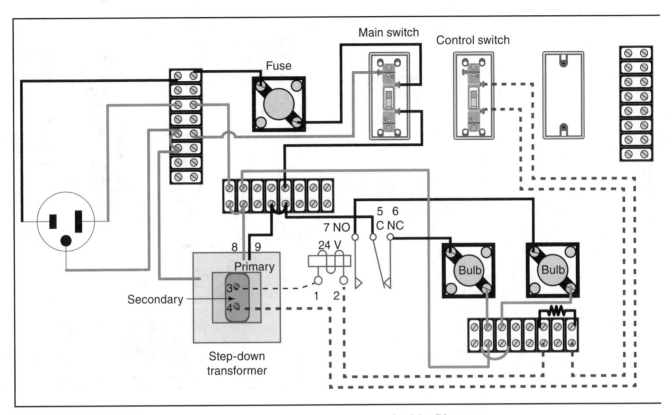

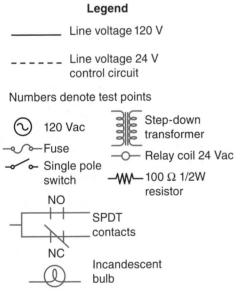

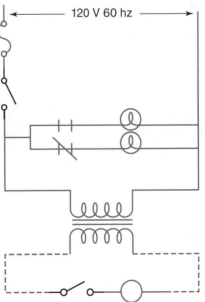

Figure 7.2-2.

8. Turn off the control switch and turn off the main switch. Remove the resistor and reconnect coil wires.

9. Wire relay contacts and light bulb sockets. See **Figure 7.2-3**.

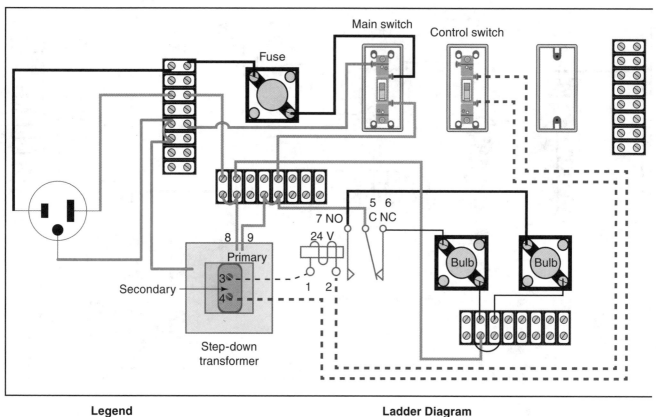

Legend

—————— Line voltage 120 V

- - - - - Line voltage 24 V
control circuit

Numbers denote test points

⊙ 120 Vac

⌇⌇ Step-down transformer

—⌒⌒— Fuse

—○— Relay coil 24 Vac

—⚬⚬— Single pole switch

NO
SPDT contacts
NC

Incandescent bulb

Ladder Diagram

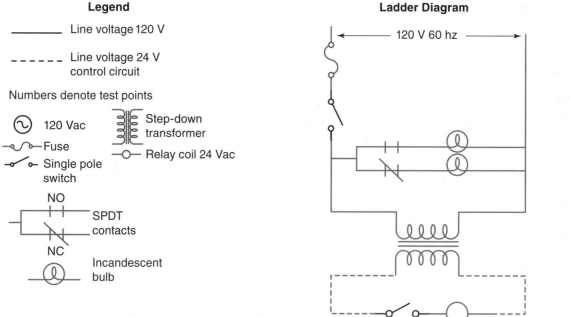

Figure 7.2-3.

Goodheart-Willcox Publisher

10. Turn on the main switch only. Observe which light turns on and which terminal is supplying power to it. Record terminal test point number and contact name based on the specific relay used.

Continued ▶

11. Turn on the control switch and observe sound made by the closing armature. Observe which light turns on and which terminal is supplying power to it. Record terminal test point number and contact name based on the specific relay used.

12. Remove the ten wire wrap from clamp jaw. Measure the current draw of each light and record.

13. The light bulbs will draw enough current for the inductive clamp to measure.

Note: Leave the lab board wired for the next lab.

Lab Questions

Use your digital multimeter or other electrical instruments to answer the following questions.

1. In procedure Step 1, was there continuity between the NO contacts when the coil was energized? What about when the coil was not energized?

_____

_____

_____

2. What is the coil holding current?

_____

3. Calculate the holding current VA. Multiply the current times the voltage.

_____

4. What is the inrush current?

_____

5. Calculate the inrush current VA. Multiply the current times the voltage.

_____

6. How much greater is the inrush current than the holding current?

_____

_____

7. What is the voltage across the energized coil (holding voltage)?

_____

8. What percentage of the rated voltage is the holding voltage? Use the equation *holding voltage/rated voltage* × 100.

_____

9. What is the applied coil voltage when the resistor is placed in series with the coil?

_____

10. How much did the applied voltage drop when the resistor was added?

_____

11. What percentage of the rated voltage is the holding voltage (with resistor)?

_____

12. How did the lower voltage affect the coil operation (sound and current draw)?

_____

_____

_____

13. Did the light bulbs alternate lighting up as the control switch was turned on and off?

_____

_____

14. What is the current draw of the light bulbs?

_____

Contactors

This lab will examine the operating characteristics of a contactor and allow the student to measure voltage.

Lab Introduction

Contactors are larger versions of relays. The contacts are normally open except for an added auxiliary set of contacts that could be normally open or closed. This lab uses a 120 Vac coil contactor that would be used to supply power to a large motor. The control circuit on the lab board is only 24 V. When the control switch is turned on, the relay coil energizes, and the normally open contacts close. The NO contact supplies 120 V to the contactor coil. This is how low voltage controls, such as thermostats, can control high voltage motors.

Equipment

- 2- or 3-pole contactor with 120 Vac coil. Contacts will not be used to power loads in this lab.

 Procedure

1. Disconnect the bulb socket that is connected to normally open contacts and connect to the contactor coil. Use quick contact terminals. Connect the white wire to the other coil terminal, **Figure 7.3-1**.

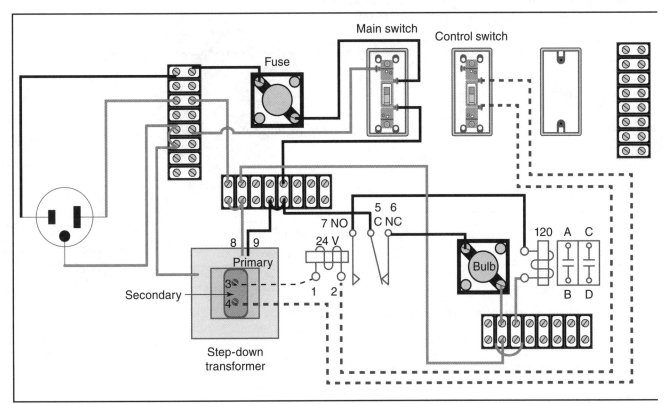

Figure 7.3-1.

Goodheart-Willcox Publisher

Continued▶

Legend

—————— Line voltage 120 V

- - - - - Line voltage 24 V
control circuit

Numbers denote test points

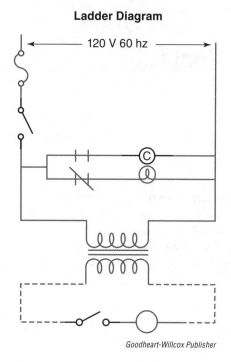

Ladder Diagram

120 V 60 hz

Goodheart-Willcox Publisher

120 Vac

Fuse

Single pole switch

Step-down transformer

Relay coil 24 Vac

Contactor coil 120 Vac

NO

SPDT contacts

NC

Incandescent bulb

Figure 7.3-1. (Continued)

2. Connect power plug.
3. Observe the bulb connected to the normally closed contact is on.
4. Turn on the control switch. The light should turn off and contactor armature should pull in.
5. Measure the contactor coil voltage.
6. The normally open contacts should be closed. Measure the contact resistance through each pole—test points A to B and C to D.

Lab Questions

Use your digital multimeter or other electrical instruments to answer the following questions.

1. What is the coil voltage when the coil is energized?

_____

_____

2. Is the coil voltage within the operating range for your contactor?

_____

_____

3. With the coil energized, what is the resistance for the normally open contacts?

_____

_____

Know and Understand

_____ 1. Which does *not* increase magnetic field strength?
   A. Iron core.
   B. Higher current flow.
   C. Magnetic wire insulation.
   D. Number of turns of wire.

_____ 2. What is the purpose of the compression spring used in some electromagnetic devices?
   A. Helps the magnetic field retract a plunger faster.
   B. Returns the plunger back to rest when the coil is de-energized.
   C. Slows down the plunger as the coil is de-energized.
   D. De-energizes the coil.

_____ 3. A collapsing magnetic in a coil will ____.
   A. induce a back emf
   B. increase current flow
   C. increase applied voltage
   D. decrease coil resistance

_____ 4. What happens when a dc powered solenoid is turned off?
   A. Power to the coil is removed.
   B. The stored magnetic energy in the coil induces a back emf.
   C. The coil continues to operate briefly.
   D. Solenoids cannot operate with dc.

_____ 5. In HVACR applications, solenoids are mostly used to ____.
   A. open electrical contacts
   B. control fluid flow
   C. close electrical contacts
   D. increase or decrease voltage

_____ 6. Reduced inductive reactance will cause ____.
   A. decreased current flow
   B. decreased applied voltage
   C. increased back emf
   D. increased current flow

_____ 7. Which of the following is *not* true about solenoids?
   A. Can be powered by dc.
   B. Can be powered by ac.
   C. Can operate 15% below rated voltage.
   D. Can operate 10% above rated voltage.

_____ 8. *True or False?* An electromagnetic coil's resistance should be zero Ω.

_____ 9. When a relay coil is energized, ____.
   A. normally closed contacts close
   B. normally closed contacts open
   C. normally open contacts open
   D. all contacts will close

_____ 10. Relays are rated for the following *except* ____.
   A. resistive load current
   B. inductive load current
   C. coil voltage
   D. contact resistance

_____ 11. Relays are for the following *except* ____.
   A. small motors
   B. low voltage control circuits
   C. large motors
   D. loads drawing less than 30 A

_____ 12. *True or False?* A relay coil rated for 24 Vac can be powered with 24 Vdc.

_____ 13. *True or False?* Contactor contacts are typically rated above 30 A.

_____ 14. Which contact configuration is likely found with a contactor?
    A. Single-pole single-throw.
    B. Double-pole single-throw.
    C. Three sets of normally open contacts.
    D. Double-pole double-throw.

_____ 15. *True or False?* Contacts should be cleaned with a file.

_____ 16. Lower than rated voltage to a contactor or relay coil will cause ____.
    A. lower current flow through the coil
    B. chatter
    C. faster operation of the armature
    D. cost savings due to lower current

_____ 17. *True or False?* VA is the transformer rating that describes the maximum safe secondary current draw.

_____ 18. Step-down transformers ____.
    A. reduce the applied primary voltage to a lower secondary voltage
    B. reduce the applied secondary voltage to a lower primary voltage
    C. increase the secondary voltage as compared to the applied primary
    D. increase the applied primary voltage to a higher secondary voltage

_____ 19. Step-up transformers ____.
    A. reduce the applied primary voltage to a lower secondary voltage
    B. increase the secondary voltage as compared to the applied primary
    C. reduce the applied secondary voltage to a lower primary voltage
    D. increase the applied primary voltage to a higher secondary voltage

Critical Thinking

1. What happens when a 24 Vdc rated relay coil has 24 Vac applied?

_____

_____

_____

_____

2. What is an advantage of using a relay? Why can you not use a switch to turn on the load instead?

_____

_____

_____

_____

8 Capacitors

DiversiTech Corporation

Chapter Outline

8.1 The Capacitor and Capacitance
    8.1.1 Types of Capacitors
8.2 Effects of DC and AC on Capacitors
    8.2.1 Capacitive Reactance
8.3 Series and Parallel Capacitors
8.4 Capacitor Failures
8.5 Capacitor Testing and Replacement

Learning Objectives

After completing this chapter, you will be able to:

- Define a capacitor and capacitance.
- Discuss types of capacitors and their construction.
- Explain capacitor ratings.
- Differentiate how capacitors work in ac and dc powered circuits.
- Explain capacitive reactance and how it is calculated.
- Calculate series and parallel capacitors.
- List capacitor failure causes.
- Explain how to test a capacitor.

Technical Terms

capacitance	dual capacitor	nonpolarized
capacitive reactance (X_c)	electrolytic	polarized
capacitor	farad (F)	run capacitor
dielectric	microfarad (mfd)	start capacitor

Introduction

The previous chapter discussed the rising and collapsing of a magnetic field to induce current toward a power source. The magnetic flux builds up during the first quarter of a cycle in a coil. Then during the second quarter, the magnetic field collapses and induces current back towards the power source. This continues for the second half of the sine wave.

    A capacitor stores an electric charge and releases the charge similar to the coil. Capacitors are used in electronic controls and single-phase motors found in the HVACR equipment and test instruments. This chapter will define how capacitors work with dc and ac power, how they are constructed, and how they are tested.

8.1 The Capacitor and Capacitance

A *capacitor* is a component that has the ability to store energy in the form of electrical charge. It is made of two conducting plates that produce a potential difference, or voltage, across them. The plates can measure over 1′ in length, so the three layers are rolled into a round or oval container. See **Figure 8-1**. A capacitor also has an aluminum product and an insulator between the plates, which is called the ***dielectric***. Capacitors are rated by charge capacity and voltage.

Capacitance is the electrical property of a capacitor. It measures a capacitor's ability to store a charge. The unit for capacitance is the ***farad (F)***, where one farad is the charge of one coulomb with the potential difference of one volt.

> **Pro Tip**
>
> ## Farads
>
> A farad is extremely large, so in most cases, the microfarad is used. Relatively speaking, a farad's surface area is in terms of square miles versus square feet for the typical values used in HVACR applications. A ***microfarad (mfd)*** is one millionth of a farad, or 1/1,000,000, as denoted by the prefix *micro*. The Greek letter μ represents micro. The voltage rating is for the dielectric's breakdown voltage.

A charge stored on the plates of a capacitor can be understood best through an object producing a static charge. Common objects can produce a static charge, which can be stored on conductive plates. See **Figure 8-2**. These plates initially are neutral since they contain an equal number of electrons and protons. Recall the function of static electricity, which is the buildup of charge on an object or surface.

Figure 8-1. Capacitor plates and dielectric are formed into a ply-roll and placed in round or oval housing. Nonpolarized capacitor symbol is shown on top. Polarized capacitor symbol is shown on bottom with circuit board type capacitors.

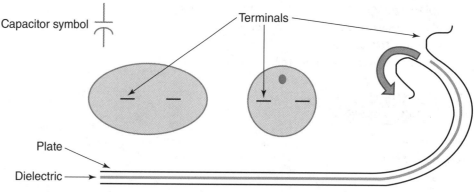

Capacitor symbol

Terminals

Plate

Dielectric

Aluminum foil plates and dielectric roles into round or oval housings

Nonpolarized capacitors

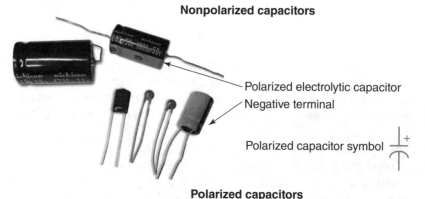

Polarized electrolytic capacitor

Negative terminal

Polarized capacitor symbol

Polarized capacitors

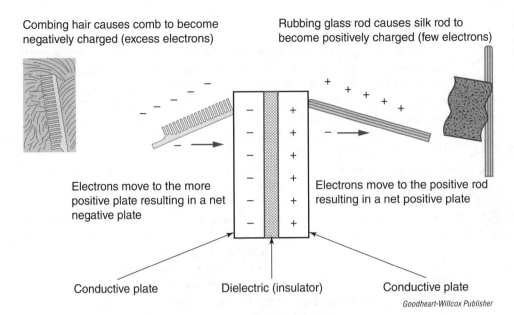

Combing hair causes comb to become negatively charged (excess electrons)

Rubbing glass rod causes silk rod to become positively charged (few electrons)

Electrons move to the more positive plate resulting in a net negative plate

Electrons move to the positive rod resulting in a net positive plate

Conductive plate

Dielectric (insulator)

Conductive plate

Goodheart-Willcox Publisher

Figure 8-2. Generated static charges are transferred to conductive plates.

A comb gains electrons as it is brushed across hair. The comb has a negative net charge. Rubbing a glass rod with a silk cloth removes electrons from the rod. The rod has a positive net charge. These charges are stored in conductive plates.

When the net-negative comb contacts the conductive plate, electrons flow from the comb to the more positive plate. The plate then contains excess electrons, and it therefore has a negative net charge. When the positive glass rod contacts the other plate, plate electrons flow into the more positive rod. This leaves the plate with fewer electrons and a net positive charge. The insulator placed between the plates restricts conduction between the plates. While the opposite charges are retained in the plates, a difference in potential exists across the plates, **Figure 8-3**. The current flows through the load until the plate charges are neutralized.

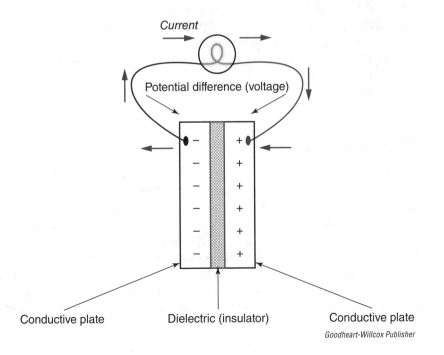

Current

Potential difference (voltage)

Conductive plate

Dielectric (insulator)

Conductive plate

Goodheart-Willcox Publisher

Figure 8-3. Stored charge flows through light bulb and illuminates bulb until plate charges are neutralized.

8.1.1 Types of Capacitors

There are two construction types for single-phase motor capacitors: start capacitors and run capacitors. A **start capacitor** assists a motor in starting up. Motor start is typically less than one second, so the start capacitor is only energized for that short period. Under normal operation, the start capacitor does not heat up. This allows for large capacity values packaged in small plastic housings since they do not have to dissipate much heat. Some start capacitors are housed in metal housings used in special applications. Start capacitor values range from about 100 μF to 600 μF.

A **run capacitor** is energized while the motor is operating, so it must dissipate the heat generated by current flow. A run capacitor is packaged in a metal container that contains nonconductive oil. The plates conduct heat to the oil, and the oil conducts heat to the metal housing. The cooler ambient air then absorbs heat from the metal housing. Run capacitor values are only about 3 μF to 60 μF, but they require a larger housing due to the cooling oil. Two capacitors with different capacity values that share one common plate are called **dual capacitors**. These have three terminals and are commonly found in residential air conditioners.

Large-value capacitors are known as **electrolytic**. They can be either polarized or nonpolarized. Capacitors used for single-phase motors are **nonpolarized**, which means they are without positive or negative polarity. This works since they are used with ac. Nonpolarized electrolytic capacitors are used to start some single-phase motors. Capacitors that are **polarized** are found in circuit boards for primarily dc use. Polarized capacitors must be wired correctly with respect to the power source because they can be connected only one way in a circuit due to their polarity. The positive lead must be wired to the positive potential and the negative terminal to the negative potential. Reversing the polarity damages the capacitor dielectric. These capacitors store values into the thousands of μF and are used in inverter motors, which are discussed in future chapters.

Safety features are built in to run and start capacitors to prevent overheating. The start capacitor has a vent to let out built-up vapor pressure caused by overheating of the dielectric. See **Figure 8-4**. Without the vent, the housing can rupture. There are two methods used for the run capacitor. A weak spot is made by using low-temperature solder to cover an opening on the top surface of the capacitor, **Figure 8-5**. Without the solder covered vent, overheating can cause the casing to rupture as the oil expands. The second method uses pressure-sensitive internal contacts that open as pressure builds up and turns off power to the capacitor.

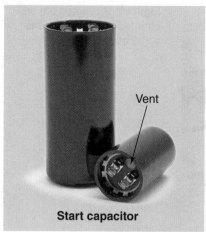

Vent

Start capacitor

DiversiTech Corporations

Figure 8-4. Start capacitor vent to release pressure due to overheating the capacitor plates.

Terminals

Solder over vent

Run capacitor

Goodheart-Willcox Publisher

Figure 8-5. Run capacitor low-temperature solder melts and releases pressure if capacitor overheats.

8.2 Effects of DC and AC on Capacitors

Capacitors can be charged with either dc or ac. For dc, the capacitor plates are initially neutral, which means there is no net charge. See **Figure 8-6A**. The positive battery terminal potential is more positive than the top capacitor connected to it. The negative battery terminal potential is more negative than the bottom plate connected to it. When the charging switch is closed, electrons flow from the top plate to the positive battery terminal. At the same time, electrons flow from the negative battery terminal to the bottom plate. Initially, there is a heavy inrush current flow causing the capacitor plates to charge. As the capacitor charge builds up, a voltage is developed across the capacitor plates. The current flow slows down as the voltage across the capacitor increases. Once the voltage across the capacitor equals the battery voltage, the current flow stops.

Essential Electrical Skills for HVACR: Theory and Labs

Current does not flow if there is no potential difference between points, **Figure 8-6B**. A capacitor does not allow current flow after it has been fully charged. This inherent characteristic of capacitors is used to block dc signals and allow ac signals to pass through in electronic applications. Capacitors are also used as short term batteries and timer circuits. In **Figure 8-6C**, the charging switch is opened, and the discharge switch is closed. The capacitor discharges the stored charge through the light bulb. The current flow rate depends on the voltage across the capacitor and the resistance of the load. The amount of time the light glows depends on the capacity of the capacitor.

In ac, recall that voltage and current rise and fall when current flows in one direction during a half cycle. Then the two values also rise and fall in the second half of the cycle when current flows in the opposite direction. Voltage and current are in phase. They both peak and cross the zero line at the same time. When a capacitor is used in an ac circuit, the current leads the voltage by 90° in a pure capacitive circuit. This is due to the inrush currents to and from the capacitor plates. The plates begin to charge up before voltage is developed across the plates and the supply voltage is at zero.

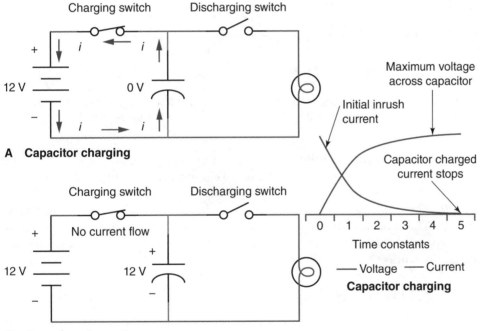

A **Capacitor charging**

B **Capacitor charged**

Figure 8-6. A—Charging switch is closed and capacitor rapidly begins charging – see inrush current in the side graph. The rate of charge decreases as the plates reach full charge. Current stops when fully charged. Voltage across the capacitor ramps up as the capacitor charges. B—Capacitor is fully charged and current has stopped. Voltage across the capacitor is the same as battery voltage. C—Charging switch is opened and Discharge switch is closed. The capacitor discharges through the load until the capacitor plates are neutralized.

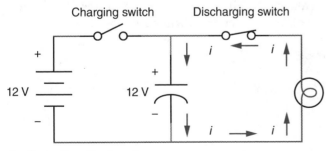

C **Capacitor discharging**

Goodheart-Willcox Publisher

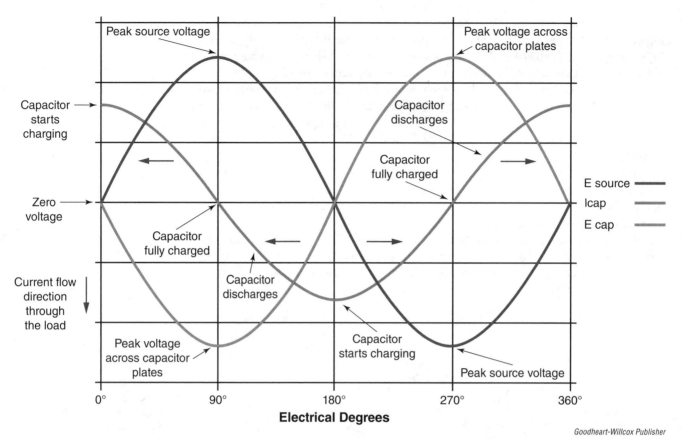

Figure 8-7. Current charges and discharges to and from the capacitor in the same direction from 0 to 90° then current charges and discharges to and from the capacitor in the opposite direction from 180° to 360°. The cycle repeats.

At 90°, a capacitor is fully charged, and the voltage across the capacitor and the source voltages have reached their maximum (peak) value. See **Figure 8-7.** At this point, the capacitor discharges as the two opposing voltages fall to zero at 180°. The capacitor begins charging up again as the voltages rise to peak value at 270°. The capacitor then discharges and voltages fall to zero at 360°. The cycle then repeats. The above pure capacitive ac circuit was used to explain how the capacitor works.

8.2.1 Capacitive Reactance

In practical circuits, there is pure resistance and inductive reactance that changes the phase angle between current and voltage. Capacitors oppose the change in voltage potentials as they charge and discharge. The opposition that results in resistance to current flow is called ***capacitive reactance (X_c)***. Capacitive reactance is measured in ohms. This concept will be explored in the Lab 8.2, *Capacitive Reactance*.

Capacitive reactance can be measured using the following equation:

$$X_c = \frac{1}{2\pi FC}$$

where
F = frequency
C = capacitance in μF
π = 3.14

As frequency or capacitance increases, the resistances decreases. For example, consider a 35 μF capacitor operating at 60 Hz. What would be its resistance?

$$X_c = \frac{1}{(2 \times 3.14 \times 60 \times 0.000035)}$$

$$= 76\,\Omega \text{ (rounded to whole number)}$$

Current does not flow through the dielectric, unless a dielectric breakdown voltage is applied. **Figure 8-8** illustrates how current flows through a load when a capacitor is used. Note the current reverses each half cycle and flows to and from the plates while flowing through the load but not through the dielectric.

8.3 Series and Parallel Capacitors

Capacitors can be wired in a series, parallel, or combination. Total capacitance can be determined in each arrangement. See **Figure 8-9**. This is useful if the required replacement capacitor value is not available in your parts inventory.

Capacitors in series use the reciprocal equation of resistors in parallel:

$$C_t = \frac{1}{(1/C_1 + 1/C_2 + 1/C_3 + 1/C_{nth})}$$

The total capacitance is less than the smallest series capacitance.

Total capacitance for two capacitors, or needing only the combined total of two when more are present, can also be calculated. The equation is as follows:

$$C_t = \frac{C_1 \times C_2}{C_1 + C_2}$$

A final equation used to calculate the same capacitor values in series can be used:

$$C_t = \frac{\text{capacitance}}{\text{number of capacitors}}$$

Consider the example where there are three 6 μF capacitors in series. To solve for total capacitance, we can calculate:

$$C_t = \frac{6\,\mu F}{3}$$

$$= 2\,\mu F$$

Note that the total 2 μF is less than the smallest 3 μF capacitor.

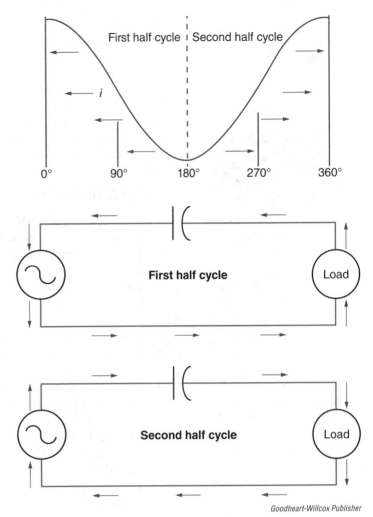

Goodheart-Willcox Publisher

Figure 8-8. Current cannot flow through the dielectric. The charging and discharging as current direction changes causes current flow through the load and power source.

Series capacitors decrease in value because only the outer plates conduct

Parallel capacitors increasing in value because the surface adds up

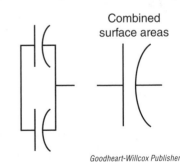

Goodheart-Willcox Publisher

Figure 8-9. Only the outer plates conduct when capacitors are wired in series while the inner plates behave like a dielectric. Capacitor values add up in parallel as their surface areas add up.

For capacitors in parallel, they can be added up just like resistors in series. They also use the equation:

$$C_t \ = \ C_1 + C_2 + C_3 + C_{nth}$$

Capacitors can also be arranged in series and parallel combinations just like resistors in a complex circuit—although this is not very practical in the field. There is a product that contains six capacitors in one housing that can be configured for up to 200 values of capacitance. The manufacturer provides jumper wires to connect terminal and produce a specific value. See **Figure 8-10**.

Goodheart-Willcox Publisher

Figure 8-10. Six capacitors in one housing allow for multiple values of capacitor values by arrangement of series, parallel, and complex combinations.

8.4 Capacitor Failures

Several failures can occur with capacitors. Overheating is a major cause of dielectric deterioration. Sources of excess heat include high voltage, voltage spikes, excess current, and ambient temperature above a capacitor's operating temperature. Gradual deterioration shortens the lifespan of the capacitor. Manufacturers provide lifespan under-normal-operating conditions in terms of hours-of-operation.

Overheated run capacitors can be detected visually. Pressure builds up as the oil expands and swells the metal casing. Oil can leak out of the terminal seals. A start capacitor is designed to operate for a fraction of a second and removed from a motor's starting circuit. Energizing the start capacitor for longer times causes overheating of the dielectric and failure.

Leakage current is a specification for the amount of current that flows between plates due to an imperfect dielectric. It is typically in the microampere range, thus is generally insignificant. However, as the dielectric decays, leakage current increases, which causes a decrease in capacity. This occurs since more current flows through the capacitor and the dielectric instead of being stored. Sudden dielectric failure can also occur if the voltage across the capacitor exceeds the breakdown voltage specification. The capacitor shorts out and becomes a conductor.

The internal plate to terminal connection can also fail prematurely due to factory defect or current surges. This results in an open capacitor, which means no continuity. A shorted capacitor allows current flow between the plates and acts as a conductor. A leaking capacitor allows some level of current flow between the plates due to a dielectric fault.

Pro Tip

Discharging a Capacitor

Improperly discharging a capacitor by shorting its terminals can damage the internal connections between the plates and the base of the external terminals due to the current surge. This can also be dangerous to the technician because a spark is produced by the short circuit current.

8.5 Capacitor Testing and Replacement

A capacitor can be first tested for functionality by performing the following steps. First power must be turned off, and the capacitor must be properly discharged. Place a 20,000 Ω 2 W resistor across the capacitor terminals. This controls current flow from one plate to another while not generating excess heat. Use the capacitor multimeter test function or stand-alone tester. Most HVACR meters have a capacitor test function.

Set the mode (function) switch to capacitor. The mode is indicated by the capacitor symbol. Be sure to read the owner's manual as there are subtle differences between meters.

Essential Electrical Skills for HVACR: Theory and Labs

Connect the meter leads to the terminals. The display may show dashing lines while performing the test. The display will show the capacity value in μF, **Figure 8-11**.

The measured value is then compared to nameplate value to determine if the capacity value is within the operating range. An open or shorted capacitor will be displayed as a value in the picofarad range in most meters. Knowing the cause of failure will aid in troubleshooting. The ohmmeter function will determine opened or shorted state by displaying *OL* for open and near 0 Ω for shorted. A potentially good capacitor can initially show a low-value resistance followed by *OL*. This will be performed in Lab 8.1, *Evaluating Capacitors*.

If the capacitor shows the varying resistance value, it needs to be tested further in the live circuit. The ohmmeter method was used to test capacitors when capacitor testers were expensive stand-alone test instruments. The low cost and miniaturizing of electronics allowed for incorporating this test function in most HVACR multimeters.

When replacing a capacitor, refer to the manufacturer's specifications because a wrong capacitor may have been installed by another technician. Never use a capacitor with a lower voltage rating.

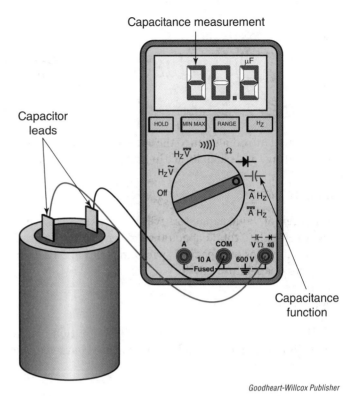

Goodheart-Willcox Publisher

Figure 8-11. A capacitor value is measured with the capacitor test function of a multimeter.

Summary

- Capacitors store an electric charge. They are made of two conducting plates separated by a dielectric.
- Capacitors are rated by voltage and charge capacity in μF.
- Capacitance measures a capacitor's ability to store a charge.
- Run capacitors use metal housings that contain oil to dissipate heat produced in plates. They use pressure-sensitive contacts or low-temperature solder to protect against housing rupturing in case of overheating.
- Start capacitors are packaged in plastic housings as they do not generate significant heat. They use a vent to protect against housing rupturing in case of overheating.
- Once a capacitor is charged with dc, current flow stops. Capacitors do not pass current in dc circuits.
- Since current alternates direction with ac, capacitors pass current in ac circuits. Current does not flow through the dielectric.
- Current and voltage are out of phase when capacitors are used. In a pure capacitive circuit, current leads voltage by 90°.
- Capacitors oppose voltage changes, and cause a resistance to current flow. It is called capacitive reactance X_c.
- Overheating is a major cause of capacitor failure.
- Capacitors should be discharged before testing with a 20,000 Ω 2 W resistor.
- Discharging a capacitor by shorting its terminals could result in injury and damage to the capacitor.
- Capacitors are tested with the capacitor function of the multimeter.
- Replacement capacitors should have the same specifications as the original.
- Total capacitance of capacitors in series will be less than the smallest capacitor.
- Total capacitance of capacitors in parallel will be sum of the individual capacitors.

Evaluating Capacitors

This lab will introduce you to the capacitor test function on a digital multimeter. You will be guided through the steps in evaluating total capacitance in both series and parallel.

Lab Introduction

Two different value capacitors will be checked using the multimeter's capacitor test function. These values will be used to calculate total capacitance in series and in parallel. The calculated values will then be compared to actual measured values of capacitors wired in series and then in parallel. An ohmmeter is used to prove the theory behind charging a capacitor with dc.

Equipment

- Digital multimeter with capacitor test function
- 2—Run capacitors 5 mfd and 10 mfd (discharged capacitors—not in the circuit)
- 1—Alligator clip jumper wires

Procedure

1. Test the capacitor one value and record.
2. Test the capacitor two value and record.
3. Calculate the total capacity for series connection and record.
4. Calculate the total capacity for parallel connection and record.
5. Set the meter mode switch to resistance. Observe the display as you probe the capacitor terminals. Describe observation.
6. Switch the probe leads to the capacitor terminals as shown in **Figure 8.1-1**. Describe observation.

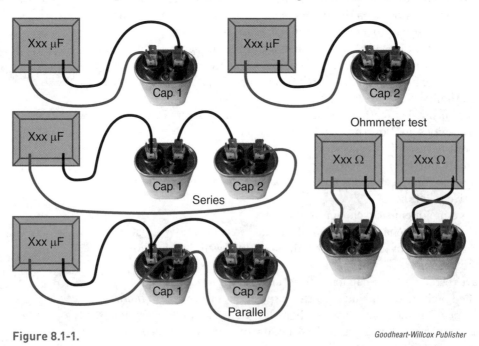

Figure 8.1-1.

Goodheart-Willcox Publisher

Lab Questions

Using capacitor one, record the following results.

1. What is the rated value of capacitor one?

   _____

2. What is the given tolerance?

   _____

Pro Tip	Run Capacitor Tolerance

Run capacitors typically have a ± 6% tolerance. So the actual value can be 6% less than the rated value or 6% more than the rated value.

3. What is the measured value of capacitor one?

   _____

4. Is the measure value within the tolerance range?

   _____

Repeat Steps 1 – 4 for capacitor two and record results.

5. What is the rated value of capacitor two?

   _____

6. What is the given tolerance?

   _____

7. What is the measured value of capacitor two?

   _____

8. Is the measure value within the tolerance range?

   _____

9. Did the ohmmeter display show a value at first and then *OL*? If not, reverse the leads and try again. The meter should display a low resistance at first to indicate that the capacitor is charging. Once the capacitor is charged, current flow stops and display shows *OL*.

   _____

   _____

10. When switching leads in procedure Step 6 did the ohmmeter show the capacitor charging each time the leads were switched?

    _____

Lab 8.2

Capacitive Reactance

This lab will explore capacitive reactance and demonstrate how it affects current flow.

Lab Introduction

This lab will show how capacitive reactance adds resistance to current flow. The 5 mfd and 10 mfd capacitors will be wired in series to a 60 W light bulb. Current measurements will be taken to evaluate the circuit.

Equipment

- Safety glasses and gloves
- Digital multimeter with inductive clamp
- 2—Run capacitors 5 mfd and 10 mfd (discharged capacitors not in circuit)
- Quick connect terminals for capacitors
- 14 or 16 AWG wire
- 15 kΩ to 20 kΩ 2 W resistor any tolerance
- Long-nose pliers

 ## Procedure

1. Wire the lab board using **Figure 8.2-1** as a guide. Do not include the capacitor in the circuit. Wire the bulb in series with the main switch.

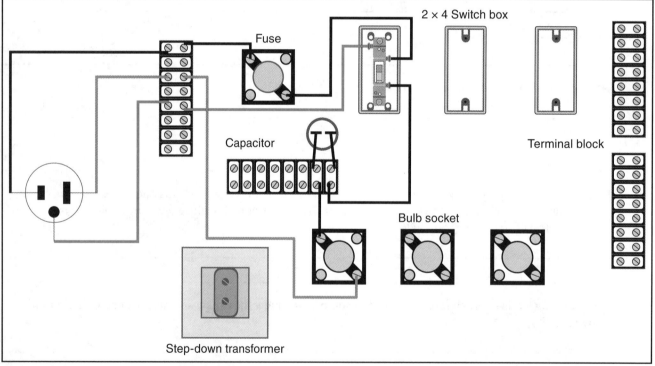

Figure 8.2-1.

Goodheart-Willcox Publisher

2. Prepare the discharge resistor by bending leads so they jump across the terminal strip that the capacitor is connected to. Hold the resistor body with long-nose pliers and wear gloves and eye protection.
3. Get your instructor's approval to plug in and turn on the circuit.
4. Observe the bulb illumination.

Continued ▶

5. Use the inductive clamp to measure current and record.
6. Measure the voltage across the bulb and record.
7. Turn off the power and wire in the 5 mfd or smallest value first.
8. Repeat Steps 3 through 6.
9. Set the meter to measure dc volts.
10. Turn off the lab board main switch and immediately measure dc voltage across the capacitor. Observe the decreasing dc voltage value.
11. Remove the meter leads and finish discharging the capacitor with the resistor.
12. Replace 5 mfd with 10 mfd capacitor and repeat Steps 3 through 6.
13. Repeat Steps 9 through 11.

Lab Questions

Use your digital multimeter or other electrical instruments to answer the following questions.

1. What is the bulb current without the capacitor?

_____

2. What is the bulb voltage without the capacitor?

_____

3. What is the bulb current with the 5 mfd capacitor?

_____

4. What is the bulb voltage with the 5 mfd capacitor?

_____

5. What is the bulb current with the 10 mfd capacitor?

_____

6. What is the bulb voltage with the 10 mfd capacitor?

_____

7. Why is the bulb brighter and the current higher without a capacitor?

_____

8. Did increasing the capacitor value increase current flow and bulb illumination?

_____

9. Do capacitors add resistance to current flow?

_____

10. Does capacitive reactance decrease with increasing capacity?

_____

11. Calculate the X_c for the 5 mfd capacitor.

_____

12. Calculate the X_c for the 10 mfd capacitor.

_____

This lab will show that a discharging capacitor can extend current flow in a half wave rectifier circuit.

Lab Introduction

A half wave rectifier uses one diode to rectify ac into dc. This dc is not pure and looks like a half of a sine wave. Current flows in direction only for half a cycle. A capacitor charges during the first half cycles and discharges in the second half cycle. The result is current flowing in the same direction for both half cycles.

Equipment

- Digital multimeter with inductive clamp
- Safety glasses and gloves
- 1–Run capacitor 20 mfd to 30 mfd value (discharged capacitors—not in the circuit)
- 1—1N002 rectifier diode

- Quick connect terminals for capacitors
- 14 or 16 AWG wire
- 15 kΩ to 20 kΩ 2 W resistor any tolerance
- Long-nose pliers

Procedure

1. Wire the lab board using **Figure 8.3-1** as a guide. Do not add the capacitor.
   Note the capacitor is connected with dotted lines to indicate connection is made later.

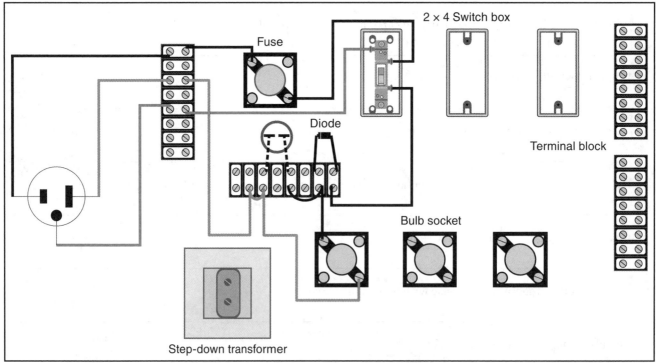

Figure 8.3-1.

Continued ▶

Schematic diagram

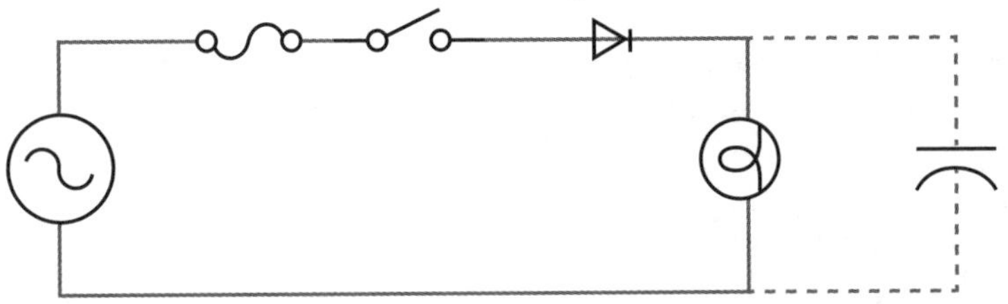

Figure 8.3-1. (Continued)

Goodheart-Willcox Publisher

2. Get your instructor's approval to plug in and turn on the circuit.
3. Observe the bulb illumination.
4. Measure the dc voltage across the bulb and record.
5. Turn off the main switch.
6. Install the capacitor.
7. Get your instructor's approval to plug in and turn on the circuit.
8. Observe the bulb illumination.
9. Measure the dc voltage across the bulb and record.
10. Turn off the switch and unplug the lab board.
11. Discharge the capacitor with the resistor.

Lab Questions

Use your digital multimeter or other electrical instruments to answer the following questions.

1. What is the dc voltage across the bulb without the capacitor?

_____

2. What is the dc voltage across the bulb with the capacitor?

_____

3. Which provided the brightest illumination?

_____

4. Is the capacitor discharging the stored charge during the second half of the sine wave?

_____

Know and Understand

_____ 1. Capacitors store a charge on plates separated by a(n) _____.
  A. inductor   C. dielectric
  B. wires   D. aluminum

_____ 2. Capacitors are rated by _____.
  A. charge capacity and voltage
  B. voltage and resistance
  C. capacitance and resistance
  D. resistance only

_____ 3. The unit of measure for a capacitor is the _____.
  A. ohm   C. volt-ampere
  B. ampere   D. farad

4. What is the purpose of the voltage rating?
   A. To match the applied system voltage.
   B. To detail the minimum voltage that can operate the capacitor.
   C. To detail the maximum safe voltage that can be applied across the dielectric.
   D. To detail the capacitor operating voltage.

5. Which best describes a start capacitors use?
   A. It can operate for any length of time.
   B. It is limited to one-hour use.
   C. It has a repeated on/off operation (short cycling).
   D. It is powered for less than one second.

6. Which capacitor is housed in a metal container?
   A. Start.
   B. Flux.
   C. Run.
   D. Inductive.

7. Which is a *not* a visual sign that a run capacitor has overheated?
   A. Rust on metal housing.
   B. Oil around the terminals.
   C. Sides are swollen.
   D. Top of bottom is swollen.

8. *True or False?* Capacitive reactance is a resistance to current flow.

9. *True or False?* The correct way to discharge a capacitor is to short the terminals with a screwdriver.

10. *True or False?* A capacitor tester displays 490 picofarad. This capacitor is no good.

11. An *OL* reading when checking a capacitor means that the ____.
    A. capacitor needs further testing
    B. capacitor is open
    C. capacitor is shorted
    D. capacitor is good

12. *True or False?* A 110 V rated start capacitor may be used with a 240 V source.

13. *True or False?* A 35 µF capacitor can be replaced with a 10 µF capacitor temporarily for one day.

14. A 40 µF and 60 µF capacitor are wired in series. What is the total capacity?
    A. 100 µF
    B. 20 µF
    C. 24 µF
    D. 2400 µF

15. Three 60 µF capacitors are wired in parallel. What is the total capacity?
    A. 20 µF
    B. 24 µF
    C. 30 µF
    D. 180 µF

Critical Thinking

1. If the capacitor in Lab 8.2 was increased in value, would the bulb light up completely as when no capacitor was used?

_____

_____

_____

2. If two capacitors are wired in parallel to a 370 V rating, can one of the two replacement capacitors have a lower voltage rating than 370 V?

_____

_____

_____

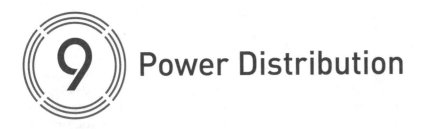

9 Power Distribution

Bohbeh/Shutterstock.com

Chapter Outline

Additional Reading

Modern Refrigeration and Air Conditioning, **21st edition**

Learning Objectives

After completing this chapter, you will be able to:

- Explain how single-phase power is used for residential power.
- Define phase conductors and neutral power wires.
- Discuss the use of grounding for safety.
- Define polarized receptacles.
- Explain impedance and how it is calculated.
- Differentiate between apparent and true power.
- Define the power factor.
- List the types of devices used for circuit protection.
- Discuss three-phase power.
- Define three-phase power line balance.

Technical Terms

apparent power
center-tapped secondary
circuit breaker
delta (Δ)
earth ground
electrode
fuse
ground-fault circuit interrupter (GFCI)

ground wire
high leg delta
hot wire
impedance
neutral wire
overcurrent
overload
phase conductor

positive temperature coefficient thermistor (PTC)
power factor
short circuit
single-phase power
three-phase power
true power
wattmeter
wye (Y)

Introduction

Chapter 6, *Alternating and Direct Current* introduced and defined the characteristics of ac and dc. You learned that ac can be transformed using a transformer while dc cannot. AC can be transformed to a high voltage and low current, thus reducing power loss. For this reason, ac is the most common form of power generation and transmission used by power companies.

Power companies generate and deliver power to residential, industrial, and commercial locations. Recall that wire has resistance, so as current flows, voltage drops. The generated ac voltage is stepped up to above 100 kV, thus lowering the current to travel through transmission, or high-tension, wires and be successfully delivered to a substation.

The high voltage and low current power are stepped down at the substation. The most common voltage phases used for HVACR are single-phase and three-phase. Single-phase is derived from three-phase. Three-phase high voltage is stepped down in substations to various voltages to supply power to industrial, commercial, and residential customers. Three-phase power is sent to industrial and commercial, while single-phase power is sent to residential and light commercial customers. The symbol for phase is represented as ∅.

This chapter will discuss power supplied for residential and commercial use. Although HVACR technicians are not licensed to perform electrical work, they must understand the characteristics of power sources. This includes understanding electrical safety, the NFPA 70 and local codes, and proper terminology on the job.

9.1 Residential Power

Single-phase power delivers one current and voltage within one cycle. See **Figure 9-1**. To deliver power to residential locations, single-phase power is sent from substations to local step-down transformers to step down the power to 240 V_{rms}. These transformers are placed on utility poles for overhead wiring locations or ground level for underground utility locations.

A step-down transformer has a single primary winding and a *center-tapped secondary* winding. The center-tapped winding is a connection point in the middle of the winding, which creates two windings in series. The secondary voltage is 240 V_{rms} across the entire winding. The center tap allows access to half the winding, which makes the top and bottom leg of the winding each 120 V_{rms} with respect to the center tap connection. The top and bottom wires connected to the secondary are called *phase conductors*. The wire connected to the center tap is the neutral conductor. The *earth ground* wire is also connected to the center tap. See **Figure 9-2**.

Most circuits for residential purposes use 120 V. In these cases, the phase conductors and the neutral wires supply power to lighting and small appliance outlets. For large appliances like HVACR systems, hot water tanks, oven/ranges, and clothes dryers, 240 V is used, which is powered by the top and bottom phase wires. Phase wires are called *hot wires*, which are ungrounded. These wires are designated as L1 and L2. The designated wire insulator color for the hot wires in residential is typically black or red. Any color except white, gray, or green can be used for hot wires according to NFPA 70.

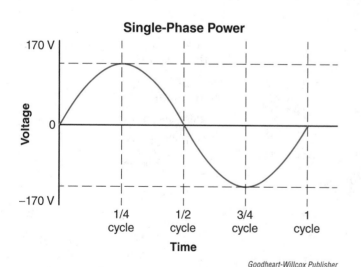

Single-Phase Power

Figure 9-1. Single-phase power provides one current within one cycle.

Goodheart-Willcox Publisher

Copyright Goodheart-Willcox Co., Inc.

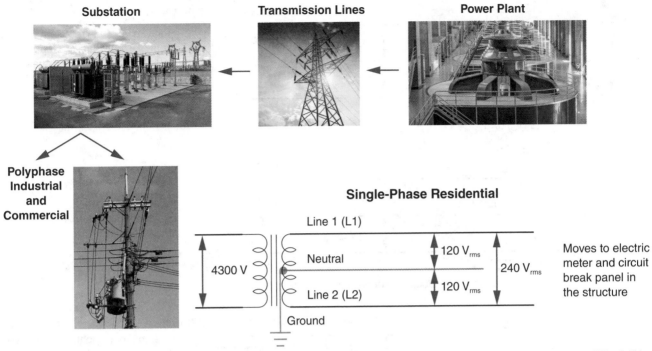

Substation

Transmission Lines

Power Plant

Polyphase Industrial and Commercial

Single-Phase Residential

Line 1 (L1)

Neutral

4300 V

$120 V_{rms}$

$240 V_{rms}$

$120 V_{rms}$

Line 2 (L2)

Ground

Moves to electric meter and circuit break panel in the structure

Figure 9-2. Power distribution is delivered from power plant to substation to end users. The local residential step-down transformer delivers power to the electric meter and circuit breaker panel. Power is sent from substations to commercial and industrial facilities.

A **neutral wire** is a current-carrying conductor that is at the same electrical potential as the ground wire. Neutral is referred to as the grounded power wire. See **Figure 9-3**. Provided that a structure is properly grounded, someone who comes in accidental contact with the neutral and earth or ground wire will not get shocked, **Figure 9-4**.

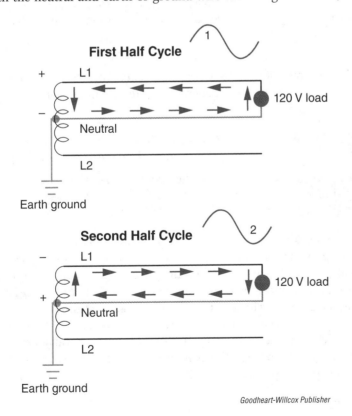

First Half Cycle

1

+ L1

120 V load

– Neutral

L2

Earth ground

Second Half Cycle

2

– L1

120 V load

+ Neutral

L2

Earth ground

Figure 9-3. Neutral is a power wire. The neutral wire completes the circuit to and from the load. During half the cycle, current flows in one direction and changes direction during the second half cycle as the electrical polarities change. The ground wire is at the same potential as the neutral since they are both connected to the secondary winding center tap.

Goodheart-Willcox Publisher

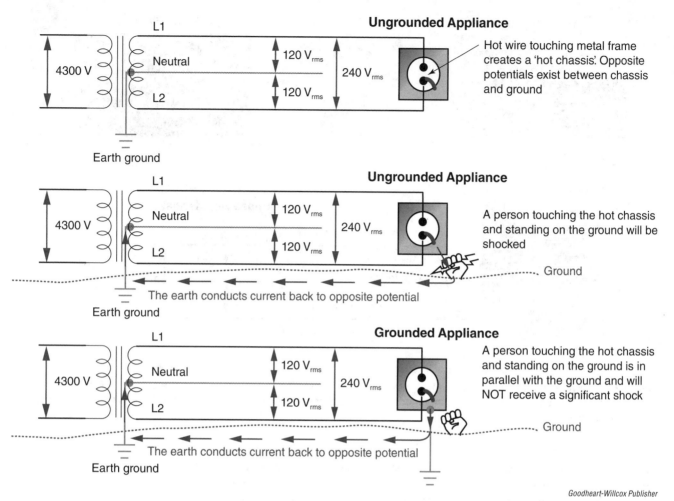

Ungrounded Appliance

Hot wire touching metal frame creates a 'hot chassis'. Opposite potentials exist between chassis and ground

Ungrounded Appliance

A person touching the hot chassis and standing on the ground will be shocked

The earth conducts current back to opposite potential

Grounded Appliance

A person touching the hot chassis and standing on the ground is in parallel with the ground and will NOT receive a significant shock

The earth conducts current back to opposite potential

Goodheart-Willcox Publisher

Figure 9-4. An ungrounded appliance, metal cabinet, will become a hot chassis when a phase wire comes in contact with surface. This will cause electric shock to a person touching it. A grounded appliance will conduct the current back to the transformer through a low resistance ground wire.

Recall that current only flows between opposite potentials. The designated wire insulator color for the neutral wire is white or gray. For internal equipment wiring, always refer to the schematic legend.

Pro Tip

Neutral Wire Terminology

The neutral wire is also referred to as the common wire. This is because all 120 V load neutral wires connect to a common bus bar. See **Figure 9-5**.

A *ground wire* is also connected at the center tap with the neutral wire. The other end of the ground wire is connected to a copper rod called an *electrode*. The electrode typically protrudes 10′ into the ground and conducts current into the ground. Grounding requirements vary depending on soil conditions, the NFPA 70 specifications, and local codes. The ground wire is for safety and not a current-carrying conductor like the phase and neutral wires. It is connected to the metal chassis of a condensing unit. If a power wire accidentally shorts to the metal housing, the ground wire conducts current through the ground back to the opposite potential (neutral potential). The ground path provides the path of least resistance, thus protecting a

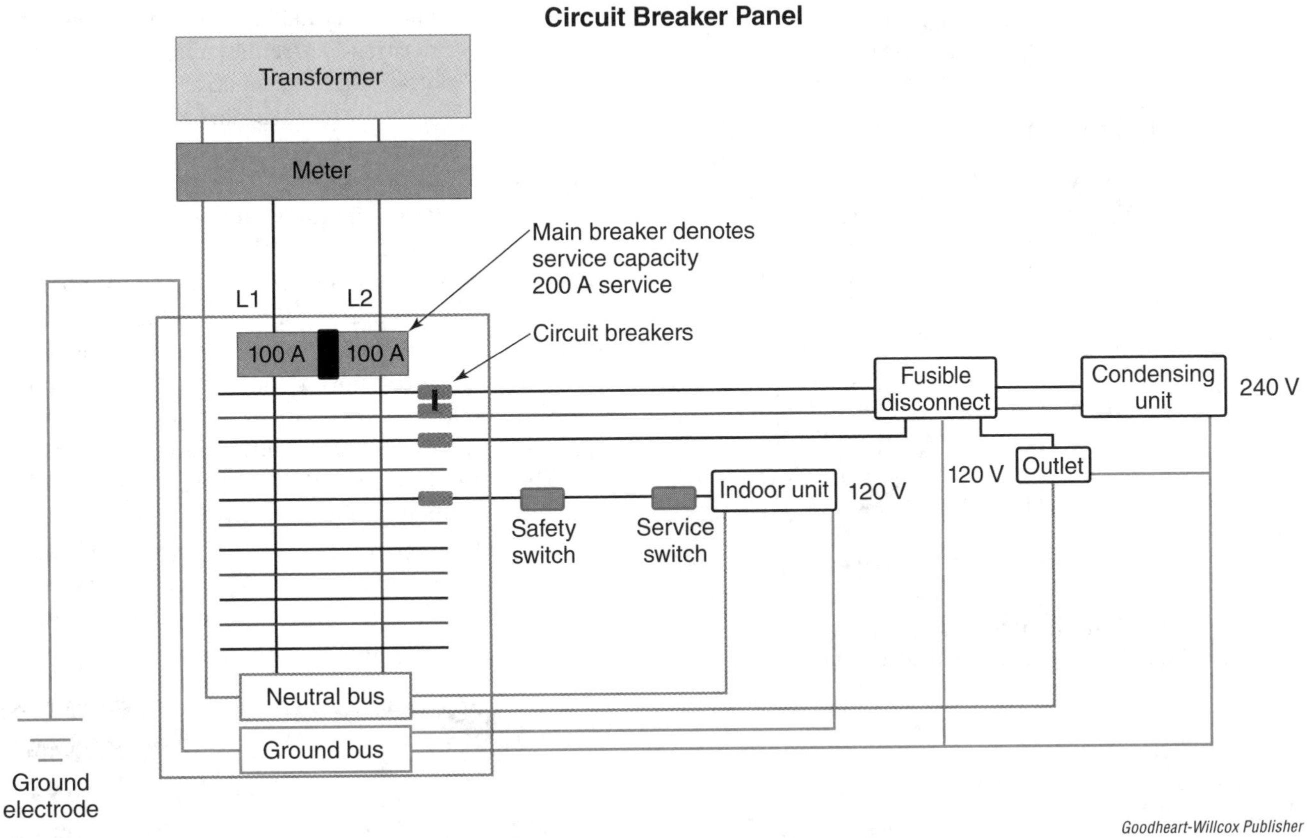

Circuit Breaker Panel
Transformer
Meter
Main breaker denotes
service capacity
200 A service
L1
L2
100 A
100 A
Circuit breakers
Fusible
disconnect
Condensing
unit
240 V
120 V
Outlet
Indoor unit
120 V
Safety
switch
Service
switch
Neutral bus
Ground bus
Ground
electrode

Goodheart-Willcox Publisher

Figure 9-5. Power from the transformer is fed to a meter and to the circuit breaker panel. The feed conductors are protected by main breakers that pass power onto individual branch circuits. The branch circuits are protected by circuit breakers and feed lighting and appliances. The neutral wire must not be switched. It must have uninterrupted continuity back to the neutral bus. The same applies to the ground wires that connects to ground bus.

person touching the metal housing. The designated wire insulator color for the ground wire is green or it can be bare, meaning it has no insulation. All metal boxes, cabinets, and conduit must be grounded. See **Figure 9-6**.

Safety Note

Proper Ground

A technician should always exercise safety. Do not assume that equipment is properly grounded. Homes predating 1970 are not grounded. Always check for proper ground.

9.1.1 AC Polarization and Ground Integrity

An HVACR technician must know how to check polarization and ground integrity. It is often assumed that the neutral wire is negative and not a power wire. Neither the neutral nor the hot wires have fixed polarities.

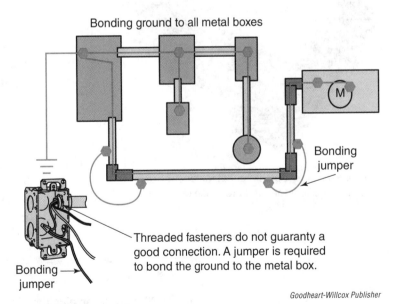

Bonding ground to all metal boxes
M
Bonding
jumper
Threaded fasteners do not guaranty a
good connection. A jumper is required
to bond the ground to the metal box.
Bonding
jumper

Goodheart-Willcox Publisher

Figure 9-6. Metal boxes, panels, cabinets and conduit must be bonded to ground. Threaded fasteners that join conduits to panels do not guaranty a good electrical connection.

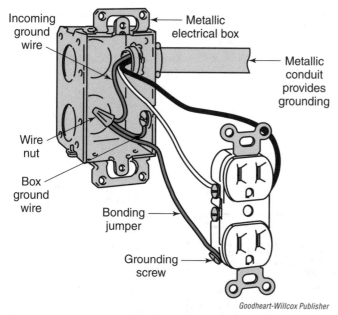

Figure 9-7. A properly bonded ground installation is shown. The ground wire is secured to the metal box and switch frame.

Incoming ground wire
Metallic electrical box
Metallic conduit provides grounding
Wire nut
Box ground wire
Bonding jumper
Grounding screw

Goodheart-Willcox Publisher

Matt Howard/Shutterstock.com

Figure 9-8. A receptacle outlet tester checks for correct polarity of conductors and presence of ground.

Polarities alternate every half cycle of an ac sine wave. The neutral wire carries current to and from the power source. The notion that the neutral is not a power wire most likely comes from the fact that in a properly grounded structure, the neutral wire does not cause shock, as does the hot wire.

Unlike battery terminals, ac polarization is not considered positive or negative. Polarization identifies the hot versus neutral wires for 120 V applications. Phase wires protected by circuit breakers feed individual circuits and the neutral wires complete the circuits back to the power source. Outlets and most plugs are polarized. It is critical that polarities do not reverse when connecting to outlets or appliances.

The outlet in **Figure 9-7** identifies polarities with a wide blade opening for the neutral wire (white) and a narrow blade opening for the hot wire. The ground wire terminal is oval as shown. A polarized plug mates to this receptacle outlet. Switches, fuses, and circuited breakers are placed on hot wires only.

The correct outlet wiring can be checked with a multimeter. The meter displays line voltage when probing the hot and neutral terminals, or the hot and ground terminals. Place a meter probe in the ground terminal. Use the other meter probe to probe the power terminals. The one that produces a line voltage display is the hot terminal. The voltage between hot and ground should be the same between hot and neutral. More than a couple of volts difference indicates loose connections in the ground path back to the circuit breaker panel or beyond. Power outlet receptacle testers are also available for quick testing of outlets, **Figure 9-8**. These do not check ground quality.

9.2 Impedance

Power used by an ac load can be calculated. First, impedance must be known. *Impedance (Z)* is the overall opposition to flow, or total resistance in an ac circuit. In other words, it is the sum of pure resistance, inductive reactance, and capacitive reactance that affects the circuit. The general equation for impedance is expressed as the following:

$$Z = R + X_L + X_C$$

where

Z = impedance (measured in ohms)
R = resistance
X_L = inductive reactance
X_C = capacitive reactance

HVACR circuits are mostly inductive or both inductive and capacitive because they use electromagnetic devices. In most cases, the inductive reactance is ten times greater than pure resistance, and so the resistive value is disregarded. Electric heating elements used for refrigeration defrost and electric heat furnaces are pure resistive circuits. In pure resistive circuits, the voltage and current are in phase. Voltage leads current in inductive circuits, and current leads voltage in capacitive circuits. Therefore, the effect of inductive reactance is countered by the use of capacitors in inductive circuits. This will be demonstrated in the lab section.

9.3 Apparent and True Power

There are two types of power taken into account for calculations. *Apparent power* is the power applied to a circuit and is calculated in volt-amperes (VA). It does not include the effects of inductive reactance or capacitive reactance. The actual power that is consumed by a circuit load is called *true power.* True power is measured in watts. In a pure resistance circuit, the voltage and current are in phase. Both voltage and current waveforms are aligned, so they peak and return to zero at the same time. This means that the load consumes all the power applied to the load. The apparent power and the true power are equal in a pure resistive circuit since the load consumes all the power. So, electrical power is calculated using the equation $P = I \times E$.

In an inductive circuit, current is returned to the power source when the collapsing magnetic field induces an emf. All the applied power is not consumed by the load, so apparent power and true power are not equal. Because the voltage and current are out of phase, the power equation cannot be used directly. The phase shift angle between the voltage and current must be factored in the equation.

An easier way to calculate the true power is to use the power factor given by motor manufacturers. The standard power equation is used as follows:

$$P = I \times E \times pf$$

where

$I \times E$ = apparent power
pf = power factor

The *power factor* is the ratio of real power absorbed by a circuit's load to the apparent power flowing in the circuit. It is a number between 0 and 1 since it is derived by dividing the true power by the apparent power. To calculate the power factor, the following equation is used:

$$Power\ factor = \frac{True\ power}{Apparent\ power}$$

The high or low power factor can tell you a lot about a motor. Consider a small 1/6 horsepower single-phase motor has the following operating characteristics:

- Applied voltage: 120 V
- Inductive clamp current draw: 3 A
- Wattmeter—true power: 171 W
- Apparent power (3 A × 120 V) 360 W
- Power factor (171 W/360 W): 0.475
- Wattmeter power factor display: 0.48

Pro Tip

Wattmeters

A **wattmeter** is an instrument that measures the true power consumed by a circuit. It also displays the power factor.

Based on the low power factor, we can determine the motor is very inefficient. The closer the power factor is to one, the more efficient the electricity is used. Utility companies charge residential and commercial customers different rates with variations in terms of paying for apparent and true power.

The motor did not use a series capacitor. Capacitors improve the power factor when used in an inductive circuit reducing the phase angle between the current and voltage. In a pure inductive circuit, the phase angle between voltage and current is 90°. A capacitor is in series with a motor winding and reduces the phase angle between voltage and current. The current through both is the same as defined by a series circuit. Because the voltage drops across the capacitor and motor winding are 180° out of phase, their sum must be added by using vectors.

Recall Kirchhoff's voltage law that the sum of voltage drops must equal the supply voltage. Either the voltage across the inductive load or the capacitor is greater than the supplied voltage. When these voltages are added by vectors, the sum is equal to the applied voltage. This is done by subtracting the larger voltage by the smaller voltage. If the inductive load is the larger voltage drop, then the circuit is inductive, and voltage leads current flow. The capacitor then decreases the phase angle between voltage and current, and therefore it increases the power factor. The reactive resistance in most HVACR motor circuits will be ten times greater or more than the pure resistance. Therefore, pure resistance is insignificant. The pure resistance is due to the conductors and coils in the circuit.

9.4 Circuit Protection Devices

Circuits must be protected from overheating, which can cause electrical fires or equipment damage. For example, a short to ground causes an uncontrolled current flow and generates enough heat to cause a fire. The current stops flowing when the wire finally breaks due to the intense heat. To prevent this type of condition, circuit protection devices are used.

A *fuse* is a one-time device that must be replaced after an overcurrent event, **Figure 9-9**. The fuse element is designed to melt and break away at specific temperatures that correspond to current. *Overcurrent* occurs when a circuit is overloaded beyond its rated current capacity. For example, the wire is rated for 20 A, but the actual current is 25 A. Damage to the load and wire insulation takes place if this current is not shut off. *Overloads* can be caused by too many loads connected to a single source. This can happen if a load is added to an existing circuit without first checking circuit capacity. Higher-than-normal voltage due to utility equipment failure increases current flow in the circuit. Internal load failure can also increase current flow.

Figure 9-9. Fuses are one-time use circuit protection devices. They must be replaced after the element has opened.

A direct short occurs when there is no circuit resistance, and current flow is uncontrolled. A **short circuit** allows all available current to flow. This current is called the short circuit available current. Short circuit available current is the amount of current a transformer can deliver in the event of a direct short. The magnitude of the current depends on transformer size and distance between the transformer and location of the short. The current is thousands of amperes.

Fuse specifications show ratings for voltage, amperage, interrupting current, and class. Fuses are categorized as fast acting, time delay, or dual acting. A fast-acting fuse opens when the current reaches its rated current. Time-delay fuses are used with inductive loads as they draw initial inrush current. A fast-acting fuse does not allow a motor to start because the fuse opens the instance the motor is turned on. It is not good practice to use an oversized fast-acting fuse. This will allow a motor to start, but it does not adequately protect the circuit. A time-delay fuse allows higher current for a specified time and for a motor to start or a solenoid plunger to fully retract. A dual-acting fuse provides time-delay and short-circuit protection. The fuse has two elements—one to allow a short duration of inrush current and the other to react to a higher current caused by a short. Note that a prolonged inrush current or overload can blow the delay element.

When a fuse element opens, it is called a blown fuse. When a fuse is blown due to a short circuit, arcing across the broken parts of the element must be stopped. If not, the available short-circuit current continues to flow across the arc. Arcing temperatures can reach 10,000°F. Fuses are packed with sand to absorb heat and designed to stop arcing.

Fuses that shut off short-circuit current in less than a half cycle before reaching maximum value are called current-limiting protective devices. The interrupting current specification is rated in thousands of amperes. Class type groups fuses by interrupting current ratings, current-limiting capability, and applications. See **Figure 9-10**. Rejection-type fuse clip holders only accept rejection type fuses. This is to prevent noncurrent limiting fuses from being used. Older noncurrent-limiting fuses will not fit into a rejection type clip. K–5 class current-limiting fuses without the rejection feature for older style fuse clips. Fuses close to the service entrance require greater interrupting current rating in the 200,000 A range and current limiting. Edison and S-type base fuses are for under 150 V only.

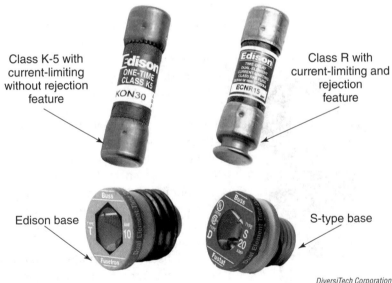

Class K-5 with current-limiting without rejection feature

Class R with current-limiting and rejection feature

Edison base

S-type base

Figure 9-10. Cartridge fuses with and without rejection features. Screw-in type base fuses are limited to circuits under 150 V.

Replace fuses with exact specifications and double check the requirements by comparing to the equipment schematic and service literature. Specific requirements for motor circuits will be addressed in Chapter 13, *Electric Motors*.

Circuit breakers are manually resettable circuit protection devices. They protect against overcurrent due to overloads and short circuits. They are rated by voltage, current, and interrupting current. Interrupting current ratings are typically 10,000 A for residential breakers. A circuit breaker trips when it opens due to an overcurrent event. They are available in one, two, or three poles.

Single-Pole Circuit Breaker

Double-Pole Circuit Breaker

DiversiTech Corporation

Figure 9-11. Single-pole breaker and double-pole breakers.

Goodheart-Willcox Publisher

Figure 9-12. Positive temperature coefficient thermistor used as automatically resettable fuse.

Single-pole is for one hot wire to feed a 120 V circuit. A two-pole is for two hot wires used for 240 V circuit, **Figure 9-11**. The three-pole is for three-phase power circuits that contain three hot wires.

Both thermal and magnetic principals are used in the operation of a circuit breaker. A short-circuit current produces a magnetic field that instantly trips the breaker contacts open. Overload and normal currents do not produce a strong enough magnetic field to open the contacts. Overload currents are handled by the thermal portion. As heat accumulates in the bimetal strip, it warps and trips the contacts. This provides a built-in delay to prevent nuisance tripping of the breaker due to short-current surges. Circuit breakers protect wiring against damage and fire but not the individual branch loads. For example, a small motor that normally draws 3 A is plugged into an outlet fed by a 20 A circuit breaker is not protected. Most motors have their own protection devices. These are covered in more detail in Chapter 13.

As discussed in Chapter 1, *Electrical Safety*, a ***ground-fault circuit interrupter (GFCI)*** compares the current flow in the hot and neutral wires. A difference of 6 mA (0.006 A) can trip the breaker. This breaker has a neutral wire that must be connected to the neutral bus bar. The GFCI breaker protects the entire branch circuit instead of only outlet receptacles when using only an outlet GFCI. In addition to the class A (greater than 5 mA leakage from hot to grounded conductor), the GFCI circuit breaker protects against overloads and short circuits. The device is manually resettable.

A ***positive temperature coefficient thermistor (PTC)***, also referred to as a *resettable fuse*, is a temperature sensitive resistor. See **Figure 9-12**. It is a solid-state device with no moving parts. As temperature increases, so does its internal resistance. PTCs are found in all facets of electronic and electric circuits to protect against overcurrent. The thermistor's resistance drops when the temperature drops. So, if the overcurrent condition is not present and the thermistor has cooled, current flow resumes. More information about thermistors is covered in Chapter 12, *Switches, Electronic Components, and Sensors*.

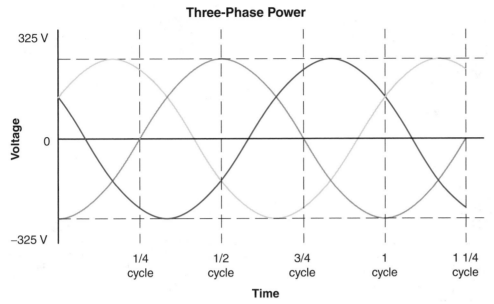

Three-Phase Power

325 V

Voltage

0

−325 V

1/4 cycle 1/2 cycle 3/4 cycle 1 cycle 1 1/4 cycle

Time

Figure 9-13. Three-phase power supplies three currents 120° out of phase within one cycle.

9.5 Commercial Power

Three-phase power is used in commercial applications that require more power than residential. ***Three-phase power*** consists of three currents separated by 120 electrical degrees within one cycle, **Figure 9-13**. Three-phase power is more efficient than single-phase, especially for motors. The power is more consistent and almost resembles dc. Higher voltages result in less current, so less wasteful heat is produced. The lower currents allow for thinner wires thus material cost savings.

There are two popular three-phase configurations for three-phase devices. These configurations are the wye and delta. See **Figure 9-14**. Both have three terminals, which are connected to three power leads that supply each set of windings with one phase of three-phase power. The ***wye (Y)*** configuration has three hot wires connected to a neutral point, while the ***delta (Δ)*** has its three-phases connected like a triangle. Each is available with different voltages. The 208 V wye and 240 V delta are illustrated in this section.

A 208 V wye four-wire transformer secondary can supply 208 V three-phase, 208 V single-phase, and 120 V single-phase. See **Figure 9-15**. Note a neutral wire connection at the center of the three-phase windings. This allows for a 120 V single-phase potential between the neutral wire and any phase wire.

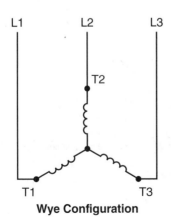

L1   L2   L3

T2

T1   T3

Wye Configuration

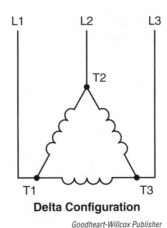

L1   L2   L3

T2

T1   T3

Delta Configuration

Figure 9-14. The wye (Y) and delta (Δ) configurations.

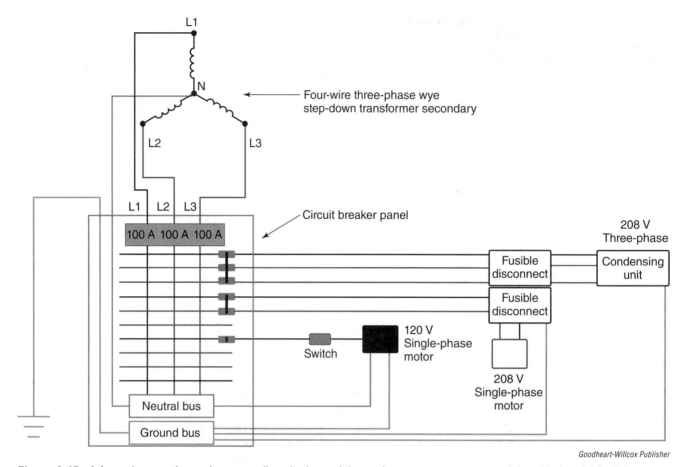

Figure 9-15. A four-wire wye three-phase supplies single- and three-phase power to commercial and industrial facilities.

The potential between any two-phase (hot) wires is 208 V single-phase. The use of all three-phase wires supply 208 V three-phase power to a three-phase load. A three-phase load requires three hot wires to operate. A higher voltage four-wire wye supplies 480 V three-phase, 480 V single-phase, and phase to neutral 277 V.

The 240 V delta four-wire transformer secondary supplies single-phase and three-phase power. See **Figure 9-16.** The delta does not have a neutral point like the wye configuration. In order to produce 120 V, one of the windings is grounded in the center. In the illustration, the ground is between L2 and L3.

Figure 9-16. Three-phase high leg delta system available voltages.

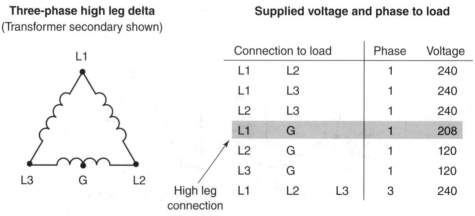

Three-phase high leg delta
(Transformer secondary shown)

Supplied voltage and phase to load

Connection to load			Phase	Voltage
L1	L2		1	240
L1	L3		1	240
L2	L3		1	240
L1	G		1	208
L2	G		1	120
L3	G		1	120
L1	L2	L3	3	240

Essential Electrical Skills for HVACR: Theory and Labs

This provides a 120 V potential between L2 to ground and L3 to ground. L1 produces an unstable higher voltage about 208 V with respect to ground. In this case, L1 is the called the **high leg delta** since it produces an unstable high voltage in delta three-phase power sources. The high leg must be identified with orange paint or marker at distribution panels per NFPA 70.

9.5.1 Measuring Three-Phase

To measure three-phase requires measuring the three combinations between the three hot wires as follows: L1 and L2, L1 and L3, and L2 and L3. Three-phase loads require three hot wires to operate. Current is measured for each phase with an inductive clamp multimeter. A 3 ∅ load will be rated for one current and not three individual ones. Each phase should draw the same current. Measuring power for 3 ∅ motors is covered in Chapter 13, *Electric Motors*.

Since 3 ∅ power supplies three separate currents, these currents should be equal in amplitude. This however would be an ideal condition. In practice, there will be variation between the currents and therefore variations in the phase-to-phase voltages. Three-phase power must be balanced or it can burn out a 3 ∅ motor. The imbalance between phases must be less than 2%. The following method is used to determine three-phase power imbalance:

1. Calculate the following phase measurements: L1 and L2, L1 and L3, and L2 and L3.
2. Determine the average of the phase measurements.
3. Determine the maximum deviation from the average.
4. Calculate using the following equation:

$$\% \ phase \ deviation = \frac{Max. \ deviation \ from \ avg.}{Average \times 100}$$

Summary

- Power plants transmit power through high voltage low current conductors to substations.
- Substations step down the voltage and distribute to local step-down transformers. Then the voltage is lowered for use in residential and commercial use.
- Single-phase power delivers one current of ac per cycle for residential use.
- Three-phase power delivers three out-of-phase currents per cycle for commercial and industrial use. Three-phase provides more power and is more efficient.
- Residential voltages are 240 V across the phase wires and 120 V across any phase wire and the neutral wire.
- A neutral wire and ground wire are connected to the center tap of the utility step-down transformer secondary. Therefore, neutral and ground are at the same potential.
- A neutral wire is a power wire and completes a circuit to the power source.
- A ground wire is not a power wire and does not flow normally. It is used for safety.
- A ground wire conducts current back to the transformer in case of a short to ground condition.
- The polarity of the neutral and hot wire alternates every half cycle.
- The hot or phase wire can be any color except white, gray or green. The most common in residential is black and red.
- Only hot wires should contain switches, fuses, and circuit breakers.

- The neutral wire should be continuous back to the neutral bus. It can be spliced appropriately but never disrupted by a switching device.
- Outlet receptacles, light fixtures, and all loads must be polarized. This means that the hot and neutral wires have specific connection points.
- Resistance to ac current flow is called impedance because there are three types of resistance. Impedance (Z) is the sum of pure resistance plus inductive reactance plus capacitive reactance.
- Since in HVACR circuits, the three types of resistances are in series, the current is the same but the voltage drops are out of phase.
- Apparent power is the power applied to a circuit, while true power is the power consumed by the circuit.
- The power factor is a number between 0 and 1 since it is derived by dividing the true power by the apparent power.
- True power can be calculated if the power factor is known.
- The power should be close to 1 to increase power use efficiency. Capacitors in series with inductive loads improve the factor.
- Circuits must be protected from overcurrent due to overloads and short circuits.
- A fuse protects against overcurrent by melting a conductive element.
- Fuses are rated by voltage, current, and interrupting short-circuit current.
- Fuses are categorized as either fast acting, time delay, or dual acting.
- Fuses that shut off short-circuit current in less than a half cycle before reaching maximum value are called current-limiting protective devices.
- Circuit breakers are manually resettable circuit protection devices.
- They are rated by voltage, current, and interrupting current. Interrupting current rating is typically 10,000 A for residential breakers.
- Circuit breakers protect wiring against damage and fire but not the individual branch loads.
- The ground-fault circuit interrupter (GFCI) compares the current flow in the hot and neutral wires. A difference of 6 mA (0.006 A) will trip the breaker.
- In addition to class A (greater than 5 mA leakage from hot to grounded conductor) protection, the GFCI circuit breaker will protect against overloads and short circuits.
- A PTC is an automatically resettable device known as a resettable fuse.
- Three-phase power is used for commercial and industrial.
- Three phase consists of three currents separated by 120 electrical degrees within one cycle.

Lab 9.1 Checking Residential Power

This lab introduces residential power to receive a better understanding of how it is produced and delivered in a circuit.

Lab Introduction

Residential power delivers 240 V between hot wires and 120 V between hot and neutral. A multimeter will be used to measure the above voltages. A receptacle outlet will be wired to the lab board and tested for correct polarity and grounding. A circuit breaker panel or disconnect box or 240 V receptacle will be needed for this part of the lab. The 240 V section must be performed in your institution lab room and under supervision of your instructor. A 3 ∅ panel or disconnect box may also be used to measure single-phase 208 V or 240 V. The separate feed feature of the outlet will be used to show circuit control. The student will connect wires to a plug to safely test the outlet. A receptacle will be added to the end to produce an extension cord that will used to check current with the inductive clamp for plug-in loads. Recall that only one conductor can be used to measure current with the inductive clamp.

Equipment

- Safety glasses
- Lab Board
- 1—15 A, 120 V receptacle outlet
- 14 AWG black, white, green wire
- 1—15 A, 125 V three-wire plug
- 1—15 A, 125 V three-wire connector
- Classroom circuit breaker panel or disconnect box with available 240 V or 208 V (3 ∅) and 120 V. (A 240 or 208 V receptacle outlet may be used.)
- Optional—terminal strip jumper clips
- Electric outlet tester

Procedure

1. Refer to **Figure 9.1-1**. Locate and cut the hot side separate feed fin. This is done to isolate the top and bottom outlets. Do not cut the fin on the neutral side.

Tamper resistant

Nontamper resistant

Cut the fin for separate feeds. Never cut the neutral side fin.

For Lab 9-1 cut the hot fin. The top outlet is controlled by the switch. The bottom is always hot and turned off only when main power is off.

Tamper resistant feature does not allow an object to be inserted into only one side. Objects have to be inserted simultaneous as with a power plug.

Figure 9.1-1.

Goodheart-Willcox Publisher

2. Refer to **Figure 9.1-2**. Securely attach appropriate wire color 14 AWG wire to the plug terminals. Make sure the wire length reaches between the switch box test points on the terminal strip. The extension cord should not be less than 1′.

Wire plug according to manufacturer's instructions

Figure 9.1-2.

Goodheart-Willcox Publisher

Continued▶

160

Copyright Goodheart-Willcox Co., Inc.

3. Follow manufacturer instructions. After this lab, a connector will be added to the other side to make an extension cord with separated wires for measuring current with the inductive clamp.

4. Wire the lab board per lab **Figure 9.1-3**. Use terminal strip jumpers if available.

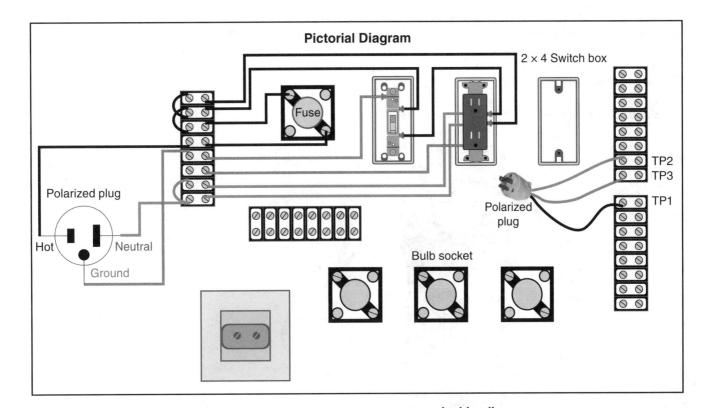

Pictorial Diagram

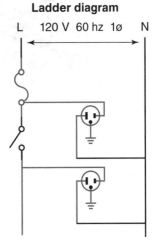

Ladder diagram

Figure 9.1-3.

Goodheart-Willcox Publisher

5. Plug in the polarized plug to the top receptacle outlet.

6. Check for shorts to ground with multimeter as with previous labs.

7. Leave the single-pole switch in the off position.

8. Get instructor approval before plugging in lab board. If approved, plug in board.

9. Measure the voltage across TP1 and TP2 and record value. Voltage should be zero.

Continued▶

10. Turn switch on. Measure between TP1 and TP2. Record value.
11. Measure between TP1 and TP3. Record value.
12. Measure between TP2 and TP3. Record value.
13. Remove plug from top outlet and insert plug into bottom outlet.
14. Turn off single-pole switch.
15. Measure between TP1 and TP2. Record value.
16. Test outlet polarity and presence of ground with plug tester and record result. Follow manufacturer's instructions.
17. After lab completion, install the three-wire connector to complete the extension cord. See **Figure 9.1-4**.

Extension cord to check current with inductive clamp for corded appliances

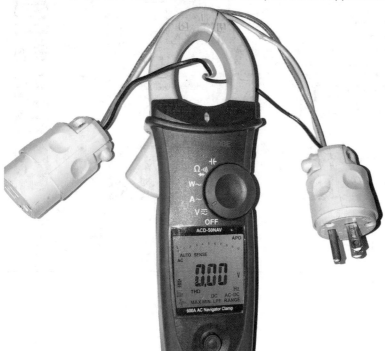

Figure 9.1-4.

Lab Questions

1. What is the voltage from Step 9?

_____

2. What is the voltage from Step 10?

_____

3. What is the voltage from Step 11?

_____

4. What is the voltage from Step 12?

_____

5. What is the voltage from Step 15?

_____

6. Explain voltage reading from Step 15.

_____

_____

_____

_____

7. Based on the outlet voltage measurements, is the polarized outlet wired correctly?

_____

_____

8. Is there a proper ground or is it compromised?

_____

_____

9. Does your outlet test conclusion agree with the outlet tester result?

_____

_____

This lab will walk you through calculating the power delivered through a circuit and determining the power factor.

Lab Introduction

To receive the most from this lab, a wattmeter should be used. However, the lab can be performed without the watt-meter. Current and voltage will be measured to calculate the apparent power delivered to a relay coil. The true power is measured with a wattmeter. The power factor can then be calculated. Most wattmeters display power factor as well. The power factor will be improved by adding a capacitor in series with the relay coil.

Equipment

- Safety glasses
- Multimeter
- Wattmeter
- Lab Board and connecting wires
- 1 – 3 to 5 µf run capacitor (only 24 V will be used) (do not use larger capacitor as coil may burn out)
- 1 – General purpose relay (Honeywell R8222 or equivalent)

☰ Procedure

1. Assemble circuit on the Lab Board per lab **Figure 9.2-1** using the jumper wire shown.

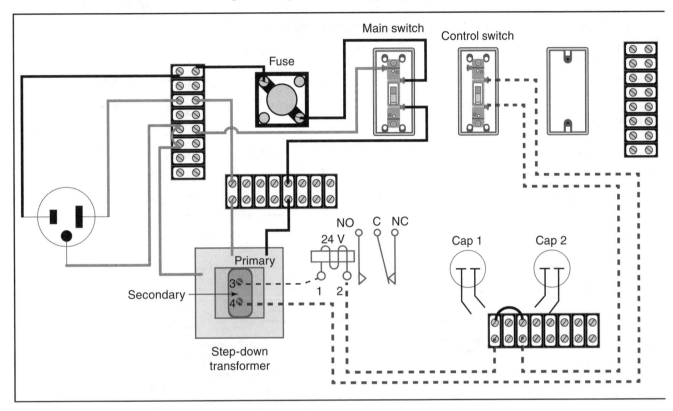

Legend

———— Line voltage 120 V

- - - - - Line voltage 24 V
control circuit

Numbers denote test points

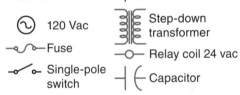

⊘ 120 Vac

⌐⌐ Fuse

⁄⁀ Single-pole switch

⊰⊱ Step-down transformer

—○— Relay coil 24 vac

⊣⊢ Capacitor

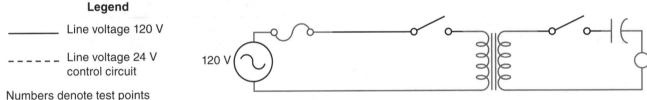

Figure 9.2-1.

Goodheart-Willcox Publisher

2. Check for shorts to ground with multimeter as with previous labs.
3. Get instructor approval before plugging in lab board.
4. Turn on power and control switch.
5. Measure the voltage at the step-down (applied voltage) and record.
6. Measure the current with the inductive clamp. Note if using a high dc resistance coil, use the ten-loop method to measure current.
7. Multiply the current by the voltage to find the apparent power and record.
8. If available, measure the true power with a wattmeter and record. If not proceed to Step 10.
9. Find the power factor by dividing true power by apparent power and record.
10. Turn off control switch and main switch.
11. Remove jumper from terminal switch and connect the capacitor.

Continued▶

12. Turn main and control switches on.
13. Measure the secondary voltage and record.
14. Measure the current with the inductive clamp.
15. Multiply the current by the voltage to find the apparent power and record.
16. If available, measure the true power with a wattmeter and record.
17. Find the power factor by dividing true power by apparent power and record.
18. Properly discharge the capacitor.
19. Turn off switches and unplug the Lab Board.

Lab Questions

Use your digital multimeter or other electrical instruments to answer the following questions.

1. What is the applied voltage from Step 5?

_____

2. What is the current draw from Step 6?

_____

3. What is the apparent power from Step 7?

_____

4. What is the true power from Step 8?

_____

5. What is the power factor from Step 9?

_____

6. What is the secondary voltage from Step 13?

_____

7. What is the current draw from Step 14?

_____

8. What is the apparent power from Step 15?

_____

9. What is the true power from Step 16?

_____

10. What is the power factor from Step 17?

_____

11. How much did the current increase when the capacitor was added?

_____

12. What is the percentage of power factor improvement with the capacitor?

Percent increase = (No cap power factor - with cap power factor) × 100

_____

Lab 9.3 Checking Three-Phase Voltage

This lab will require three-phase power in the lab facility.

Lab Introduction

The lab must be performed under the supervision of an instructor. A breaker panel, disconnect or three-phase ground outlet can be used. Three-phase voltages will be measured with a multimeter to identify the type of system used in your lab facility and available voltages. Line imbalance will be calculated by using the worksheet provided in this lab. System grounding will be checked by measuring voltage between phase wires and ground. All metal junction boxes, breaker panels, and disconnect switch boxes should be bonded to ground.

Equipment

- Approved safety glasses and electrically insulated gloves
- Multimeter
- Worksheet Figure 9.3-1

 ## Procedure

1. Put on personal protective equipment.
2. Ensure meter leads are installed and working. Set mode selector is set for ac volts.
3. Get approval to turn power on. Proceed with instructor supervision.
4. Measure phase-to-phase voltages and record. L1 and L2, L1 and L3, and L2 and L3.
5. Measure phase wires to Neutral and record. L1 and N, L2 and N, and L3 and N.
6. Measure phase wires to ground wire. L1 and G, L2 and G, and L3 and G.
7. Measure phase wires to ground unpainted metal surfaces. L1 and G, L2 and G, and L3 and G.
8. Fill out line imbalance worksheet. See **Figure 9.3-1**.

1. Phase measurements

					round to whole number
A)	L1	and	L2		
B)	L1	and	L3		
C)	L2	and	L3		
			sum =		divide by 3 = average _____

2. Average of measurements

Enter the difference between phase measurements and the average: A and Avg _____ B and Avg _____ C and Avg _____

The largest difference is the maximum deviation from the average _____

3. Maximum deviation from average _____

4. Calculate: *% phase deviation from avg. / Average × 100*

% phase deviation = _____ / _____ × 100 = _____ %

Figure 9.3-1.

Lab Questions

Use your digital multimeter or other electrical instruments to answer the following questions.

1. Are the phase-to-phase measurements in Step 4 close in value?

_____

2. Are the phase-to-phase measurements in Step 5 close in value?

_____

3. Which wire is the high leg?

_____

4. Are the phase-to-phase measurements in Step 6 close in value?

_____

5. Are there different values between grounded components?

_____

6. What type of three-phase was used?

_____

7. Is the system imbalance below 2%?

_____

Know and Understand

_____ 1. What should the voltage be across L1 and L2 in a residential power source?
  - A. 120 V
  - B. 208 V
  - C. 240 V
  - D. 277 V

_____ 2. Which color is *not* allowed for a phase or hot wire?
  - A. Red.
  - B. Orange.
  - C. Black.
  - D. Gray.

_____ 3. What is the difference in potential between the neutral and ground wires?
  - A. 120 V
  - B. 240 V
  - C. 24 V
  - D. 0 V

_____ 4. Which statement is *not* true about the neutral wire?
  - A. Neutral wire is not a power wire.
  - B. It is the grounded conductor.
  - C. It completes the load circuit.
  - D. Polarity changes every half cycle.

_____ 5. *True or False?* Circuit breakers, fuses, and switches can used on the hot wires, but not the ground wire.

_____ 6. Total ac resistance is called _____.
  - A. pure resistance
  - B. inductive resistance
  - C. reactive resistance
  - D. impedance

_____ 7. In a pure resistive circuit, which statement is *not* correct?
  A. $P = I \times E$
  B. Current and voltage are in phase.
  C. True power is greater than apparent power.
  D. True power equals apparent power.

_____ 8. *True or False?* Capacitors improve the power factor when used in an inductive circuit.

_____ 9. Power factor is calculated by _____.

  A. $\dfrac{apparent\ power}{true\ power}$

  B. $P = I \times R \times true\ power$

  C. $\dfrac{true\ power}{apparent\ power}$

  D. $P = I \times R$

_____ 10. If the power factor is known, how is it used to find true power?

  A. $P = \dfrac{I \times E}{power\ factor}$
  B. $P = I \times Z \times power\ factor$
  C. $P = I \times E \times power\ factor$
  D. $P = I \times R \times Z$

_____ 11. In an inductive circuit, _____.
  A. true power is greater than apparent power
  B. apparent power is greater than true power
  C. current leads voltage
  D. $P = I \times E$

_____ 15. Which condition is *not* an example of an overload?
  A. Too many loads connected to the same branch circuit.
  B. There is a temporary voltage increase.
  C. An internal load failure results in increased current flow.
  D. The insulation on a hot wire wears and the bare wire contacts a grounded surface.

_____ 13. A fuse that allows inrush current without blowing but still protects against overloads and shorts to ground is the _____.
  A. time delay   C. fast acting
  B. dual acting   D. inrush acting

_____ 14. Which is *not* true about rejection type fuses?
  A. Have time delay features.
  B. Interrupting short-circuit current above 50,000 A.
  C. Automatically resettable.
  D. Protect against overloads.

_____ 15. *True or False?* Circuit breakers can protect all loads.

_____ 16. Which statement is *not* true about circuit breakers?
  A. Contain both thermal and magnetic elements.
  B. Interrupting current of at least 10,000A.
  C. Protect against short circuits.
  D. Reset automatically.

_____ 17. *True or False?* GFCI circuit breakers can protect against overloads and short circuits.

_____ 18. Which solid-state device acts as an automatically resettable fuse?
   A. PTC thermistor.
   B. Dual acting.
   C. Fast acting.
   D. Reusable.

_____ 19. Which voltage is *not* available with a four-wire wye system?
   A. 240 V single-phase
   B. 120 V single-phase
   C. 208 single-phase
   D. 208 three-phase

_____ 20. *True or False?* Three-phase line imbalance should be below 2% between phases.

Critical Thinking

1. The voltage across a condensing unit feed is 240 V. The voltage across one hot leg and an unpainted metal part of the condenser cabinet measures 90 V. Is the unit properly grounded? What should you look for?

_____

_____

_____

_____

2. If two 370 V rated capacitors are wired in parallel and one fails, can it be replaced with a lower rated value as with capacitors wired in series?

_____

_____

10 How Electric Meters Work

Goodheart-Willcox Publisher

Learning Objectives

After completing this chapter, you will be able to:

- Describe the operation of an analog meter.
- Explain the purpose of a shunt.
- Discuss meter configurations to measure voltage, current, and resistance.
- Describe range settings.
- Describe the operation of a digital meter.
- Explain how to read digital meter displays.
- Explain the digital meter configurations to measure voltage, current, and resistance.
- Describe additional built-in digital meter functions.
- Describe special function meters.
- Explain meter specifications.

Technical Terms

analog meter	automatic ranging	meter movement	shunt
analog signal	digital meter	microprocessor	test lead
analog to digital converter (A/D)	duty cycle	noncontact voltage (NCV)	volt-ohm-milliammeter (VOM)
	infinity	phantom voltage	
audible continuity test	megohmmeter	scientific notation	wattmeter

Goodheart-Willcox Publisher

Figure 10-1. An analog meter is known as a volt-ohm-milliammeter (VOM).

Introduction

A multimeter has been discussed and used to take measurements in prior chapter labs. This chapter will examine a meter's internal operation and each of its meter functions. Both analog and digital meters have been used in the HVACR field. The operating characteristics of the original analog meter are similar to those for the digital meter, but its mechanical complexities have been replaced by electronics. Today, the digital multimeter is most commonly used due to its high accuracy and usability. It contains software-driven microprocessors that allow the meter to hold multiple functions.

Understanding the internal operation of these meters is critical in troubleshooting electrical circuits and components as well as determining when a meter malfunctions. A technician also must be familiar with the safety features and capabilities of a meter, which can be understood by interpreting the owner's manual specifications.

10.1 The Analog Meter

An **analog meter** is an electromechanical device that measures current flow to determine the basic electrical quantities of voltage, current, and resistance. The current measurement on an analog meter is displayed by a pointer, or needle, that moves due to an internal magnetic coil. The magnetic coil and pointer are known as the **meter movement**. The meter movement is calibrated by the meter designer to determine voltage, current, and resistance through Ohm's law principles. The meter designer uses calculated resistance values that are correlated to the amount the pointer travels. The amount of travel indicates the amount of current measured. Each meter function requires a specific circuit configuration to control the meter movement.

A **volt-ohm-milliammeter (VOM)** is an analog meter that combines the three basic functions in one unit. See **Figure 10-1**. The correct range, or the maximum value to be measured, must be selected, **Figure 10-2**. For example, the 50 V range has a maximum value of 50 V, so it cannot be used when measuring 120 V—the 250 V range must be used. If not, the movement can become damaged. See **Figure 10-3**.

The following sections describe how measured current is translated into current, voltage, and resistance based on the mode and range settings in a typical VOM.

10.1.1 Current Measurement

The analog meter measures ac and dc. With ac, it is first rectified to dc to prevent inductive effects. The meter itself can only handle a minute amount of current, so the meter is calibrated where a small amount of current measured correlates to the actual amount of current. To prevent a large amount of current to pass through

Current ranges

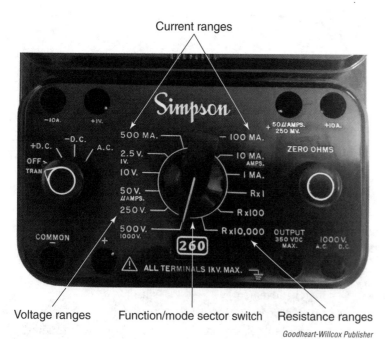

Voltage ranges   Function/mode sector switch   Resistance ranges

Goodheart-Willcox Publisher

Figure 10-2. The mode switch selects the functions and corresponding ranges.

120 V on 250 V range

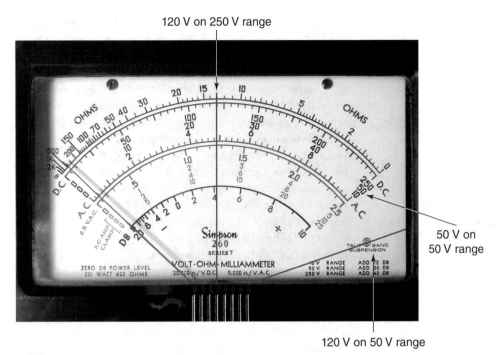

Figure 10-3. A voltage measurement using two ranges is shown. The measure 120 V the 250 V range must be selected. If the 50 V range is selected, the pointer will be off scale and the movement can become damaged.

50 V on
50 V range

120 V on 50 V range

Goodheart-Willcox Publisher

the meter, a shunt is used. A ***shunt*** is a resistor that produces a low-resistance path for current to pass through another part of the circuit. This prevents the device from overheating. Analog ammeters typically have four or more current ranges that use internal and external shunts. Shunts are available in the hundreds of amperes.

The magnetic coil is wired in parallel to the shunt. See **Figure 10-4.** In this example, we see if the coil has a maximum capacity of 1 mA (0.001 A), then shunt 1 is used for a 10 mA range. When a 10 mA current is measured, 90%, or 9 mA, flows through the shunt.

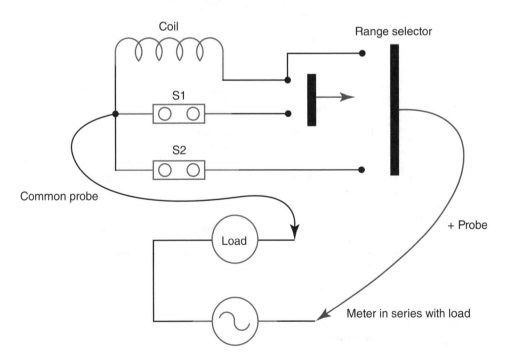

Figure 10-4. The meter is configured to measure current.

90% of current flows through shunt 1 and 10% flows through the coil.
99% of current flows through shunt 2 and 1% flows through the coil.

Goodheart-Willcox Publisher

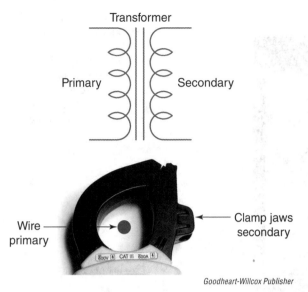

Figure 10-5. An inductive clamp is used to measure current without having to place the meter in series with load.

Goodheart-Willcox Publisher

One percent flows through the coil and deflects the needle to the 10 mA on the display. Shunt 2 is used for a 100 mA range. A 100 mA measured current causes 1%, or 0.001 A, to flow through the coil. The remaining 99% flows through shunt 2.

Current measurement requires an in-line connection. This means the meter must be connected in series with the load. The inductive clamp is the safest and most preferred method to check large currents since a direct connection is not required. See **Figure 10-5**. The conductor being measured acts as a primary winding in a transformer. The rising and collapsing magnetic field induces a current into the clamp. The clamp acts as a transformer secondary and passes this current into the meter movement. An add-on inductive clamp can be connected to many VOM meters, or combination units are also available.

10.1.2 Voltage Measurement

Voltage can be measured by placing the VOM in parallel with the power source. A resistor is placed in series with the measured voltage and the meter coil. Like current measurements, ac voltage is rectified to dc voltage. See **Figure 10-6**.

In this example, the coil has a dc resistance of 1000 Ω, so a series resistor of 49,000 Ω is required to limit the coil current to 0.001 A when 50 V is measured in the 50 V range. The 50 V is the maximum scale value and is the maximum voltage that can be measured. Higher voltages cause movement damage. To use the 250 V range, a 249,000 Ω resistor is required to limit current to 0.001 A. If the meter is placed across 250 V, then 249 V is dropped across the series resistor, and 1 V is dropped across the coil. The pointer is at full deflection when 0.001 A flows through and a corresponding voltage drop of 1 V.

Regardless of the number of ranges, the coil current and voltage drop across a coil does not exceed its limits. The current does not exceed 0.001 A, and the voltage drop does not exceed 1 V for the movement coil in this example.

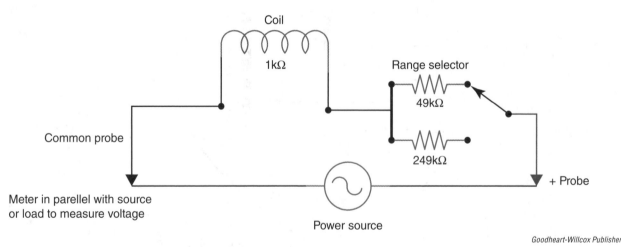

Goodheart-Willcox Publisher

Figure 10-6. A voltage measurement using two ranges. The meter is configured to measure voltage. Note only two ranges are shown for simplicity in this figure, but actual meters have multiple range selections.

Maximum Voltage Ratings

10.1.3 Resistance Measurement

The same meter movement used for voltage and current is used for the resistance measurement. This means there are the same coil current limits. The meter must be connected in parallel to the component. The resistance function configures the meter to supply power to the component being tested. The meter's internal battery supplies the power for current to flow, **Figure 10-7**. The typical battery (dc) voltage is between 1.5 V and 9 V.

When using the resistance function, the resistor under test must form a series circuit with the meter movement coil and shunt. Test probes are then connected together. At this point, the pointer must fully deflect to show 0 Ω. The total coil circuit resistance made up of the coil, variable resistor, and fixed resistor limits coil current to 0.001 A. The variable resistor, named the zero ohms adjustment, compensates for battery voltage loss. If 0 Ω is not displayed when test probes are connected together, the zero adjustment is turned until the pointer is over the zero mark. A weak meter battery does not allow full pointer deflection.

When the test leads are connected to a component, current flows from the battery through the parallel branches formed by the coil circuit and shunt resistor. Then it flows through the component under test and continues back to battery positive battery terminal. See **Figure 10-7**. A 1000 Ω resistor is measured by using the R × 1 and R × 100 range selections on the meter to show how the range setting affects pointer deflection.

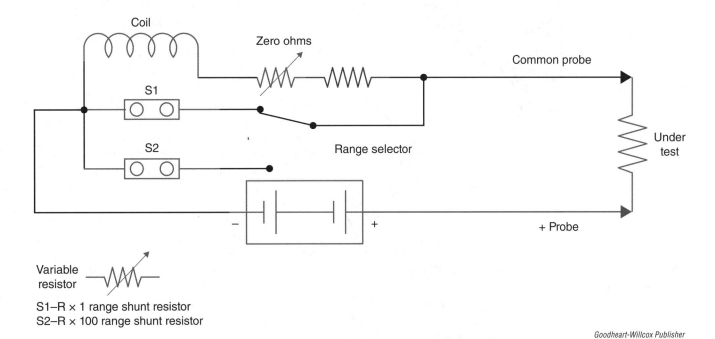

S1–R × 1 range shunt resistor
S2–R × 100 range shunt resistor

Goodheart-Willcox Publisher

Figure 10-7. A resistance measurement using two ranges. The meter is configured to measure resistance.

Figure 10-8. Selecting the R × 100 results in better accuracy to measure the 1000 Ω resistor.

R × 100 range

R × 1 range

1000 Ω resistor measured using R × 1 and R × 100 ranges

Goodheart-Willcox Publisher

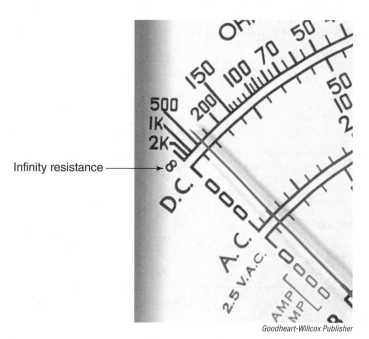

Infinity resistance

Goodheart-Willcox Publisher

Figure 10-9. Resistance greater than the meter's maximum capacity is displayed as infinity.

The R × 1 range means most of the current flows through the shunt, leaving only a fraction of current to flow through the coil. As a result, pointer deflection is minimal. See **Figure 10-8**. In this example, the pointer is on the 1k mark. This is a small region with many increments, which can cause more inaccurate readings. The R × 100 range is a better choice for measuring a 1000 Ω resistor. With this range, less current flows through the shunt, thus allowing more current through the coil and more pointer deflection. Notice the pointer is about 60% of travel and points to 10. Because the increments are farther apart, there is higher accuracy. Since the R × 100 scale is used, the marked value 10 is multiplied by 100 and results in 1000 Ω.

All meters have a maximum value of resistance that can be displayed. The meter shown has a maximum value of 20,000,000 Ω per manufacturer's specification. The 20,000,000 is indicated by the 2k mark and using the R × 10,000 range (2000 × 10,000 = 20,000,000). Left of the 2k mark is the *infinity* symbol (∞) which indicates that the meter is out of range. A resistance greater than 20,000,000 would be displayed as infinity. See **Figure 10-9**.

10.2 The Digital Meter

A *digital meter* is a device that displays electrical quantities as a numerical value rather than a position on a meter. Digital pertains to only two states: zero and one. Digital electronics convert the analog information, or the *analog signal*, in long strings of zeros and ones through a device called the ***analog to digital converter (A/D)***. The processed information is then passed on to a ***microprocessor***, which is the central processing unit and contains all the internal functions. The microprocessor uses stored calibrated tables and Ohm's law calculations to generate the information on the display module for the individual to read.

A digital meter has many advantages:

- It has higher accuracy.
- It has both automatic and manual ranging.
- It measures frequency and duty cycle.
- It has an easy-to-read display with back lighting.
- There are multiple functions built-in, including the capacitor test, diode test, audible continuity, built-in inductive clamp, inrush current, maximum and minimum value recording, and long-term recording.
- It has remote access to data, such as Bluetooth, Wi-Fi, and internet.
- It includes data storage.

Digital meter functions are selected through a mode switch. Many meters can have additional switches for different features. This is based on the meter's brand and specific model. See **Figure 10-10**. Like analog meters, digital meters use ranges to protect the meter electronic components from excessive current.

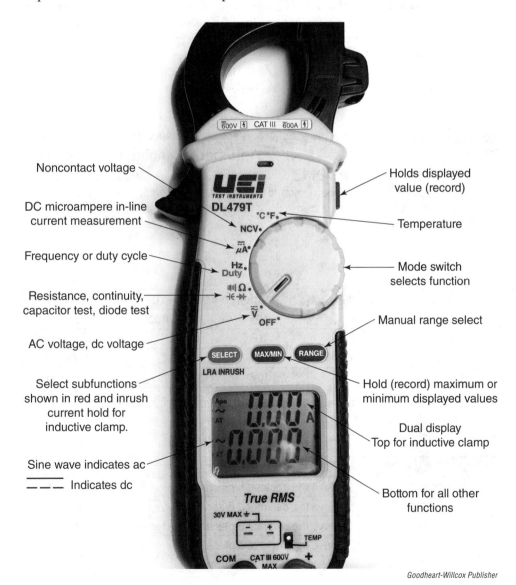

Figure 10-10. A digital multimeter's various selection switches.

Noncontact voltage

DC microampere in-line current measurement

Frequency or duty cycle

Resistance, continuity, capacitor test, diode test

AC voltage, dc voltage

Select subfunctions shown in red and inrush current hold for inductive clamp.

Sine wave indicates ac

Indicates dc

Holds displayed value (record)

Temperature

Mode switch selects function

Manual range select

Hold (record) maximum or minimum displayed values

Dual display
Top for inductive clamp

Bottom for all other functions

Goodheart-Willcox Publisher

Most meters use an *automatic ranging* feature, so user selection is not required. The meter analyzes the measured current and function used and then selects the range to be displayed. These meters also allow manual range selection.

Pro Tip

Multimeter Functions and Terminology

The digital multimeter contains several meter functions combined in one piece of test equipment. When voltage is measured, it is common to say that the voltmeter is used. This applies for using the term ohmmeter when measuring resistance and ammeter when measuring current. The meter function selector switch, also called mode selector switch, is positioned to select the type of measurement to be taken.

10.2.1 Reading the Display

Most portable digital meters display only four characters, or digits, with 1999 being the typical maximum display. Because values ranging from 0.000001 to 40,000,000 (this upper value may vary among meter models) can be measured on some meters, prefixes are used in front of a unit of measure to express these values. Prefixes are used for A, V, Ω, and F. The prefixes use *scientific notation*, which is used to represent a large or small number. It expresses a number by a power of ten in increments of three. For example, 10^3 and 10^6 are expressed in this notation, which mean 1000 and 1,000,000 respectively. The table shown in **Figure 10-11** shows various prefixes and their corresponding basic units. Often, it is necessary to convert prefix values to basic units, **Figure 10-12**.

Pro Tip

Automatic Ranging and Prefixes

The user must be aware of prefixes when reading the display. Most meters are auto-ranging, which cause the device to display values using prefixes. Always take the time to read the value correctly. If a display shows 0.457 kΩ instead of 457 Ω, ensure the prefix is noted.

Figure 10-11. Prefixes denote specific unit values.

Prefix	Symbol	Notation	Basic Unit	Example
mega	M	10^6	1,000,000	1 MΩ
kilo	K	10^3	1000	1 kΩ
milli	m	10^{-3}	0.001	1 mA
micro	μ	10^{-6}	0.000001	1 μA , 1 μF
nano	n	10^{-9}	0.000000001	1 nF

Goodheart-Willcox Publisher

Figure 10-12. Convert a basic unit by moving the decimal point to a new place value. The prefix is then dropped.

Display	Convert to Basic Unit
0.457 MΩ	Move the decimal 6 places to the right to yield 457,000 Ω
0.457 kΩ	Move the decimal 3 places to the right to yield 457 Ω
42.34 mA	Move the decimal 3 places to the left to yield 0.042 A
4.234 μA	Move the decimal 6 places to the left to yield 0.000004 A *

* Rounded to the millionth place. The 234 portion of the value is insignificant.

Goodheart-Willcox Publisher

10.2.2 Current Measurement

Most HVACR multimeters' in-line current measurement capacity is limited to the dc μA range. However, multimeters used for HVACR direct digital controls can measure in-line ac μA, dc μA, and mA ranges. Some multimeters can measure up to 10 A ac and dc in-line currents by using larger shunts. Currents larger than 10 A are not common for portable meters. The inductive clamp (or clamp-on meter) is the safe and simple way to measure the larger current draw of motors.

Figure 10-13 illustrates the meter configuration for measuring current. The meter is in series with the load, so that load current flows through the meter. The red probe connects to the power source. Some of the current flows through the fuse and feeds the analog to digital converter, while the remaining larger current flows through the shunt and to the load.

When the inductive clamp is used, the small-induced current is first amplified and then sent to the A/D. It is processed the same as in-line current measurement.

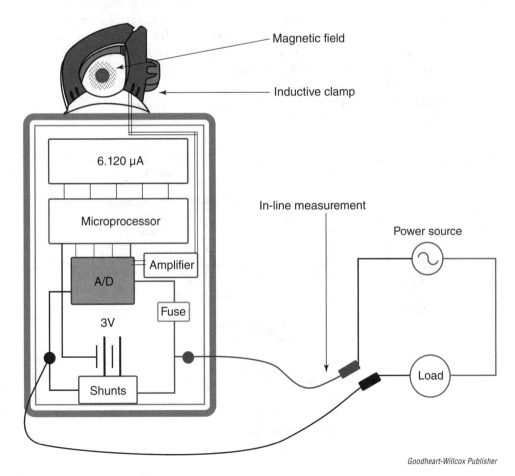

Figure 10-13. A digital multimeter configured to measure current by in-line or inductive clamp method.

Goodheart-Willcox Publisher

Currents below 0.5 A do not induce a strong enough current into the inductive clamp. The ten loop coil method must be used, which was first introduced in Lab 7.1, *Inductive Reactance and Transformer Evaluation*. The typical HVACR clamp measures up to 600 A, although greater capacities are available.

When using the inductive clamp, only one wire can be placed inside the jaws. The wires that have opposite current flow cause the magnetic fields to cancel out, resulting in no induced current in the clamp. Even if the current flows in the same direction, the reading can be unreliable. The rule is one wire only. Avoid placing the clamp near strong magnetic fields. The magnetic field generated by transformers, solenoids, relays, and contactors alter the measurement.

10.2.3 Voltage Measurement

When the meter is configured for voltage measurement, current flows through a high resistance, **Figure 10-14**. A high resistance in the millions of ohms limits the current for the delicate electronic components. The series resistor passes current through the fuse to feed the A/D. The digital information is fed to the microprocessor. There, the current and type ac or dc, selected by the user, is processed and sent to the display unit.

Digital voltmeters use an insignificant amount of power when connected to a circuit. They are high impedance measuring devices, and thus, they do not become part of the circuit under test. This is different from the low-impedance analog voltmeter.

When measuring voltage, the meter probes are placed in parallel across a voltage source, load, or switch. Note: Voltage is measured across, and current is measured through loads, per industry terminology. Current flow through the meter is shown in **Figure 10-15**. When the meter's probes are across the source voltage, current flows only through the meter. For the meter testing, voltage across the open switch current flows from the source through the meter and load back to the source to complete the circuit.

In **Figure 10-15A**, the meter's internal 10 MΩ resistor is in series with the load's 1 kΩ resistance. In a series circuit, the largest resistance drops the most voltage. In this case, the meter drops (and displays) 119.99 V while the much smaller load resistance drops 0.01 V. In **Figure 10-15B**, most of the current flows through the closed switch. Because there is contact resistance, a minute amount of current flows through the meter and the meter displays 1.2 V. The switch resistance is in series with the load, so the load drops the remaining 118.8 V.

Figure 10-14. The digital multimeter is configured to measure voltage. Frequency and duty cycle can also be measured while using this function with some meters.

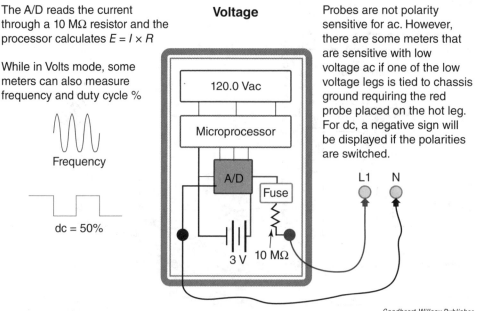

The A/D reads the current through a 10 MΩ resistor and the processor calculates $E = I \times R$

While in Volts mode, some meters can also measure frequency and duty cycle %

Frequency

dc = 50%

Voltage

120.0 Vac

Microprocessor

A/D

Fuse

3 V   10 MΩ

L1   N

Probes are not polarity sensitive for ac. However, there are some meters that are sensitive with low voltage ac if one of the low voltage legs is tied to chassis ground requiring the red probe placed on the hot leg. For dc, a negative sign will be displayed if the polarities are switched.

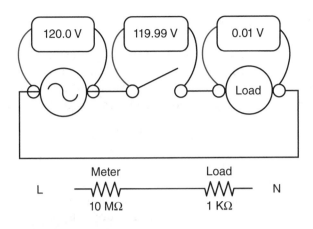

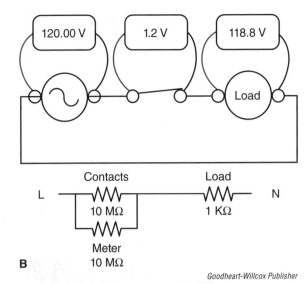

Goodheart-Willcox Publisher

Figure 10-15. A—Voltage is measured across an open switch. All the current flows through the meter, and it measures close to line voltage. B—Voltage is measured across a closed switch. Most of the current flows through the closed switch. A very small current flows through the meter resulting in a close to zero volts displayed.

Safety Note

Mode Switches

Ensure that mode switch is set to voltage. Checking voltage in other functions can damage the meter, which, in turn, may injure the user.

Pro Tip

Phantom Voltage

When measuring voltage on an unpowered circuit, the meter may display a voltage. The value can be high enough to cause confusion of its validity. This is called *phantom voltage*. It is caused by capacitance developed from nearby live wires. The unpowered wires act like capacitor plates. Some meters offer an optional feature to filter out phantom voltage.

10.2.4 Resistance Measurement

The meter is configured to supply power to the component under measurement, **Figure 10-16.** Current flows from the negative battery terminal through the component under test and back to the meter's A/D converter and shunts. The analog battery voltage is also digitized. This information is processed and sent to the display unit.

Fundamentally, the microprocessor is calculating Ohm's law ($R = E/I$). In the figure illustration, the meter battery voltage is 3 V. Since 1 A flows through the resistor, the resistor value is 3 Ω. As with current measurement, shunts are with resistance measurement to reduce current flow through the A/D.

Connecting the probes together should yield 0 Ω. Several factors can generate small resistances. It is common to see a reading of 0.1 Ω to 0.2 Ω. These factors are test lead wire resistance and contact resistance of the test leads to the meter and internal meter calibration.

Resistance

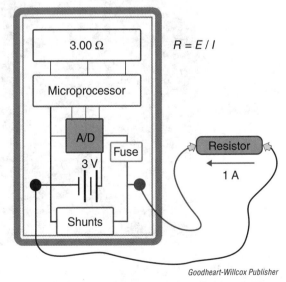

$R = E/I$

Goodheart-Willcox Publisher

Figure 10-16. The digital meter is configured to measure resistance.

10.3. Additional Multimeter Functions

The capability to miniaturize electronic components has made way for using specialized microprocessor integrated circuits that house multiple meter functions in one test instrument. The most common functions found in a test instrument are audible continuity, diode, capacitance, frequency/duty cycle, and noncontact voltage.

10.3.1 Audible Continuity Test

The **audible continuity test** is a quick way to verify continuity, or continuous flow current, between two points. This could be used for a length of wire or between switch contacts. Internally, the meter is configured similar to the resistance function. The meter supplies battery power to the component under test. The returning current is sampled by the A/D and sent to the processor. If the resistance of the component tested is less than a specific value, then the audible buzzer sounds off. If the resistance is greater, there is no sound but the resistance is displayed. After a certain value of resistance, the display will show *OL*.

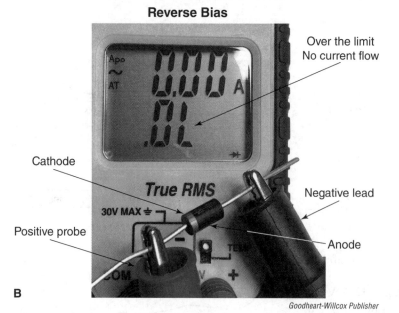

Forward Bias

Voltage drop

Diode symbol

Cathode

True RMS

30V MAX

Positive lead

Negative probe

Anode

A

Reverse Bias

Over the limit
No current flow

Cathode

True RMS

30V MAX

Negative lead

Positive probe

Anode

B

Goodheart-Willcox Publisher

Figure 10-17. A—The diode is forward biased, and the meter displays voltage drop across the diode. B—The diode is reverse biased, and the meter displays *OL*.

> **Pro Tip**
>
> ### Resistance Values
>
> The owner's manual specifications must be checked to find the resistance value when the audible alarming feature will not sound. A typical value is between 40 Ω and 60 Ω. When this resistance value is exceeded, the meter displays the ohmic value but up to a specified amount. A typical value is about 200 Ω and the display will show *OL*. The technician should verify the condition of the component with the ohmmeter before making a troubleshooting decision.

10.3.2 Diode Test

Recall from Chapter 6, *Alternating and Direct Current* that diodes allow current to flow in one direction from cathode to anode. Diodes are used to rectify ac to dc power. When the diode test function is selected, the meter configures itself to supply battery power to the diode. The red positive probe is connected to the diode's anode, while the black negative probe is connected to the cathode. See **Figure 10-17**.

Since the diode is forward biased, current flows through a good diode, and the meter displays the voltage drop across the diode. The value is between 0.4 to 0.7 V. *OL* indicates a faulty diode. The probe connections to the diode are reversed so that the diode is wired in reverse bias. Current shows no flow and is indicated by *OL* on the display. If current flows and the meter displays a voltage drop, the diode is faulty.

10.3.3 Capacitor Test

The meter is configured to charge the capacitor under test and record the charging time. Recall that current stops flowing when the capacitor is charged. The meter then discharges the capacitor and that time is recorded. The microprocessor uses the charge and discharge times to calculate capacity, which is displayed as a value in microfarads. An open or shorted capacitor is displayed by *OL*. The capacitor test takes several seconds to complete. Larger capacitors take longer. Some meters display dashes while the test is in progress. The displayed value should be compared to the capacitor nameplate value.

> **Safety Note**
>
> ### Properly Discharge Capacitors
>
> Always properly discharge capacitors with a 20,000 Ω 2 W resistor before testing.

> **Pro Tip**
>
> ### Checking Prefixes
>
> Check the prefix displayed. Be aware of displayed values in the nanofarad (nF) range when the measured capacitor value should be in the microfarad range.

10.3.4 Frequency and Duty Cycle

While the meter is set in voltage mode, the A/D and processor analyze the measured signal to derive the frequency of the ac voltage. Most meters require pressing a switch to engage this feature. **Duty cycle** is displayed as a percentage of on-versus-off time for a square waveform used in pulse width modulation. This means the power is turned on and turned off for specific time periods to control the operation of a load. This topic will be explored in Chapter 12, *Switches, Electronic Components, and Sensors.*

10.3.5 Noncontact Voltage (NCV)

The **noncontact voltage (NCV)** feature checks for the presence of voltage without using the test probes. Instead, it uses magnetic induction to sense voltage. An antenna is typically placed at the inductive clamp jaw, **Figure 10-18**. A live wire induces a current in the antenna. The meter annunciates the live wire by sounding a buzzer and a lamp. Most meters use a dedicated push button for this function.

> **Safety Note**
>
> ### Always Check Actual Voltage
>
> The NCV is for quick reference only. Always check actual voltage with the voltmeter function. Ensure that power is off before working on the circuit.

10.4 Test Leads

Test leads are the conductors connected to the meter to make measurements. The test leads contain plugs to connect to the meter and probes to connect to the test points. Test leads must be very flexible since they are bent and twisted during normal use. While test leads typically use 18 AWG stranded wire, it is made of about 190–40 AWG strands.

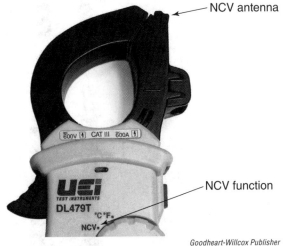

Goodheart-Willcox Publisher

Figure 10-18. NCV – noncontact voltage antenna

Figure 10-19. Check leads by connecting probes together and moving wire in circular motion while observing display. A break in the lead wire is indicated by flashing random numbers displayed.

Test leads eventually fail. They must be checked with the resistance function by using alligator clips to connect leads together. While observing the display, hold one probe and move the wire in a circular motion at the strain relief area. See **Figure 10-19**. The displayed value should remain constant. Repeat this test for each connection. A flickering display reveals a break in the wire. Visually inspect lead insulation for cracks and damage. Make sure test lead plugs securely connect to meter jacks.

Probe tip shields are used to reduce the amount of exposed metal, **Figure 10-20**. This is to reduce arc flash risk. If the metal probes are in close proximity, current can arc across them due to a high potential difference. Insulated alligator clips allow the technician to leave a meter connected while evaluating other parts of the system.

Figure 10-20. Various test lead probe attachments.

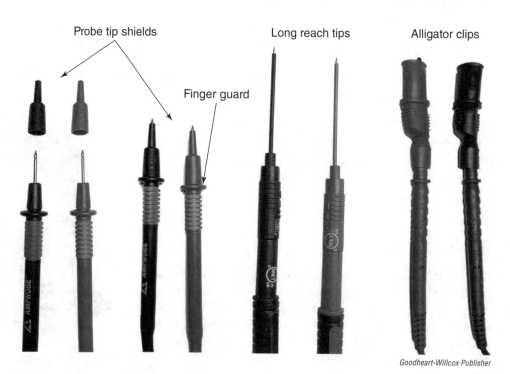

Another useful feature is to attach one alligator clip to the neutral wire while using the other probe tip to check for power at various points throughout a circuit. There are many test leads rated for different voltages available today.

10.5 Specialty Meters

Specialty meters, such as wattmeters and megohmmeters, are typically single-type meters. Some meters include these functions in one unit together with traditional multimeter functions.

A **wattmeter** is used to measure true power that is consumed by a system. The wattmeter function is added to some multimeters, or it can serve as a standalone unit. The nonportable standalone unit is typically panel mounted for stationary use. The wattmeter uses an inductive clamp to measure current, while source voltage is measured with the meter probes. See **Figure 10-21**. The meter analyzes the phase shift between voltage and current and their amplitudes to derive the true and apparent power. The display shows the true power and power factor.

A **megohmmeter**, also known as a *megger*, tests the resistance of electrical insulators. See **Figure 10-22**. For HVACR applications, these meters supply 500 to 1000 Vdc to the wire under test. A high potential difference is required to test insulation values in the millions of ohms. The manufacturer's instructions and the owner's manual must be carefully followed when using this meter. Details such as the specified time that the high voltage is applied to the test circuit must not be exceeded or the circuit can become damaged. Pass/fail insulation resistance values depend on meter voltage and test time duration.

10.6 Specifications

Every meter has specifications that can be found in the owner's manual. It is critical that these specifications are understood. Meter specifications are used to define the following features:

- **Capacity limits.** The amount of voltage, current, resistance, and other functions a meter can measure.
- **Accuracy.** How close a measurement is compared to a reference point.
- **Resolution.** The smallest increment that can be measured.
- **Precision.** Indicates the repeatability of the same measurement. See **Figure 10-23**.

Pro Tip

Refresh Rate

A meter's refresh rate is a factor considered for its precision. This rate describes how often a signal is updated on a display. A typical value is three times per second. Because current is a signal, this is sampled by the A/D and processor several times per second.

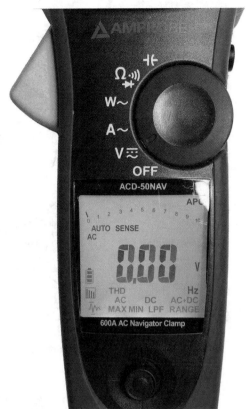

Goodheart-Willcox Publisher

Figure 10-21. Wattmeter built in to a multimeter.

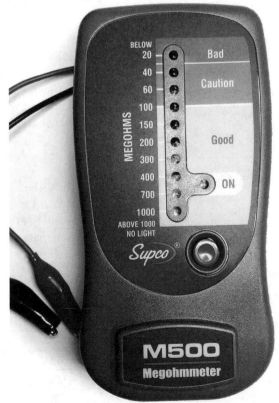

Goodheart-Willcox Publisher

Figure 10-22. The megohmmeter (megger) checks insulation integrity.

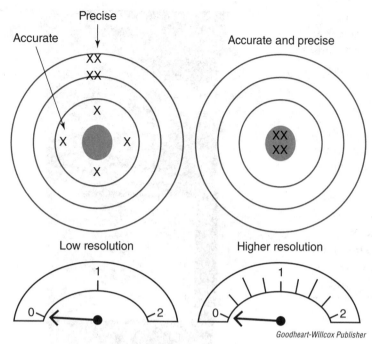

Figure 10-23. Accuracy, precision, and resolution examples.

Goodheart-Willcox Publisher

There are also specifications for how the measured values on a meter are displayed. A meter display is rated by the number of digits and counts. Digits indicate the amount of numbers it can appear on the display, while counts indicate the highest value it can display. Some examples are the following:

- **3 1/2 : 2,000.** A 3 1/2 display has 4 digits, but the most significant digit can only be a 0 or 1. Therefore, the display range is 0.001 to 1999. This is also called a 2000 count display.
- **3 5/6 : 6000.** A 3 5/6 display has 4 digits but the most significant digit is between 0 and 5. Therefore, the display range is 0.001 to 5999. This is also called a 6000 count display.
- **4 3/4 : 40,000.** A 4 3/4 display has 5 digits but the most significant digit can only be a 3. Therefore, the display range is 0.0001 to 39999. This is also called a 40,000 count display.

Consider measuring voltage, and the display shows 120.0 V. The 600 V range must be used, and the least significant digit is 100 mV. This means the display shows one decimal in the tenth place. The displayed information and resolution detail values are plugged into a formula to derive the range of values that the measurement can be.

$$\text{Range} = \pm (0.01 \times 120.0) + (0.100 \times 3)$$
$$= 1.5 \text{ V}$$

Thus, the error in the measurement is ±1.5 V. This means the actual voltage is between 118.5 and 121.5 V. This is providing the temperature is between 32 and 104°F, and relative humidity is below 75% for the specific meter used in this example. The error rate increases when the temperature and humidity are outside the given parameters.

Figure 10-24. Meter specifications for classifying accuracy.

Range	Resolution	Accuracy	O/L Protection
600 mV	0.1 mV		
6 V	1 mV		
60 V	10 mV	± (1.0 % + 3 digits)	1000 V$_{rms}$
600 V	100 mV		
750 V	1.0 V		

Goodheart-Willcox Publisher

Voltmeters are rated by category and voltage for safety. The category rating system was developed to protect the meter and user from destructive spikes (transients in the power system). Most HVACR meters are rated for category 3 and 600 V. The Cat (category) ratings are 1 through 4. Cat 4 has the greatest transient protection and voltage ratings above 600 V. Some Cat 3 have a 1000 V rating, but the transient protection is limited to Cat 3. Cat 2 is limited to single-phase power distance from circuit breaker panels where large transients are less likely. Cat 1 is for electronics work where transients are not likely to happen. Cat type, voltage rating, and certifying testing laboratory information are found on the meter body, **Figure 10-25**.

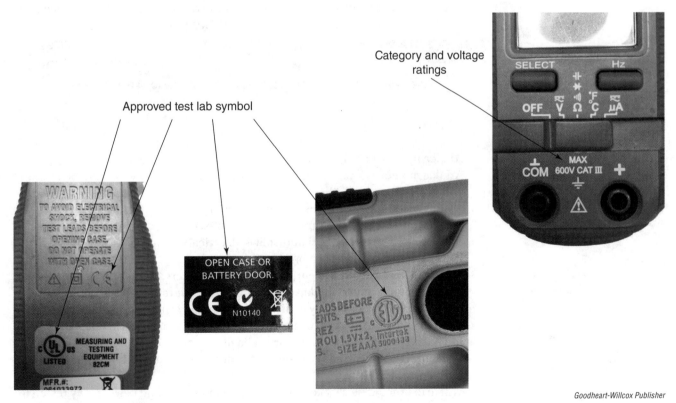

Goodheart-Willcox Publisher

Figure 10-25. Category rating and certifying laboratories.

10.7 Features

There are many added features that a digital meter can have based on its make. These additional features include the following:

- **Data hold.** Captures and records the value displayed. This value is displayed until the feature is turned off or the meter is turned off.
- **Max hold.** Captures and records the maximum value displayed while taking a measurement within a specified amount of time. This value is displayed until the feature is turned off or the meter is turned off.
- **Min hold.** Captures and records the minimum value displayed while taking a measurement within a specified amount of time. This value is displayed until the feature is turned off or the meter is turned off.
- **Inrush or Lock Rotor Amperage.** Captures and records the initial current draw of an inductive load. The inrush current happens too quickly for the human eye to see. Therefore, the meter must capture the value within a fraction of second and record it. The range must be set correctly before using this feature. This value is displayed until the feature is turned off or the meter is turned off.
- **Back lit display.** Allows user to see the display in low-light conditions.
- **Work light.** Used to light up low-light areas and make measurements safely.
- **Automatic range select.** The best resolution for the measurement taken is displayed. Modern quality meters reliably select the best range for best accuracy.
- **Manual range select.** It can select a range based on resolution preference. An example is to round value to whole numbers.
- **Temperature.** Measure temperature in Fahrenheit and Centigrade scales with the included thermocouple. Many meters allow for user calibration, which is preferred.
- **Dual display.** The inductive clamp uses a dedicated display thus allowing the display for any of the other functions to be used simultaneously.
- **True RMS.** A meter measures the power dissipated by a resistor instead of calculating the RMS value. This is more accurate since it is a measure of true power used.

Pro Tip

Owner's Manual

Refer to the owner's manual for the safe and correct way to use any feature.

Summary

- Analog meters and digital meters can only measure current.
- Analog meters use a magnetic movement that uses a pointer to move across a calibrated scale.
- Analog meters configure the internal circuitry to translate the measured current into the selected function values: volts, ohms, or current.
- A digital meter is a device that displays electrical quantities as a numerical value rather than a position on a meter.
- Digital meters are more accurate than analog meters.
- Digital meters convert analog signals to digital information. The digital information is processed by a microprocessor and sent to the display.
- Due to the miniaturization of electronics, digital meters perform many functions.
- Included in most digital meters are capacitor, diode, continuity, frequency, noncontact voltage, and recording functions in addition to the basic volt-milliamps-ohms functions.
- Test leads should be checked for damage before using and periodically tested for intermittent failure by using the ohmmeter function.
- The wattmeter measures true power and power factor.
- The megohmmeter uses high voltages to check wire insulation condition.

Lab 10.1   Multimeter Evaluation

The purpose of this lab is to acquaint the technician with the limitations, safety requirements, and features of their multimeter.

Lab Introduction

The HVACR technician must rely on the multimeter when troubleshooting electrical systems. It is impossible to accurately diagnose and find the root cause of a problem without the facts provided by the multimeter. However, the technician must know how to use the meter correctly and know the meter's limitations and accuracy. The meter is a tool, and if used incorrectly, it can be damaging to the user and system under test. The multimeter owner's manual will be required for this lab. If necessary, the manual can be downloaded from the manufacturer's website.

Equipment

- HVACR type multimeter with built-or or plug-in inductive clamp
- Test leads
- Alligator clip attachment
- Temperature probe
- Owner's manual

Procedure

1. Read the safety warnings found in the manual.

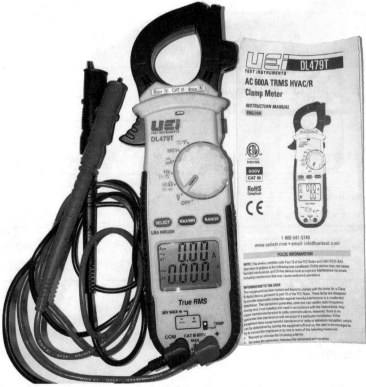

Figure 10.1-1.

Goodheart-Willcox Publisher

Continued ▶

2. Check the battery condition. Refer to the owner's manual.
3. Find and record the category and voltage rating in the manual and on the meter body.
4. Read the features section.
5. Look over the symbols section. These symbols are used in the display and switch settings.
6. Look over the overview section. This is a detailed exploded view of the meter.
7. Read the general specifications.
8. Locate the specifications for each function.
9. Read the instructions for special features. It is assumed that the user knows how to use the meter for the basic functions since the meter is designed for trained professionals.
10. Check for a section about leads and attachments.

Lab Questions

1. What are the category and voltage ratings?

_____

2. Who is/are the authorized testing laboratory?

_____

3. What is the maximum resistance value?

_____

4. What is the continuity buzzer cut-off resistance value?

_____

5. Does this meter feature True RMS?

_____

6. What is the display count and how many digits (some meters show count only)?

_____

7. What is the accuracy of ac voltage?

_____

8. Which range is used when 24 Vac is to be measured?

_____

9. What is the resolution for that range?

_____

10. What is the accuracy of dc voltage?

_____

11. What are the ranges and resolutions for the inductive clamp?

_____

12. What is the inductive clamp accuracy?

_____

13. What is the accuracy of the in-line current measurement?

_____

14. How many ranges are there for in-line current measurement?

_____

15. Does the meter measure both ac and dc in-line current?

_____

16. What is the accuracy of the displayed temperature when measuring 70°F?

_____

17. What is the accuracy for the capacitor test function?

_____

18. What is the accuracy formula?

_____

19. What is the refresh rate?

_____

20. The automatic power off feature turns the meter off when not in use. How long does this take your multimeter?

_____

Lab

10.2 Measurement and Accuracy

In this lab, you will calculate the tolerance according to the accuracy listed in the specifications for voltage, current, and resistance measurements

Lab Introduction

Tolerance is the span between the minimum and maximum values that are calculated by using the given accuracy formula. Therefore, the actual value of the number displayed can be anywhere between the range of the tolerance. The inductive clamp will be used to measure the current drawn by three parallel bulbs.

Equipment

- Safety glasses
- Lab board
- 14 AWG wire – green, white, black, or red
- 3—60 to 100 W incandescent bulbs

- 1—single-pole switch
- 1—multimeter with inductive clamp
- 1—resistor of any value

 Procedure

1. Wire the lab board per **Figure 10.2-1**.

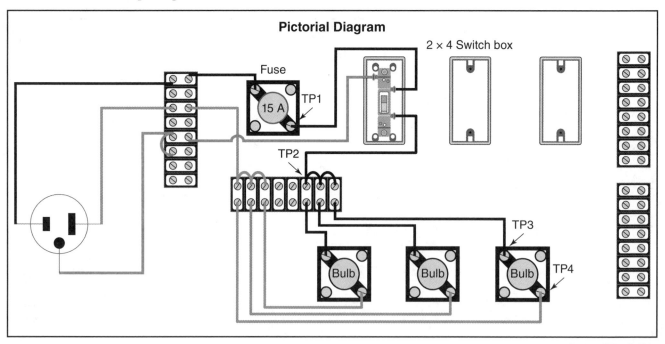

Figure 10.2-1.

Goodheart-Willcox Publisher

Continued ▶

Schematic Diagram

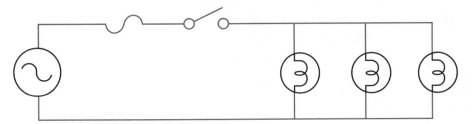

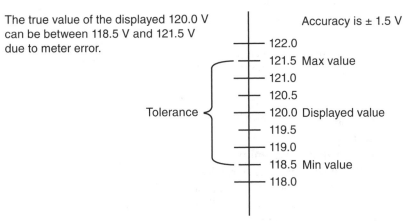

The true value of the displayed 120.0 V can be between 118.5 V and 121.5 V due to meter error.

Accuracy is ± 1.5 V

122.0

121.5 Max value

121.0

120.5

Tolerance 120.0 Displayed value

119.5

119.0

118.5 Min value

118.0

Figure 10.2-1. (Continued)

Goodheart-Willcox Publisher

2. Get your instructor's approval before powering up the lab board.
3. Plug in board into outlet.
4. Before turning the switch on, measure the source voltage between TP1 and any neutral terminal. Line voltage should be displayed. Leave one probe on TP1 and test the other neutral terminals.
5. Record the line voltage.
6. Find the accuracy for ac voltage in the specification section of your manual.
7. Apply the formula: ± (*displayed value* × *% error*) + (*resolution* × *# of digits*).
8. Perform the calculation for the measured value and record.

Pro Tip

Formula Example

Consider for example: the displayed value 240.0 V. This means the accuracy given in the specification is ± (1% + 3 digits). The percent error is 1%, so use 0.01 in the above formula.

The resolution is given in a separate column as 100 mV or is included with the range value as 600.9. In either case 0.1 is the least significant digit to be multiplied by the number of digits in the equation. So, the equation is set up as follows for the example.

± (240.0 × 0.01) + (0.1 × 3)

2.4 + 0.3 = 2.7 V

Thus, the tolerance is ± 2.7 V. The actual voltage is between 237.3 and 242.7 V.

Continued ▶

9. Turn on the switch. The three bulbs should be on. Measure the current with the inductive clamp. Remember to use only one wire. The hot wire between the switch and TP2 is a good choice. The neutral wire may be used but ensure it is common to all three bulbs. Check the accuracy for current as in Step 7 and record tolerance.
10. Turn off the bulbs.
11. Measure the resistor value and check your accuracy as in Step 7 and record.
12. With the board still plugged in and the switch in the off position, check voltage between TP1 and TP2 and record value.
13. Leave the meter connected to TP1 and TP2 (use alligator clips or have a lab partner). Use a second meter and measure voltage between TP3 (which is same point as TP2) and TP4. Record the value.
14. Remove any two bulbs and repeat Steps 11 and 12. Record the new values.
15. Remove the last bulb and measure the voltage between TP1 and TP2. Record value.
16. Replace all bulbs and turn the switch on.
17. Measure the voltage between TP1 and TP2. Record the value.
18. Measure the voltage between TP3 and TP4. Record the value.
19. Turn off the switch, and allow the bulbs to cool down about three minutes.
20. Set up the meter to measure inrush (LRA) current. Place the wire in clamp jaw.
21. Turn on the switch. Record the inrush current. Incandescent bulbs draw a more current initially due to the cold filament. Note: the current may be too small for some meters to capture. (This will be performed with a motor in Chapter 13, *Electric Motors*.)
22. Leave the lab board wired for use in Lab 10.3.

Lab Questions

Use your digital multimeter or other electrical instruments to answer the following questions.

1. What is the source voltage?

_____

2. What is the actual voltage range (tolerance) for Step 8?

_____

3. What is the actual current range (tolerance) for Step 9?

_____

4. What is the actual resistance range for Step 11 (tolerance)?

_____

5. What is the voltage between TP1 and TP2 with switch off?

_____

6. Explain the measured value?

_____

7. Explain the measured value between TP3 and TP4 in Step 12?

_____

8. With 2 bulbs removed, what is the voltage between TP1 and TP2?

_____

9. Is the voltage different between TP1 and TP2 without the two bulbs? Why?

_____

_____

_____

10. What is the voltage across TP1 and TP2 when the switch is off and the bulbs are removed? Why?

_____

_____

_____

11. What is the voltage across TP1 and TP2 when the switch is closed?

_____

12. What does the voltage value represent?

_____

13. What is the inrush current?

_____

Lab

((10.3)) Multimeter Features

This lab will allow you to explore and use the various features offered on your digital multimeter.

Lab Introduction

Most HVACR digital multimeters include continuity, capacitor, diode, noncontact voltage, and temperature test features. Refer to the owner's manual when performing this lab. All of the above features will be used in this lab.

Equipment

- Safety glasses
- Lab board (as wired from Lab 10.2)
- 1 – 1N1001 diode or equivalent
- 1 – 100 Ω resistor
- 1 – run or start capacitor
- 1 – cup for crushed ice and water to calibrate temperature probe
- 1 – HVACR multimeter

 Procedure

1. Use the same lab board from Lab 10.2. Do not plug it in. Set the meter for continuity, and touch the probes together to ensure it is working. The buzzer should sound, and the display should read close to 0 Ω. The buzzer means there is continuity—the circuit is complete. The display shows the resistance of the portion of the circuit tested.
2. Use the alligator clip on one probe and attach it to the hot terminal on the lab board plug.
3. Use the test lead to probe TP1. Note the buzzer and displayed value. Record.
4. With the switch open, probe TP2. Note the buzzer and displayed value. Record.
5. With the switch closed, probe TP2. Note the buzzer and displayed value. Record.
6. With the switch closed, probe TP3. Note the buzzer and displayed value. Record.
7. Remove the probe from the hot wire plug terminal and attach it to TP3.
8. With the switch off, probe TP4. Note the buzzer and displayed value. Record.
9. Remove the bulb corresponding to TP3 and TP4. Leave the switch off.
10. Probe TP4. Note the buzzer and displayed value. Record.
11. Turn the switch on and repeat Step 10.
12. Remove the middle bulb and repeat Step 10.
13. Remove the last bulb and repeat Step 10.
14. Replace all the bulbs. Attach the alligator clips to both test lead probes.
15. Attach the probes to hot and neutral plug terminals.
16. With the switch off, note the buzzer and displayed value. Record.
17. With the switch on, note the buzzer and displayed value. Record.
18. Connect the 100 Ω resistor in series with fuse and loads. See **Figure 10.3-1**.

Continued▶

Pictorial Diagram

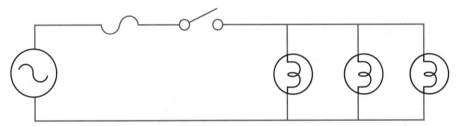

2 × 4 Switch box

100 Ω Resistor

Fuse

15 A

TP1

TP2

TP3

TP4

Bulb   Bulb   Bulb

Schematic Diagram

Figure 10.3-1.

19. With the switch on, note the buzzer and displayed value. Record.
20. Set the meter for capacitor test. Probe the capacitor. Record value.
21. Calculate the accuracy for the capacitor measurement and record. Refer to Lab 10.1, if needed.
22. Set the meter for diode test. Refer to the diode test and the owner's manual.
23. Test and record the forward bias displayed value.
24. Test and record the reverse bias displayed value.
25. Remove the resistor and place hot wire to the original position.
26. Get approval from your instructor before plugging in the board. Turn the switch off. Plug in board.
27. Set the meter for NCV function.
28. Place NCV antenna near hot terminal along the board including the bulb sockets.
29. Turn on the switch, and place NCV near bulb socket. Note the proximity between the antenna and hot terminal to cause the meter to sound and/or light an indicator lamp.
30. Set the meter for temperature and connect the thermocouple. Measure the ambient temperature and record.
31. Read the temperature calibration instructions (not available on some meters) and prepare to calibrate.
32. Crush ice and fill the cup with a 50/50 mix of crushed ice and water. Stir well. This mixture is at 32°F.
33. Place the probe in saturated mixture. Do not directly put thermocouple on ice—suspend it between ice chips. Proceed to calibrate per meter instructions.
34. After calibration, measure the ambient temperature and record.

Lab Questions

Use your digital multimeter or other electrical instruments to answer the following questions.

1. Was there continuity for Step 3?

   _____

2. Was there continuity for Step 4?

   _____

3. Was there continuity for Step 5?

   _____

4. Was there continuity for Step 8?

   _____

5. Was there continuity for Step 10? What is shown on the display?

   _____

6. Why is there continuity in Step 10 since the bulb was removed?

   _____

7. When checking continuity between TP3 and TP4, did the switch position make a difference? Why or why not?

   _____

   _____

   _____

8. Was there continuity for Step 16? What was the displayed value?

   _____

9. Was there continuity for Step 17? What was the displayed value?

   _____

10. Was there continuity for Step 19? What was the displayed value?

    _____

11. Did the buzzer sound in Step 19?

    _____

12. What was the resistance displayed in Step 19?

    _____

13. Does no buzzer sound always indicate an open circuit?

    _____

14. What is the tolerance (meter accuracy) for the measured capacitor?

    _____

15. What was the displayed forward bias value for the diode?

    _____

16. What was the displayed reverse bias value for the diode?

_____

17. What was the ambient temperature before calibration?

_____

18. What was the ambient temperature after calibration?

_____

19. How close does the NCV antenna have to be to the live source for it to work?

_____

Know and Understand

_____ 1. Analog and digital meters can only measure _____ and translate it into other units of measure.
A. voltage
B. resistance
C. current
D. Ohm's law

_____ 2. When a meter is configured to measure current, what protects the meter from excess current flowing through its delicate components?
A. Shunts in series with meter.
B. Shunts in parallel with meter.
C. Series resistors.
D. Capacitor in series.

_____ 3. When the meter is set up to measure resistance _____.
A. it uses the circuit voltage to send current to the meter
B. the measured resistance charges a capacitor in the meter
C. the measured resistance is powered by an external source
D. the meter sends current through the resistance to be measured

_____ 4. What is term used when an analog ohmmeter measures a resistance greater than its limit?
A. Ohmmed out.
B. Overload.
C. Infinity.
D. Over capacity.

_____ 5. What is not a benefit of digital meters?
A. It becomes part of a circuit due to its low impedance.
B. Automatic ranging.
C. High resolution display.
D. High internal impedance.

_____ 6. *True or False?* Digital meters are 100% accurate.

_____ 7. What does the prefix *M* represent in a digital display?
A. 1000
B. 10,000
C. 1,000,000
D. 100,000

_____ 8. What does the prefix *k* represent?
A. 000
B. 10,000
C. 1,000,000
D. 100,000

_____ 9. What does the prefix *m* represent?
A. 1000
B. 0.001
C. 0.01
D. 0.000001

_____ 10. What does the prefix μ represent?
  A. 1000   C. 0.000001
  B. 0.001   D. 0.000000001

_____ 11. What is term used when a digital ohmmeter measures a resistance greater than its limit?
  A. Infinity.
  B. O.L—over the limit, overload, over range.
  C. Ohmmed out.
  D. Over capacity.

_____ 12. What protects a meter when measuring voltage?
  A. High value series resistance.   C. Capacitors in parallel.
  B. Shunts in parallel.   D. Capacitor in series.

_____ 13. What is the typical resistance value that a continuity test buzzer stops sounding?
  A. 200–300 Ω   C. 400–500 Ω
  B. 100–200 Ω   D. 40–60 Ω

_____ 14. *True or False?* Test leads should be tested frequently.

_____ 15. True power is measured by ____.
  A. inductive clamp current multiplied by the voltage
  B. inductive clamp current plus the voltage across the load
  C. wattmeters
  D. load resistance times load voltage

_____ 16. The megohmmeter is used to measure ____.
  A. continuity   C. insulation
  B. phantom voltages   D. resistance under 1 Ω

_____ 17. *True or False?* Resolution refers to the smallest increment that can be measured.

_____ 18. What is the range of values that a 6000 count display can show?
  A. 1 to 6000   C. 0.1 to 5999
  B. 0.001 to 5999   D. 0.01 to 6000

_____ 19. *True or False?* Temperature and humidity do not affect a meter's accuracy.

_____ 20 *True or False?* Dual-display meters allow the user to check voltage and resistance at the same time.

Critical Thinking

1. If a voltage is measured across an open switch in a live circuit and the load has an equivalent resistance to the meter's internal resistance, what voltage would the meter display?

_____

_____

_____

2. Why does a megohmmeter require high voltage to detect insulation failure?

_____

_____

_____

11 Introduction to Practical Circuits

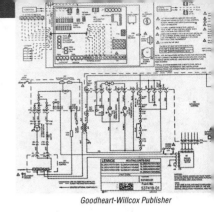

Chapter Outline

Additional Reading

***Modern Refrigeration and Air Conditioning,* 21st edition**

17.2 Electrical Diagrams
17.3 Electrical Troubleshooting Process

Learning Objectives

After completing this chapter, you will be able to:

- Discuss the three sections of an HVACR circuit.
- Define line voltage and low-voltage controls.
- Discuss ladder diagrams and how they are interpreted.
- Compare ladder diagrams, schematics, and pictorial diagrams.
- Explain how to analyze line-voltage controls.
- Explain how to analyze low-voltage controls.
- Describe how to trace voltage when checking a circuit.
- Describe the basic structure of ac circuits.
- Explain the basic structure of heating system circuits.
- Discuss the basic structure of refrigeration circuits.

Technical Terms

fusible disconnect	load side	schematic diagram
ladder diagram	low-voltage control	split ac unit
line side	packaged ac unit	
line-voltage control	pictorial diagram	

Introduction

In previous chapters, we focused on basic circuits to teach electrical fundaments. These circuits consisted mostly of resistors and light bulbs. This chapter will build upon those fundamentals and introduce how practical circuits are structured for HVACR applications. Although there are many equipment brands, the overall operation for air conditioning, heating, and refrigeration systems is similar.

This chapter will first examine the fundamental circuit structure for each of these systems. These concepts apply to any HVACR system. The circuit structure then will be translated into the different types of wiring diagrams used for installation and troubleshooting.

11.1 Circuit Structure

A circuit structure can be divided into three sections. The first section is the line that delivers the power to the circuit. The second section is the load, or loads, that consumes the power. The third section is the control that operates the load. The control (or control circuit) turns the load on and off automatically based on the HVACR system operation. A control can directly operate the load if it is in series with the load. This is called a ***line-voltage control.*** If a control only indirectly operates the load, it is called a ***low-voltage control.*** A low-voltage control requires a separate circuit since there is a different voltage used to energize a relay or contactor that closes its contacts to pass power to the higher-voltage load.

A circuit can be broken down into the line, load, and control sections when looking for the root cause of a circuit malfunction. With troubleshooting scenarios such as this, circuit schematics called ladder diagrams are used. A ***ladder diagram*** consists of vertical side rails that deliver power and horizontal rungs that make up the parallel load branches. Each rung includes all the switches that supply power to each load. This makes it easy to see how the load receives power. Since each load is on a separate rung, the ladder diagram helps a technician better trace individual load branches. This makes these diagrams clearer than other electrical diagrams like schematics and pictorial diagrams. Schematics and pictorial diagrams provide their own specific information and are still needed for troubleshooting.

An example of a ladder diagram is shown in **Figure 11-1**. The line section contains circuit protection and switching devices. There is the circuit breaker panel where the power originates in the building or house. This power is then fed to the equipment disconnect switch. If the disconnect switch has a fuse, it is called a ***fusible disconnect.*** Power is fed to the control switch that, when closed, feeds power to the load. This control switch is considered a line-voltage control since it is directly interrupting the load. Thus, the switch breaks the power wire connected to the load. The neutral wire must then have a direct uninterrupted path back to the circuit breaker panel.

Figure 11-1. Line voltage controlled load.

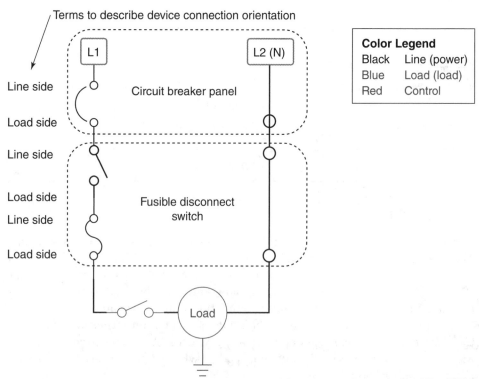

When a load requires 208 or 240 V, both hot legs must be protected and interruptible through a switching device. The control can be line voltage or low voltage. **Figure 11-2** shows a 240 V load controlled by a low-voltage control circuit. The control circuit energizes a contactor coil that activates two normally open contacts. Once contacts close, they pass power to the load. A three-phase circuit is shown in **Figure 11-3**. The line section of the circuit contains protection and switching for each hot leg. The control circuit energizes a contactor coil that activates three normally open contacts. Once contacts close, they pass power to the load.

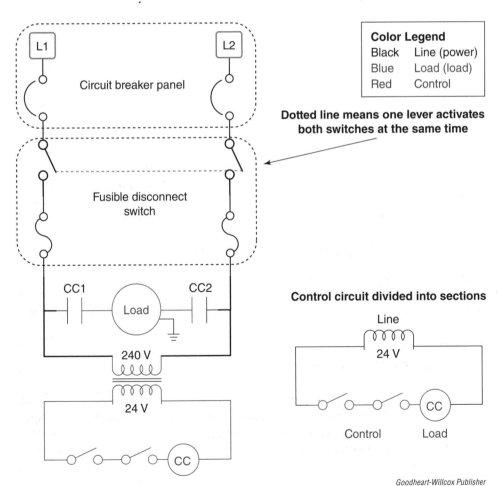

Color Legend

Black	Line (power)
Blue	Load (load)
Red	Control

Dotted line means one lever activates both switches at the same time

Control circuit divided into sections

Figure 11-2. Low-voltage control circuit to control the load.

Goodheart-Willcox Publisher

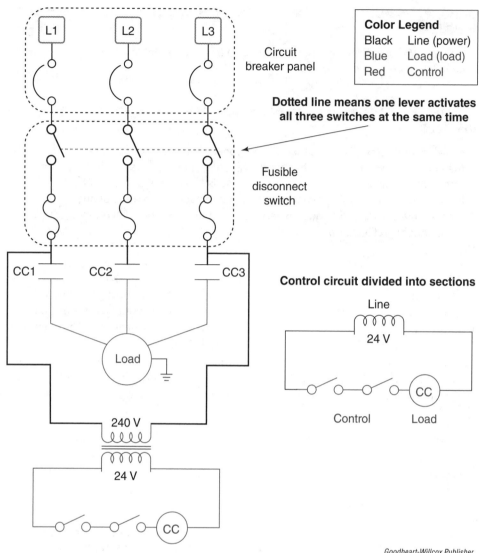

Figure 11-3. Low-voltage control circuit controlling a three phase load.

Circuit breaker panel

Color Legend
Black — Line (power)
Blue — Load (load)
Red — Control

Dotted line means one lever activates all three switches at the same time

Fusible disconnect switch

CC1 CC2 CC3

Load

240 V

24 V

CC

Control circuit divided into sections

Line

24 V

Control Load

CC

Goodheart-Willcox Publisher

The next two sections will delve into the troubleshooting process for circuits with line-voltage and low-voltage controls. You then will be introduced to how these circuits operate within air conditioning, heating, and refrigeration systems.

11.1.1 Analyzing Circuits with Line-Voltage Control

A technician must be able to interpret and understand how a circuit works in order to begin the troubleshooting process. A load in a circuit needs power to operate. This is a fundamental place to start. The following questions need to be considered by a technician:

- Where does load power originate?
- What devices interrupt the power to the circuit?
- What controls the operation of the load?

The circuit in **Figure 11-1** demonstrates how these questions can be answered. Power originates in the circuit breaker panel where one circuit breaker is used since the load requires 120 V. Therefore, the return conductor from the load to the panel is a neutral wire. The fusible disconnect switch disrupts power to the circuit manually by the operator or automatically by a fuse. The load is controlled by a line voltage switch.

If this circuit is operating normally, then power is passed through the circuit in the following steps:

1. L1 feeds the line side of circuit breaker.
2. The load side of circuit breaker feeds the line side of the disconnect switch.
3. The load side of the disconnect switch feeds the line side of the fuse.
4. The load side of the fuse feeds the line side of the control switch.
5. The load side of the control switch feeds the hot side of the load.
6. The neutral side of the load is fed by the neutral terminal in the disconnect switch.
7. The disconnect switch neutral bus terminal is fed by the circuit breaker neutral bus bar terminal.

A break in any of the wire segments or terminal connections used to wire between the above components can cause an unintentional open circuit. This prevents power from being supplied to the load. To troubleshoot this, a voltmeter can be used to trace power through the circuit. See **Figure 11-4.** In this process, a common test probe is anchored on the load's neutral side, while the red probe is moved. The red probe is placed at the load first and moved progressively toward the power source. Since the circuit is complete, line voltage is read at all locations. If all locations yield 0 V, then there is a break in the neutral wire.

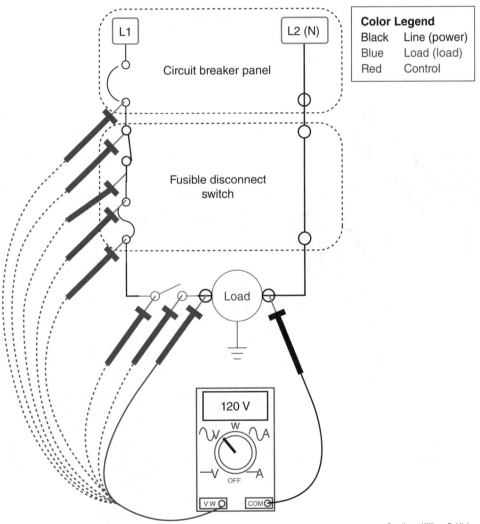

Color Legend

Black	Line (power)
Blue	Load (load)
Red	Control

Figure 11-4. This illustration outlines the method used to trace voltage throughout the circuit.

Goodheart-Willcox Publisher

In this case, the common probe must be moved up toward the circuit breaker panel to determine where the neutral wire is broken. In **Figure 11-5**, 0 V is measured at the load—both sides of the control switch and at the load side of the fuse. However, line voltage is measured on the fuse's line side. This indicates an open fuse or a bad connection at the fuse holder.

Tracing the power does not necessarily have to begin by measuring voltage across the load. Measuring voltage across the load was the appropriate starting point in this scenario because the technician knew there was power supplied to the circuit. Although all switches were turned on, the load was inoperable. A logical approach for larger circuits is to determine the section in which the interruption occurred. Another way to find the open fuse is to measure voltage across each switch, fuse, and circuit breaker in the hot side of the circuit. The open device should display line voltage.

Figure 11-5. Power is indicated on the line side of the fuse indicating the fuse is blown because 0 V is measured at the load side of the fuse.

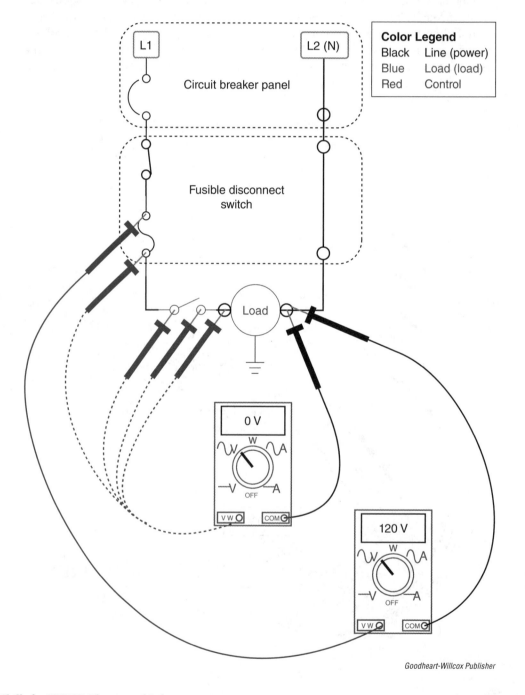

Goodheart-Willcox Publisher

11.1.2 Analyzing Circuits with Low-Voltage Control

When the control is a different voltage than the load, a step-down transformer is usually used. A step-down is required for a traditional 24 V control circuit. However, a three-phase 240 V refrigeration system that uses a 120 V control circuit may not require a step-down transformer if a neutral wire is available. A hot and neutral provide the 120 V. The control then becomes a complete circuit itself and is broken down into the three sections. The lower voltage power source is the line section. Switching devices make up the control, and the load contains one or more relays and contactor coils.

Breaking down a circuit into these three sections is used for troubleshooting. See **Figure 11-6**. In this example, all switches are closed, but the load is not operating.

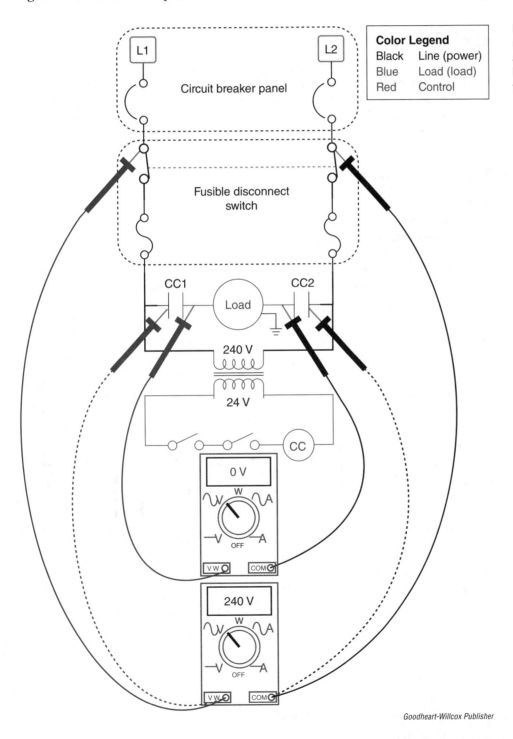

Color Legend	
Black	Line (power)
Blue	Load (load)
Red	Control

Figure 11-6. Power is not reaching the load as contacts are not passing power to the load. The problem is likely in the control circuit.

Goodheart-Willcox Publisher

The logical approach is to ensure that power is feeding the system. Because there is 240 V across the line side of the fusible disconnect switch, this means there is power present. The logical question to ask in this case is whether the power is reaching the load. Voltage across the load sides of contactor contacts cc1 and cc2 measure 0 V, so we know the load is not receiving power.

However, 240 V is measured across the line side of cc1 and cc2 contacts. Since the contactor contacts are closed by energizing the contactor coil, the problem is most likely in the control circuit. The control circuit can also be broken down into sections to locate the problem. See **Figure 11-7**. Note that the fault can be a loose or broken wire at the contact terminals, primary transformer winding terminal, or an open transformer primary winding.

> ### Pro Tip
>
> **Troubleshooting**
>
> First, make sure there is power to the equipment. This may require measuring the voltage, or it may be obvious there is power if parts of the system are operating. Next, begin testing at the most accessible locations to avoid unnecessary work. Logically isolate the section that contains the problem, and proceed to find the root cause of the problem. Be aware that more than one problem can exist.

11.2 Air Conditioner Circuits

There are three basic loads in an ac circuit. These are the compressor motor, condenser fan motor, and evaporator blower motor circuits. The compressor creates a pressure differential between the evaporator and condenser (to facilitate heat absorption in the evaporator and heat rejection in the condenser) and moves the refrigerant throughout the system. The evaporator blower forces the room air through the evaporator coil to condition the air. The condenser fan forces air through the condenser coil, so it can reject heat.

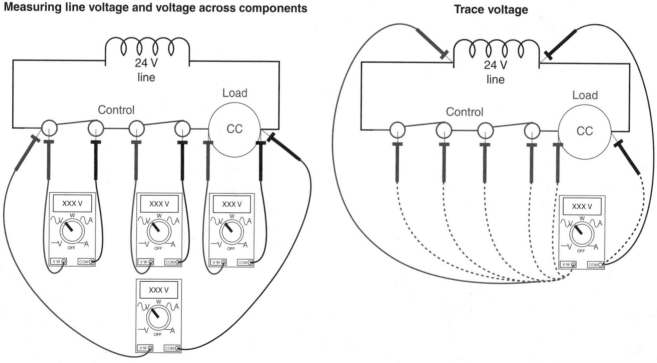

Figure 11-7. Analyzing the control circuit by measuring voltage across components.

To operate these motors, the line side contains a disconnect switch (to turn the unit on and off) and required circuit protection devices. The basic control circuit includes a thermostat that controls the operation of the loads and limit switches that protect the loads. An example of a limit switch is a high-pressure control that interrupts current to a control device or directly to the load.

The control circuit increases the complexity of the system, depending on the types of system features and capacity. Modern ac controls are either a combination of electromechanical devices and electronic circuit boards or 100% electronic circuit boards. In addition, the control section can manage heating systems and heat pumps. In either case, the control section passes power from the line section to the load section.

AC systems are available as packaged units or split units. A *packaged ac unit* contains the entire system in one housing. Window units, packaged terminal ac, and roof top units are all examples of package units. A *split ac unit* has an indoor unit that houses the evaporator and blower assembly and an outdoor unit consisting of the compressor, condenser, and fan. The outdoor unit is referred to as the condensing unit.

Figure 11-8 shows a ladder diagram for a split ac system. The cooling thermostat closes on a rise in temperature and feeds power to the relay coil (RC), which closes RC contacts that allow line power to feed the blower motor. At the same time, the thermostat feeds the Y terminal that feeds the Y terminal in the outdoor unit's control circuit. The Y terminal then feeds the CC contactor coil, which closes contacts cc1 and cc2 to pass power to the compressor motor and fan motor. Note that the blower can be turned on independent of the thermostat by switching to the blower ON position.

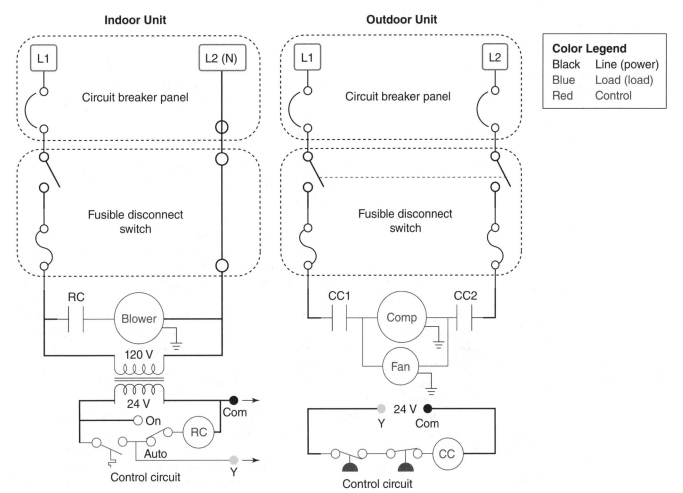

Goodheart-Willcox Publisher

Figure 11-8. A ladder diagram of a split ac system.

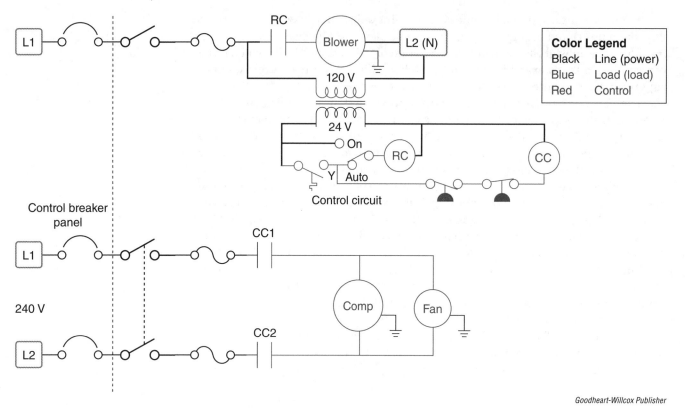

Color Legend
Black Line (power)
Blue Load (load)
Red Control

Figure 11-9. A schematic diagram of a split ac system.

Figure 11-9 shows a split ac system in a schematic diagram, which shows the entire system wiring. *Schematic diagrams* show detailed information such as wire color, AWG wire size, and terminals and connectors. They also include a legend.

11.3 Heating System Circuits

Heating systems are categorized by the type of heat distribution method and heat source. Heat is distributed through forced air, water, or steam. Heat sources are natural gas, propane, oil, heat pump, and electric.

Figure 11-10 shows forced air systems with three different heat sources represented on ladder diagrams. The typical line voltage loads in the gas-fired unit are the blower motor, inducer motor, and ignition module. The control circuit includes a heating thermostat and IFC (integrated furnace control). The IFC is an electronic circuit board, which will be covered later in Chapter 14, *Troubleshooting Printed Circuit Board Control Systems.*

The oil-fired unit uses a line voltage oil pump motor, which also supplies air for combustion by driving a blower in addition to the oil pump. The oil burner controller energizes the oil pump motor and ignition module while the blower is energized by a thermostatic controller. The main line voltage loads in the electric unit are the blower and the electric heating elements. The low-voltage control unit sequentially energizes the blower and heating elements.

The three ladder diagrams in **Figure 11-10** can be used for hydronic (water) systems by replacing the blower with a circulator pump motor. Typical residential steam systems do not require circulators or pumps. However, large steam systems may require condensate pumps to return water to the boiler. The heat pump circuit structure is the same as the air conditioner with the addition of a reversing valve solenoid, thermostat, and outdoor coil defrost controller. Note that state-of-the-art system incorporates controls in a computer-driven circuit board.

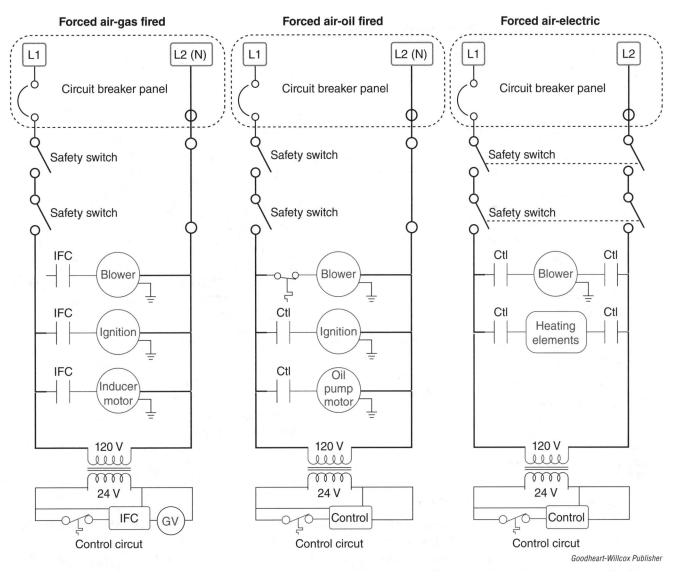

Forced air-gas fired

L1   Circuit breaker panel   L2 (N)

Safety switch

Safety switch

IFC — Blower
IFC — Ignition
IFC — Inducer motor

120 V

24 V

IFC   GV

Control circuit

Forced air-oil fired

L1   Circuit breaker panel   L2 (N)

Safety switch

Safety switch

Blower

Ctl — Ignition

Ctl — Oil pump motor

120 V

24 V

Control

Control circuit

Forced air-electric

L1   Circuit breaker panel   L2

Safety switch

Safety switch

Ctl — Blower — Ctl

Ctl — Heating elements — Ctl

120 V

24 V

Control

Control circuit

Figure 11-10. Gas, oil fired, and electric heat source forced air ladder diagrams. Components, such as sensors and safety devices, have been left out for simplicity.

11.4 Refrigeration Circuits

The main difference between ac and refrigeration systems is the evaporator temperature. Evaporator temperatures typically drop below freezing, so the condensed moister freezes. The evaporator must be defrosted by electric heating elements, hot gas bypass, or off-cycle defrosting. The refrigeration circuit starts with the basic ac components—compressor, evaporator blower, and condenser fan. The defrost feature requires a controller and sensors to start, stop, and set the duration of the defrost cycle. The controller may consist of a mechanical clock timer or a computerized circuit board.

Figure 11-11 shows a domestic nondigital controlled refrigerator pictorial diagram. *Pictorial diagrams* show the relative location of specific components and physical connections within a unit. In this case, it is a refrigerator. The terminals inside a connector are identified by numbers or letters, so the technical can locate specific test points in the circuit. Wire color and gage number are also shown. For example, the defrost control is a mechanical clock type and is in the line voltage side of the diagram. The connector and terminals are identified. This refrigerator uses a line-voltage thermostat, and therefore all controls are line voltage.

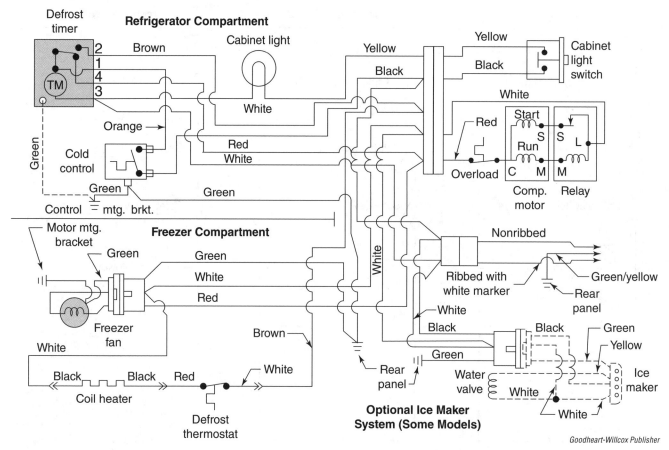

Figure 11-11. Pictorial diagram of a domestic refrigerator.

Goodheart-Willcox Publisher

Summary

- To better understand circuit operation, a circuit can be divided into three sections. The line provides the power to the circuit. The control manages the operation of the load, and it allows power to pass from line to load. The load includes one or more loads.
- Line-voltage controls pass power directly to the load. Low-voltage controls pass power indirectly to the load through a relay or contactor.
- Ladder diagrams are ideal for troubleshooting. Power lines are shown as the rails of a ladder, and each parallel is shown as the rungs of a ladder.
- A fusible disconnect is a switch assembly that contains a fuse to protect the circuit.
- Analyzing a circuit requires a logical and systematic approach. The path for each load must be clearly defined. Ask how does power get to the load.
- A technician must first ensure that there is power to the system being checked, which can be done by tracing voltage.
- There are three basic loads in an air conditioning circuit: compressor motor, condenser fan motor, and evaporator blower motor circuits.
- A packaged ac unit contains the entire system in one housing. A split ac unit has an indoor unit that houses the evaporator and blower assembly and an outdoor unit consisting of the compressor and condenser and fan.
- Schematic diagrams show details including wire color and size, terminals, and connector connection to controls and loads.
- Heating systems are categorized by the type of heat distribution method and heat source.
- The refrigeration circuit starts with the basic ac components—compressor, evaporator blower, and condenser fan. The main difference between refrigeration and ac systems is the evaporator temperature.
- Pictorial diagrams show the relative location of components within a unit.

Lab 11.1 Analyzing Circuit by Tracing Voltage

This lab will walk you through how to trace voltage through a circuit.

Lab Introduction

The lab board is shown as a pictorial diagram. It shows relative location and details about the wires. In addition, a ladder diagram is shown to simplify the understanding of how the circuit works. Both the ladder and pictorial diagrams show test points for ease of completing this lab and to show the correlation between the diagrams. Actual equipment diagrams show test point locations as required for technicians to check key locations when following service manual troubleshooting guides. Note that the terminal strips are used to test voltage across the switches instead of directly probing the switch terminals. This is done for safety as the switch terminals are close to the metal boxes. The sequence of operation for the circuit is as follows.

All switches are initially in the off state. When the main switch is closed, the power on the bulb illuminates and the primary winding is energized. For the control circuit, 24 V is available. When both control switches are closed, the relay coil is energized. Line voltage is passed to the load by the normally open contacts that close when the coil is energized.

Equipment

- Lab board
- Safety glasses
- Multimeter
- 3—single-pole switches
- 24 V transformer
- Relay 24 Vac coil, SPST only needed for lab
- 14 AWG wire black and white
- 18 AWG wire (available colors)
- 2—low wattage incandescent or LED bulbs

1. Wire the lab board per **Figure 11.1-1**. The connections made from the switches to terminal strips are for the safety of the technician and to avoid shorting out the transformer when checking voltage.

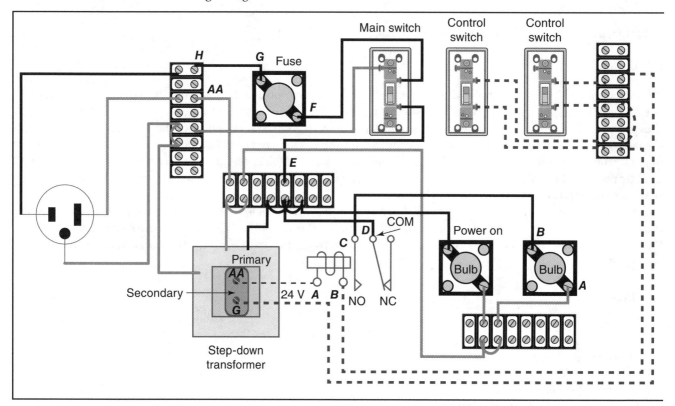

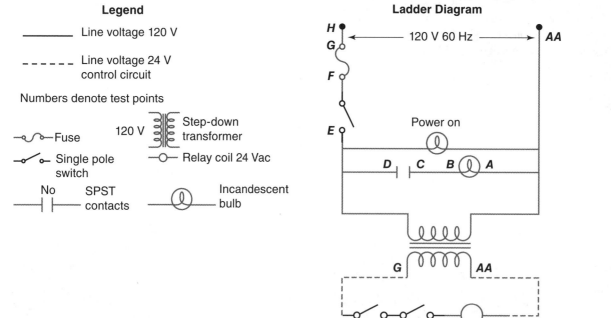

Figure 11.1-1.

Continued ▶

2. Get your instructor's approval before plugging in the board.
3. Check the multimeter operation, and set to measure ac voltage.
4. Close the main switch and observe the power on bulb.
5. Close both control switches.
6. Begin testing the circuit by tracing voltage on the high voltage side. Place one probe on test point (TP) AA and the other probe on test point (TP) B. Record the voltage. Move the probe from TP AA to TP A, and use the other probe to sequentially probe TP C through H. Record the voltage for each TP.
7. Open one control switch, and repeat Step 6. Record TP values.
8. Close the control switch, and test the control circuit. Place one probe on TP AA and the other on TP B and record voltage. Move probe from TP AA to TP A, and use the other probe to sequentially probe TP C to G and record voltage.
9. Open one control switch and repeat Step 8. Record values.

Note: Do not dismantle the lab board. The same circuit is used for Lab 11.2.

Lab Questions

1. What is the voltage between AA and B?

_____

2. What is the voltage between A and B?

_____

3. If there was a break in the wire between AA and A, what would the voltage be between A and B?

_____

4. Did all test point measurements result in line voltage for Step 6? Why or why not?

_____

_____

_____

5. Did all test point measurements result in line voltage for Step 7? Which ones did not? Why not?

_____

_____

_____

6. Which test points measured line voltage? Explain why they measured line voltage.

_____

_____

_____

7. Which diagram was easier to follow the circuit operation?

_____

8. Which diagram is required to find the actual test points?

_____

Lab

11.2 Analyzing Circuit by Line, Control, and Load Sections

This lab will use a different approach to analyze the same circuit in Lab 11.1. Voltage will be measured at various locations to identify the circuit section at fault.

Lab Introduction

Instead of sequentially checking for power, this lab will focus on strategic test points to narrow down the source of the problem.

Equipment

- Lab board
- Safety glasses
- Multimeter
- 3—single pole switches
- 24 V transformer
- Relay 24 Vac coil, SPST only needed for lab
- 14 AWG wire black and white
- 18 AWG wire (available colors)
- 2—low wattage incandescent or LED bulbs

Procedure

1. Use the same lab board from Lab 11.1. See **Figure 11.2-1**. Turn off all switches.

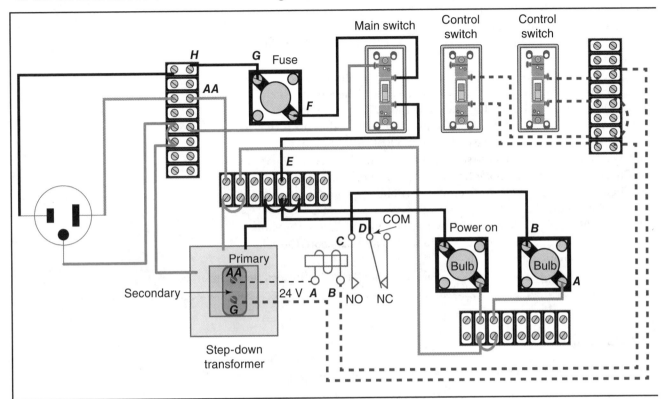

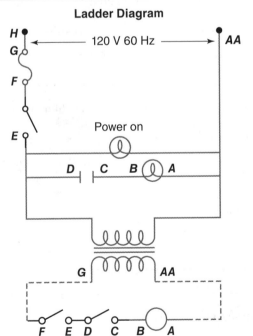

Figure 11.2-1.

Goodheart-Willcox Publisher

Continued ▶

2. Establish that power is being supplied to the circuit. Check the voltage across TP F and AA. Record the value.
3. Turn on the main switch. The power on lamp should illuminate.
4. Check the voltage across the load TP B and A. Record the value.
5. The load is fed from the NO contacts. Check the voltage across the contacts TP C and D. Record the value.
6. Remove the bulb representing the load and Step 5. Record the value.
7. Replace the bulb.
8. Close both control switches. The load bulb should illuminate.
9. Check the voltage across NO contacts TP C and D. Record the value.
10. Check the voltage across the load TP B and A. Record the value.
11. Open both control switches to simulate a problem in the control circuit.
12. Check the voltage across the coil TP A and B. Record the value.
13. Check the voltage across the secondary TP G and AA. Record the value.
14. Check the voltage across control switch TP F and E. Record the value.
15. Check the voltage across both control switches TP F and C. Record the value.
16. Close one control switch.
17. Check the voltage across control switch TP F and E. Record the value.
18. Check the voltage across control switch TP D and C. Record the value.
19. Close the other control switch.
20. Check the voltage across both control switches TP F and C. Record the value.

Lab Questions

Use your digital multimeter or other electrical instruments to answer the following questions.

1. What is the source voltage across F and AA?

_____

2. What is the voltage across the load in Step 4?

_____

3. What is the voltage across the NO contacts in Step 5?

_____

4. What is the voltage across NO contacts with bulb removed in Step 6?

_____

5. Explain the voltage value measured.

_____

_____

_____

6. Why is the voltage across the coil 0 V per Step 12?

_____

_____

_____

7. What is the voltage across the transformer secondary tps G and AA?

_____

8. What is the voltage across tps F and C?

_____

9. Explain the voltage reading for the above question.

_____

_____

_____

10. Explain the voltage reading for Step 17.

_____

_____

_____

11. Explain the voltage reading for Step 18.

_____

_____

_____

12. Explain the voltage reading for Step 20.

_____

_____

_____

The purpose of this lab is to convert a pictorial diagram into a ladder diagram.

Lab Introduction

The pictorial diagram is for a basic refrigerator that uses nondigital controls. **Figure 11.3-1** shows the pictorial diagram with helpful call outs. Note how wire splices are shown on the diagram. Two wires are typically spliced at a solderless crimp terminal that is housed in a connector body. An example is shown in **Figure 11.3-2** of the freezer fan circuit. The rails of the ladder are the hot and neutral wires that originate at the power plug. The tracing of the circuit in the pictorial diagram begins with the hot wire. The black wire goes to the cold control and exits the control as an orange wire. The orange wire connects to the defrost timer terminal 1 and exits on terminal 4 as a red wire. From there, the wire connects to the fan switch and exits as a red wire to feed the freezer fan. A white wire connects the fan to the neutral rail to complete the circuit.

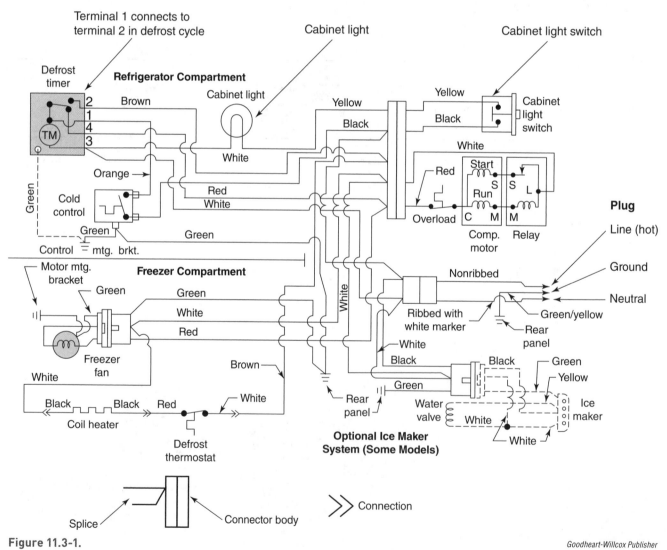

Figure 11.3-1.

Goodheart-Willcox Publisher

Name ________________ Date ____________ Class ____________

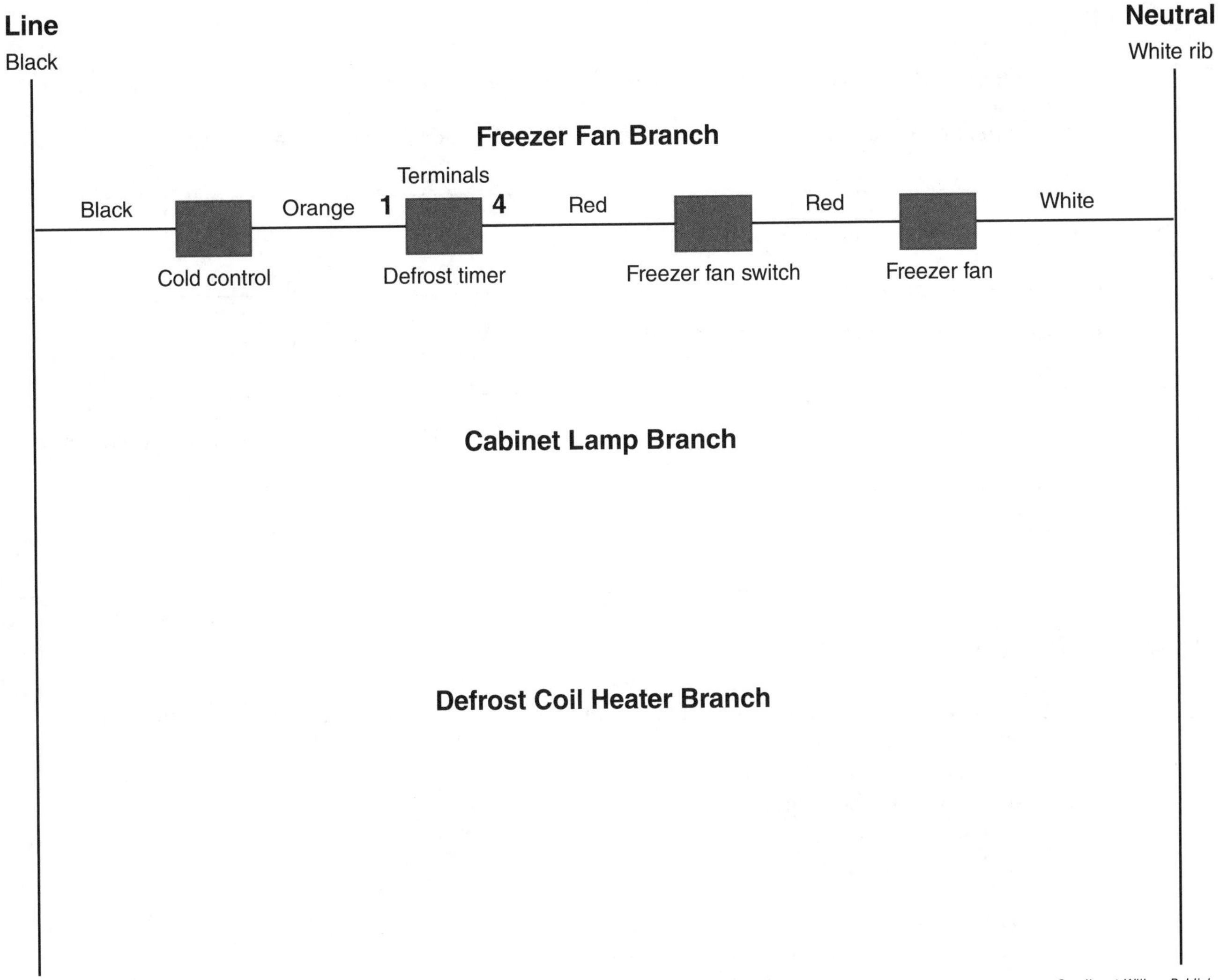
Line
Black
Neutral
White rib
Freezer Fan Branch
Terminals
Black
Orange
1
4
Red
Red
White
Cold control
Defrost timer
Freezer fan switch
Freezer fan
Cabinet Lamp Branch
Defrost Coil Heater Branch

Figure 11.3-2. *Goodheart-Willcox Publisher*

Equipment

- Pencil
- Paper
- Lab drawings

Procedure

1. Using the pictorial diagram from **Figure 11.3-1** and the freezer fan example shown in **Figure 11.3-2**, complete the ladder diagram. Begin the path with the hot wire and end the path (circuit) at the neutral wire.
2. Indicate the wire color and all components along the path as shown in the example.
3. The compressor motor can be done after studying motor in Chapter 13, *Electric Motors*.

Lab Questions

Use your digital multimeter or other electrical instruments to answer the following questions.

1. At which component does the wire color change from black to yellow for the cabinet fan circuit?

   _____

2. What is the wire color on the neutral side of the cabinet fan?

   _____

3. Is the red wire connected to the defrost thermostat hot or neutral?

   _____

4. What color does the brown wire (connected to the defrost timer terminal 2) connect to at the defrost thermostat?

   _____

5. Is the orange wire a hot wire?

   _____

Know and Understand

_____ 1. To facilitate the understanding of circuit operation, a circuit can be divided into which of the following three sections?
   A. Fuses, switches, and circuit breakers.
   B. Power, switches, and fuses.
   C. Line, control, and load.
   D. Line, power, and load.

_____ 2. A line-voltage control means that _____.
   A. the control uses a higher voltage than the load
   B. the control uses the same voltage as the load
   C. the control uses a lower voltage than the load
   D. the load does not require a control

_____ 3. Which diagram is best for understanding system operations and troubleshooting?
   A. Pictorial.   C. Line.
   B. Schematic.   D. Ladder.

_____ 4. Which diagram uses side rails for power lines and rungs for parallel branches of a circuit?
   A. Pictorial.   C. Floor plan.
   B. Schematic.   D. Ladder.

_____ 5. *True or False?* A fusible disconnect is a switch assembly that contains a fuse to protect the circuit.

_____ 6. *True or False?* The side of a switch connected towards the power source is referred to as the load side.

_____ 7. *True or False?* When a load requires 208 or 240 V, both hot legs must be protected and interruptible through a switching device.

_____ 8. *True or False?* The neutral wire must be a direct uninterrupted path back to the circuit breaker panel.

_____ 9. Which statement is incorrect about control circuits?
   A. They protect against load from overcurrent.
   B. They pass power from line to load.
   C. They control the operation of the load.
   D. They can be either line or low voltage.

_____ 10. *True or False?* Power source interruption to a load can be found by measuring voltage along the power source path with respect to a fixed opposite potential point.

_____ 11. *True or False?* A step-down transformer is required when a 24 V control circuit is used.

_____ 12. What is the first step in troubleshooting a system?
   A. Check voltage at loads.
   B. Start voltage checks at the furthest points in the system.
   C. Jumper out safety switches.
   D. Establish if there is power to the equipment.

_____ 13. Which is *not* a basic load in an ac system?
   A. Compressor.
   B. Evaporator blower motor.
   C. Condenser fan motor.
   D. Defrost heating elements.

_____ 14. Which statement is *not* true of a split ac system?
   A. It contains an evaporator and blower unit.
   B. It contains a compressor, condenser, and condenser fan motor.
   C. All components are housed in one unit.
   D. The indoor unit contains the low-voltage control circuit transformer.

_____ 15. In the diagrams shown in this chapter, the Y terminal feeds power directly to the _____.
   A. compressor motor
   B. condenser fan
   C. contactor coil
   D. compressor and fan motors

_____ 16. Which diagram includes circuit details without showing relative location of components within a system?
   A. Pictorial.
   B. Schematic.
   C. Ladder.
   D. Floor plan.

_____ 17. Which is *not* a typical load in a heating system?
   A. Inducer motor.
   B. Ignition system.
   C. Blower motor.
   D. Reversing valve solenoid.

_____ 18. Which control is used in refrigeration system, but not in ac systems?
   A. Compressor.
   B. Evaporator.
   C. Defrost.
   D. Thermostat.

_____ 19. Which diagram shows the relative location of components and specific descriptions?
   A. Pictorial.
   B. Schematic.
   C. Floor plan.
   D. Ladder.

_____ 20. Which diagram shows the clearest path for current flow between the power source and the load?
   A. Ladder.
   B. Pictorial.
   C. Schematic.
   D. Floor plan.

Critical Thinking

1. 240 V is measured across the line side of normally open contacts that are in the closed position. However, the load is not working and voltage across the load side of the contacts measure 0 V. Does this mean that the load is always at fault?

_____

_____

_____

2. Why do HVACR equipment manufacturers supply both ladder and schematic and/or pictorial diagrams?

_____

_____

_____

12 Switches, Electronic Components, and Sensors

Goodheart-Willcox Publisher

Learning Objectives

After completing this chapter, you will be able to:

- Define a switch and its different types.
- Explain the terms used to describe switch characteristics.
- Explain the operation of various switches.
- Define automatically-activated switches.
- Discuss switch specifications.
- Describe the purpose of electronic components.
- Explain the terms used to describe electronic components.
- Explain the operation of a diodes, diacs, triacs, SCRs, and transistors.
- Describe the two types of resistors.
- Explain resistor specifications.
- Discuss how sensors used in HVACR.
- Define the types of transducers used in HVACR.
- Define a latching circuit.
- Explain the function of a lockout relay.

Technical Terms

active component
bimetal
bipolar transistor
break
break-before-make (BBM)
cadmium sulfide cell
current-interrupting capacity
current transducer
diac
diode
doping
float switch
flow switch
full-wave rectifier
half-wave rectifier
light emitting diode (LED)
limit switch
lockout relay
maintained
make
make-before-break (MBB)
momentary
N-type material
NPN transistor
P-type material
passive component
PNP transistor
pole
potentiometer
resistor
rheostat
sensor
silicone-controlled rectifier (SCR)
solid-state
stop/start station
temperature transducer
thermistor
thermocouple
throw
transducer
triac
wetting current
Zener diode

Introduction

Earlier in this textbook, switches were introduced. These simple switches were used to demonstrate how power is passed to the circuit or to control the operation of a load. This chapter will cover switches more in depth, including specific types, how they are used, and how they are displayed in circuit diagrams. Also, switch nomenclature and specifications will be covered. Later, electronic components used for switching applications will be explored. It will also cover sensors once an understanding of mechanical and electronic switching devices is developed.

12.1 Switches

Switches are mechanical devices that can be divided into two groups: manual and automatic. Manual switches require physical human activation, while automatic types are activated by a system's physical properties, such as temperature and pressure. An automatic device activation can also be mechanical, by a moving lever. Power is passed through the automatic type by physical contacts or through electronic components.

Switches are available in many configurations. Configuration pertains to their internal switching action arrangements. See **Figure 12-1.** Two terms used to describe their configuration are poles and throws. A ***pole*** refers to how many sources of power feed the switch, while a ***throw*** indicates how many terminals power can be passed to. In a single-pole single-throw (SPST) switch, current from one power source feeds a load when the switch is closed. In contrast, a single-pole double-throw (SPDT) can pass one source of power to two different terminals. Power can be passed to either load A or load B, but not both at the same time.

An SPDT is often called a selector switch, or a three-way switch. A double-pole single-throw (DPST) disrupts two power sources, or two circuits, by activating one throw. The dotted line indicates there is only one lever. The switch can be considered two SPST switches contained in one body. In the example shown in **Figure 12-1**, a 240 V load is powered by two hot legs—L1 and L2. The double-pole double-throw (DPDT) can switch two power sources to either the 240 V load A or load B. A three-phase load requires a three-pole single-throw switch. The three-pole double-throw switch supplies power to either load A or load B.

Note that in both three-pole switches, there is only one lever, which is indicated by a single dotted line. The momentary SPST normally open (NO) push button switch contacts close when the button (throw) is pushed down. When pressure is removed from the button, the spring returns the button to its normally open state. The momentary SPST normally closed (NC) push button switch contacts open when the button is pushed down.

Multiple throw switches allow connection from one source to multiple paths. They are available in rotary or slide types. The term used to describe when contacts are closed is ***make***. When contacts are opened, it is considered ***break***. Switches with more than one throw are typically designed as ***break-before-make (BBM)***. This is shown in the four-throw slide switch in **Figure 12-1**. When the slide is moved away from the first position, the circuit is broken. However, with the ***make-before-break (MBB)*** switch, both positions one and two are connected to the source terminal during the transition between the two positions. For example, an MBB switch is required when the circuit must not be in an open state while transitioning between two positions.

Figure 12-2 shows a miniature SPST switch arranged in a dual-in-line package (DIP). These are found in circuit boards to customize a system's operating profile.

The above mechanical switches require manual activation. They can also be activated automatically in response to physical movement caused by changes in temperature, pressure, fluid flow, fluid level, or position of a moving part.

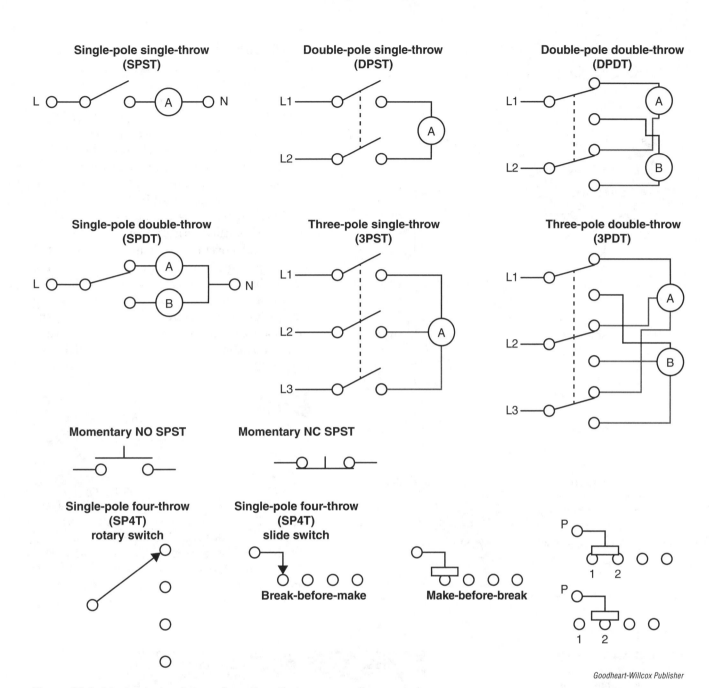

Figure 12-1. Mechanical switch configurations that are manually operated.

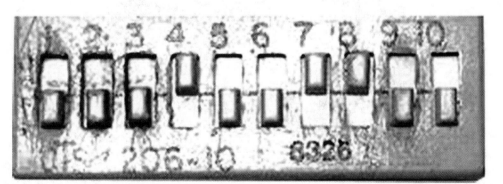

Goodheart-Willcox Publisher

Figure 12-2. Dual-in-line packaged miniature switches found on circuit boards.

Goodheart-Willcox Publisher

Low pressure switch

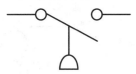

Closes on a rise in pressure

High pressure switch

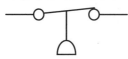

Opens on a rise in pressure

Flow switch

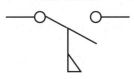

Closes with fluid flow

Float switch

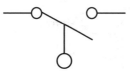

Closes with rising liquid level

Heating thermostat

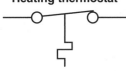

Opens on a rise in temperature

Cooling thermostat

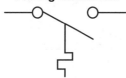

Closes on a rise in temperature

Reed switch

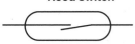

Closes with magnetic field
within proximity

Limit switch

Closes upon movement

Goodheart-Willcox Publisher

Figure 12-3. Mechanical switches that are automatically activated be an external force.

See **Figure 12-3.** All the switches respond to an external force that either closes or opens the contacts. There are other configurations available based on system requirements—for example, an SPDT. Pressure switches respond to rising or falling pressure that act on a diaphragm that, in turn, open or close the switch contacts.

There are many types of switches. *Flow switches* are activated by air or liquid movement. A *float switch* is activated by liquid level. Flow and float switches are available as NO, NC, or both. Temperature-activated switches respond to the physical movement of a bimetal, **Figure 12-4.** The warping movement of the bimetal moves the switch throw to open or close the contacts. The Reed switch contacts open or close in response to a magnetic field. A *limit switch* closes or opens in response to movement. For example, when a louver opens, it activates a limit switch. See **Figure 12-5.**

Figure 12-4. The bimetal is made of copper and steel. Copper has a higher expansion rate than steel so when heated, the copper expands and causes the bimetal to warp outward. If it is cooled by removing the same amount of heat energy that was added, the bimetal returns to the reference position shown. If the bimetal is cooled below the reference temperature, the copper contracts faster than the steel causing the bimetal to warp inward.

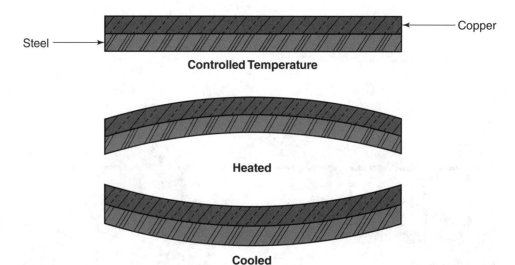

Steel

Copper

Controlled Temperature

Heated

Cooled

Goodheart-Willcox Publisher

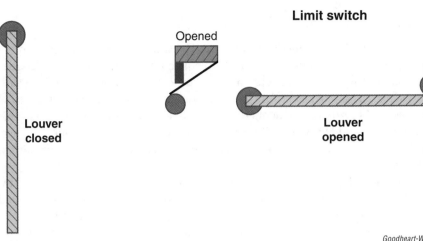

Limit switch

Opened   Closed

Louver closed

Louver opened

Figure 12-5. When the louver opens it pushes the switch plunger to close the switch contacts.

Goodheart-Willcox Publisher

Switch specifications must be considered when using or selecting a switch. They include the following characteristics:

- Resistive and inductive load current capacity
- Horsepower
- Voltage
- Contact material
- Current-interrupting capacity
- Wetting current
- Configuration (poles and throws)
- Activation type (toggle, rocker, rotary, slide, push button)
- Action (maintained or momentary)
- Operating temperature
- Application
- Life cycle

In order to provide the current demand of HVACR motors, switch contacts are typically made of silver or silver-cadmium alloy. These alloys are great current conductors and resist surface corrosion and oxidation. Switch contacts must dissipate heat energy generated by the current during operation and while making and breaking. For this reason, they are sized according to current and voltage capacity. Contacts must make and break quickly to reduce arcing. The ***current-interrupting capacity*** is the maximum current that can be stopped when the contacts open. Greater current continues to flow by arcing across the open contacts. Low-energy contacts in switches and relays used in control circuits are affected by the wetting current specification. The ***wetting current*** is the minimum current that can break through the contact's surface film. Contacts develop a film that acts as an insulator and so a required amount of current is needed to burn through this film. Application notes denote the typical use for a particular switch. The life cycle shows the number of expected activation cycles under normal use.

Safety Note

Replacing a Switch

When replacing a switch, replace it with the same or equivalent part number. This ensures that the new switch meets the original design intent requirements. Never replace with a switch of a lower current and voltage capacity.

12.2 Electronic Components

While HVACR technicians do not design or repair electronic circuit boards, they must understand the basic operation of certain electronic components. This knowledge is key when troubleshooting modern systems that are controlled by electronic circuit boards and sensors. Modern electronic components have replaced the vacuum tube and are called *solid-state* components or devices. They have no moving parts. This makes them more reliable because there are no components that wear out from use. Heat, shock, and electrostatic discharge are still conditions that must be considered and can be destructive.

Most solid-state devices use silicon or germanium as a base material. Impurities are added to change their current-conducting characteristics. This process is called *doping*. It is used to produce a positive-type (or P-type) material and a negative-type (or N-type) material that are the fundamental building blocks of solid-state electronics. The *P-type material* is doped to have a deficit of free electrons, while the *N-type material* is doped to have a surplus of electrons. Solid-state devices are also called *semiconductors* due to their ability to conduct current in varying amplitude as opposed to a fixed value. These devices can be used for switching but with the added benefit of modulation. That means not just on and off, but they can also supply variable power to a load.

Electronic components are categorized as active or passive. *Active components* regulate current flow based on an external signal or special condition applied. These include diodes, diacs, triacs, SCRs, and transistors. *Passive components* do not require an external signal for operation. These components include capacitors, inductors, transformers, which were discussed in previous chapters, and resistors.

12.2.1 Diodes

The most basic electronic component is the diode. A *diode* is a two-layer device made of one layer of positive and one layer of negative doped material. A diode is made of positive-type material called the anode and negative-type material called the cathode. The diode conducts current when it is forward biased and does not conduct when it is reverse biased. This makes it ideal to rectify ac into dc, **Figure 12-6**. There are different rectifiers depending on the number of diodes used.

A *half-wave rectifier* has only one diode and conducts only for half a cycle of the sine wave. A *full-wave rectifier* uses two diodes and conducts during the entire cycle but with reduced voltage as it requires a center-tapped secondary. The full-wave bridge conducts during the complete cycle with full secondary voltage. Diodes are rated by current and peak inverse voltage (PIV). If the PIV is applied to a diode when in reverse bias, it then conducts current. For example, in the half-wave rectifier shown, if voltage is increased during the negative half cycle to the PIV value, this causes diode and circuit malfunction in a standard diode. A standard diode cannot conduct current in both directions.

However, this characteristic is used to make a *Zener diode* that is used for voltage regulation. It is heavily doped to produce different values of PIV, also known as breakdown voltage. A typical application circuit is shown for a 5.1 V Zener diode. The purpose of this diode is to provide the load with a stable voltage even if the supply voltage varies within a specified range and load current varies. A *light emitting diode (LED)* is made from different materials that emit light when forward biased.

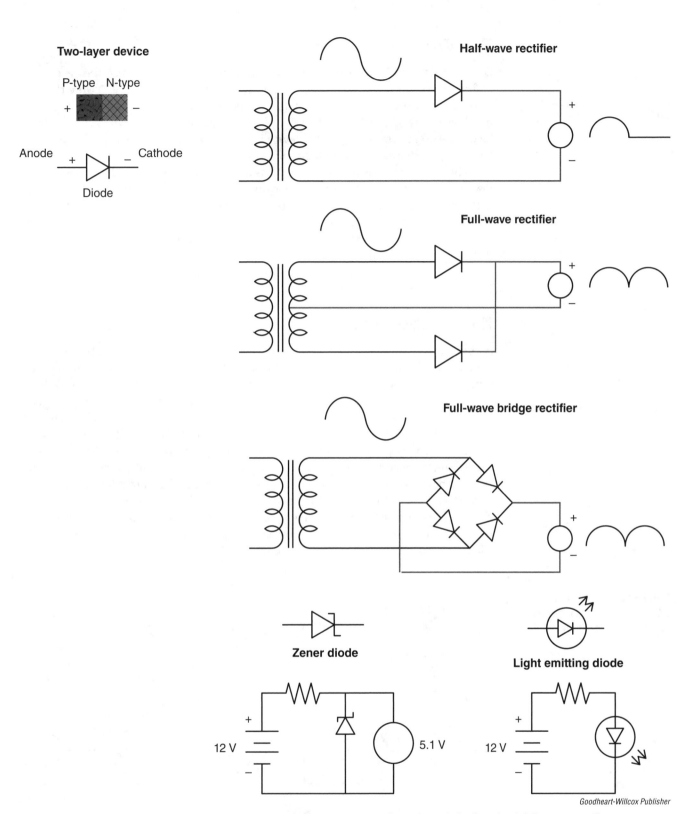

Figure 12-6. One diode is used for a half-wave rectifier where current flows for only half cycle. A full-wave rectifier uses two diodes and allows conduction of the full cycle, but the voltage is half of the secondary voltage. The full-wave bridge rectifier conducts for a full cycle and provides the full secondary voltage. The Zener diode is reverse biased and will conduct when the peak inverse voltage is reached. The diode is used as a voltage regulator. The LED emits light when it is forward biased.

12.2.2 Diacs, Triacs, and SCRs

Diacs, triacs, and SCRs are devices classified as thyristors because they are made of four layers. A ***diac*** conducts ac in both directions. See **Figure 12-7**. Note the diac symbol has two diodes in opposite directions. However, the voltage must reach the breakover voltage level for conduction to begin. Conduction stops when voltage drops below the breakover voltage. This device is ideal for providing the proper voltage to a load that is sensitive to low voltage conditions.

The ***triac*** is similar to a diac, but it has a control gate. The added gate allows for a low-voltage control signal (pulsed or steady) to turn on conduction in both directions without having to reach a breakover voltage. The triac continues to conduct if the gate signal is removed. This means the only way to stop conduction is to drop current below the holding current specified for the device. A ***silicone-controlled rectifier (SCR)*** conducts current in one direction like the rectifier diode—hence the name. It has a gate to control current flow through the cathode and anode like the triac. Current continues to flow after the removal of gate current.

12.2.3 The Bipolar Transistor

A ***bipolar transistor*** is a three-layer device that consists of a collector, emitter, and base. For current to flow between the collector and emitter, there must be base-to-emitter-current present. The two types of bipolar transistors are NPN and PNP. See **Figure 12-8**. An ***NPN transistor*** requires a forward bias current to flow from the emitter to the base. This means the base must be more positive than the emitter, which is called the bias current. The bias current allows current flow between the collector and emitter, providing the collector is a more positive emitter. The transistor is used as a switching device by using the maximum bias current allowed for the device. In turn, the collector-to-emitter current can be at a maximum value. The transistor functions as an amplifier when the bias current varies from a minimum to a maximum value. Collector-to-emitter current is proportional to the base to emitter current. A small base-to-emitter current causes the collector-to-emitter to conduct a larger current. So, a small signal input to the base results in a larger signal output at the collector or emitter junction.

Figure 12-7. Diac, triac, and SCR symbols and how the devices control loads.

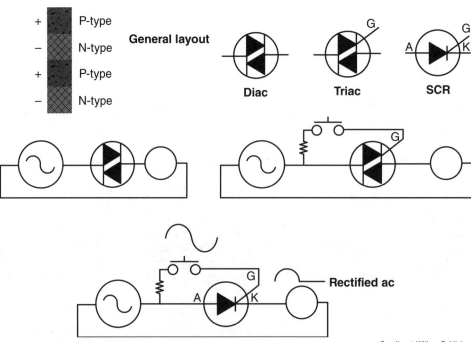

Goodheart-Willcox Publisher

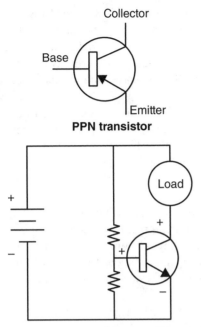

Collector

Base

Emitter

PPN transistor

Load

+

−

+

+

−

NPN transistor

Load

−

+

−

−

+

Goodheart-Willcox Publisher

Figure 12-8. Transistor types and symbols. Transistors are used as switching devices.

A **PNP transistor** operates the same as the NPN except that the polarities are changed. The base must be more negative than the emitter, and the collector must be more negative than the emitter. It can be used as a switching device or an amplifier.

12.2.4 Resistors

Resistors limit current flow. They are available as either variable or fixed. There are two types of variable resistors—the rheostat and the potentiometer. See **Figure 12-9.** The **potentiometer** is a three-terminal device with an adjustable center contact, called the wiper, and two fixed terminals. The **rheostat** is a two-terminal device with one fixed terminal and an adjustable wiper. The potentiometer is equivalent to two variable resistors in one device. Terminal *A* and the wiper is one resistor and terminal *B* and the wiper is the second variable resistor. As the wiper is moved, the values of the two resistances change where one has more resistance than the other.

On the top view of **Figure 12-9A**, the wiper is not centered and is closer to terminal *A*. This results in a smaller resistance in series with the left bulb and more resistance in series with the right bulb. Therefore, the right bulb is dimmer than the left bulb. In **Figure 12-9B**, a potentiometer wiper centered is shown, so both bulbs are in series with equal resistance and

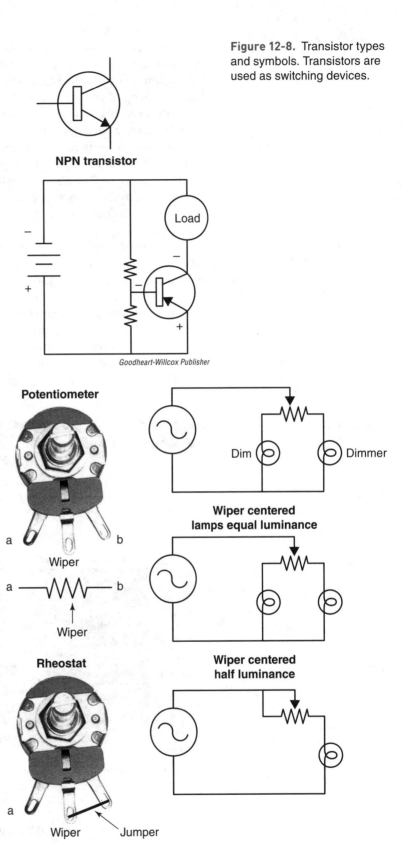

Potentiometer

a   b

Wiper

a —/\/\/\— b

Wiper

Dim   Dimmer

**Wiper centered
lamps equal luminance**

**Wiper centered
half luminance**

Rheostat

a

Wiper   Jumper

a —/\/\/\—

Wiper

Goodheart-Willcox Publisher

Figure 12-9. Potentiometer and rheostat symbol in circuit application

result in equal luminance. All components have tolerance, and it is unlikely that the two bulbs are identical in terms of filament resistance. In reality, one lamp will be slightly brighter than the other. This is where a potentiometer is beneficial. Rather than centering to potentiometer, it is adjusted until both lamps are equally bright. The bulbs were used to illustrate the operation of the device, but in practice, it is used to control current to two control circuits.

The rheostat controls the current flow to only one load, as shown in the illustration. A rheostat can be made by placing a jumper between the wiper and a fixed terminal as shown. It is also available as a two-wire device. These devices are rated the same as fixed resistors. The rated resistance value is from one fixed terminal to the other. For the two-terminal rheostat, it is the fixed terminal to the extreme travel of the wiper.

Resistors are rated by ohmic value, tolerance, and wattage. Recall that tolerance indicates how much the actual value is allowed to deviate from the rated value per the manufacturer. Typical resistor tolerance is ± 5%–10%, although precision resistors are typically ± 0.1%–2 %. Resistors vary in physical size, according to wattage. They must be able to dissipate heat, thus greater current flow requires greater surface area for heat transfer to the ambient air.

Many resistor bodies are too small to include print information. Therefore, a system of colored bands is used instead, **Figure 12-10**. The pictured resistor has a value of 1000 Ω and a tolerance of ± 5% that is derived as follows:

Figure 12-10. A resistor band chart.

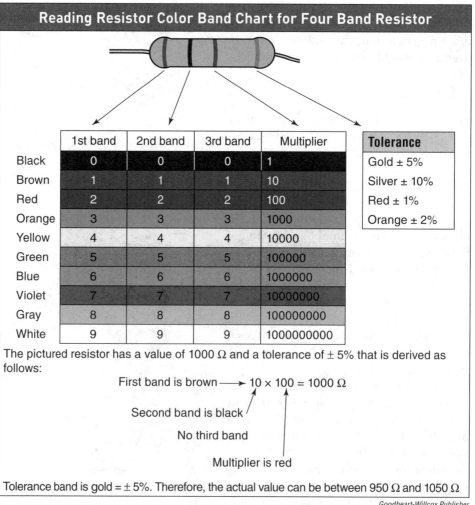

Wattage is based on the physical size of the small resistors in a circuit board and those used in start capacitor bleed resistors. Typical values of 1/4 W through 3 W for carbon resistors range in size from about 1/4″ to 1″ in length. Power resistors are rated in hundreds of watts, are usually wire wound, and are big enough to print the rating. See **Figure 12-11**. If the wattage rating is exceeded, the resistor can overheat and possibly ignite. For example, if a 10 Ω 1/4 W resistor is connected to a 3 V source, the current draw is 0.3 A. However, the power is 3 V × 0.3 A = 0.90 W. The consumed power exceeds the 1/4 W rating. Therefore, the resistor would burn. In cases where the potential exists for current to increase beyond design intent, special flame-retardant resistors are used.

12.3 Sensors and Transducers

Sensors are devices that sense, or detect, a change in physical quantities, such as temperature. There are a wide variety of sensors. Some are specific to industries, such as medical, food processing, manufacturing, and HVACR. They all respond to changes in a specific quantity that results in either activating a set of contacts or producing a signal that is proportional to the detected change. Sensors can be mechanical, electronic, or a combination of both.

Transducers change one form of energy to another. For example, a pressure transducer changes mechanical energy to electrical energy. The pressure sensed by the transducer is converted to an analog electrical signal. The words transducer and sensor are used interchangeably in the case where a sensor produces an analog signal compared to simply opening or closing a set of contacts. A transducer can also receive an analog electrical signal called an input and convert it to mechanical energy called the output. A sensor cannot perform this function.

The following are common HVACR sensors and transducers:

- Bimetal
- Thermistors
- Thermocouple
- Cadmium Sulfide Cell
- Transducers (temperature, pressure, or current)

A **bimetal** is made from two different metals that are bonded together. The different expansion rates chosen for the materials cause the bimetal to warp in one direction when heated. See **Figure 12-12**. When the bimetal is cooled to the initial temperature, it returns to the original shape. If it cooled further, it warps in the opposite direction. This warping provides movement that is used to activate a switch.

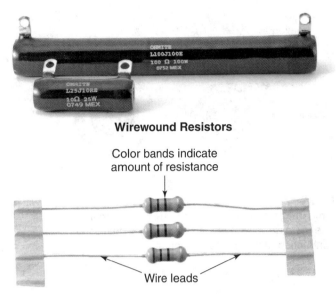

Wirewound Resistors

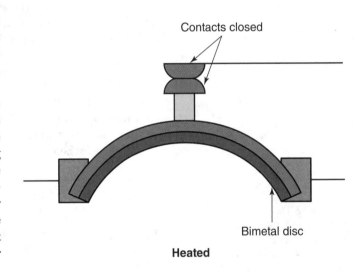

Color bands indicate amount of resistance

Wire leads

Carbon Composition Resistors

Goodheart-Willcox Publisher

Figure 12-11. Two types of resistors.

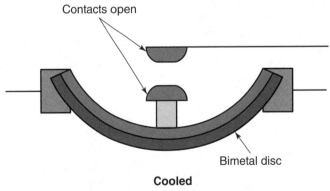

Contacts closed

Bimetal disc

Heated

Contacts open

Bimetal disc

Cooled

Goodheart-Willcox Publisher

Figure 12-12. When the bimetal warps, it closes a set of contacts. When the bimetal is cooled, it returns to the original shape and the contacts open.

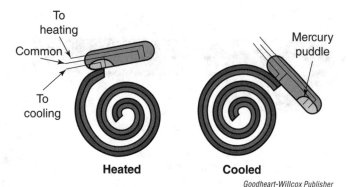

To heating
Common
To cooling

Mercury puddle

Heated   **Cooled**

Goodheart-Willcox Publisher

Figure 12-13. The coiled bimetal expands and contracts with temperature change. This will cause the mercury bulb to tilt. The mercury will move toward the two terminals and allow conduction, thus closing the control circuit.

Bimetals are manufactured in different shapes depending on the amount and rate of movement required. For example, **Figure 12-13** shows a coil shape with a mercury bulb thermostat.

A ***thermistor*** is a resistor that responds to temperature changes by altering its ohmic value. There are two types used in HVACR applications: the positive temperature coefficient (PTC) and the negative temperature coefficient (NTC). The resistance of a positive temperature coefficient increases with temperature rise. Its operating characteristic is designed to have a low resistance up to a specific temperature that, when reached, causes the resistance to greatly increase. See **Figure 12-14A**. The PTC is used to stop current flow when temperature rises due to internal heat produced by overcurrent or external ambient heat. Once the PTC cools, the resistance drops to the operating value.

> **Pro Tip**
>
> **PTC and NTC**
>
> In HVACR, the terms PTC and NTC are typically used without the word thermistor attached. Both are thermistors where one has a negative temperature coefficient, and the other has a negative temperature coefficient.

The NTC resistance decreases as temperature rises. The negative temperature coefficient is used to measure temperature since it has a linear operating curve. For example, a commonly used NTC has a value 10 kΩ at about 70°F. See **Figure 12-14B**. The manufacturers provide tables that equate resistance to temperature. These tables are used when testing NTC thermistors.

Figure 12-14. A—A PTC thermistor's resistance increases rapidly when the temperature rises to a specific value. B—A NTC thermistor's resistance increases gradually as the temperature drops. The operating curve is near linear. For this reason, it is used to measure temperature.

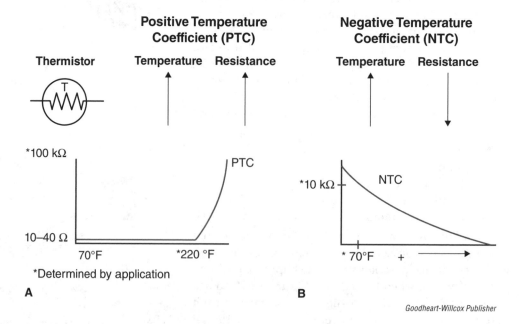

Goodheart-Willcox Publisher

Thermocouples are made of two dissimilar metals that are joined together. When these two metals are heated, a potential difference is developed across the metals. There are many types to accommodate specific usage requirements for extreme temperatures and harsh environments. Different combinations of pure metals and alloys produce different voltages and stability of the temperature to voltage measurements. Type K and Type J are common thermocouples found in HVACR. A Type K thermocouple is typically used to measure temperature. It is found in many electronic thermometers. A Type J thermocouple is used to provide standing pilot flame in gas-fired equipment, such as furnaces. A thermocouple generates 20–30 millivolts dc when the proper pilot flame is present. If the controller does not detect this voltage, then the gas supply is shut off. Thermocouples have a hot junction where the metals are joined and the heat source is sensed. The ends of the wires that attach to the hot junction are called the cold junction. The cold junction is where the produced voltage can be checked. The cold junction is connected to the controller. See **Figure 12-15**.

A *cadmium sulfide cell* is a photoconductive device that reacts to changes in light. In darkness, its resistance is greater than 20,000 Ω, and, in light, it is less than 1600 Ω. It is used to detect the flame in oil-burner heating equipment. This sensor is a safety device that sends a signal to the oil-burner controller. If the burner is firing, the flame should produce the correct level of light, so the cadmium (cad) cell can send a corresponding current to the controller. A low current causes the burner to shut down because the sensor does not see a flame.

A transducer senses a physical quantity and then converts that quantity to an electrical signal that is calibrated to correspond to the sensed value. For example, consider a pressure transducer that is attached to a 300 psig port on a refrigerant line. The pressure acts on a strain gage that produces a small current that is amplified.

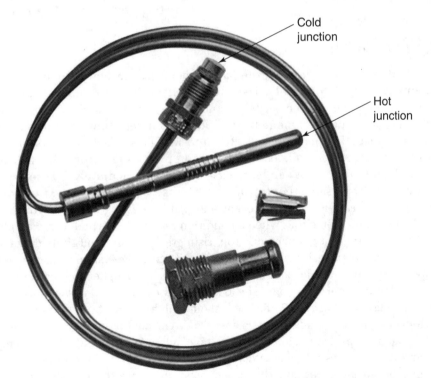

Cold junction

Hot junction

Figure 12-15. The thermocouple has a hot junction to sense heat and a cold junction that connects to a controller.

White-Rodgers Division, Emerson Climate Technologies

The amplified signal is calibrated to output a specific voltage that corresponds to pressure. A typical transducer output voltage is 3 Vdc to represent the 300 psig. Transducer output voltages vary, but some popular outputs are 0 V to 4.5 V and 0 V to 5.0 V. These voltages can represent different pressure ranges. For the example outlined, 1 V represents 100 psig. The pressure transducer provides an analog signal that corresponds to actual system pressures as opposed to the digital on and off action of a pressure limit switch.

The transducer's output is connected to a microprocessor-based controller. Some transducers output a current between 4 mA to 20 mA instead of voltage. The type of output depends on the controller requirements. The **temperature transducer** works the same. The output voltage or current corresponds to a specific temperature. The **current transducer** senses current by magnetic induction the same as the inductive clamp on the multimeter. The small current produced by the current-carrying wire is processed in the same manner the other transducers. The output of the current transducer is either a voltage or current. Typical transducers use a three-wire system, **Figure 12-16**.

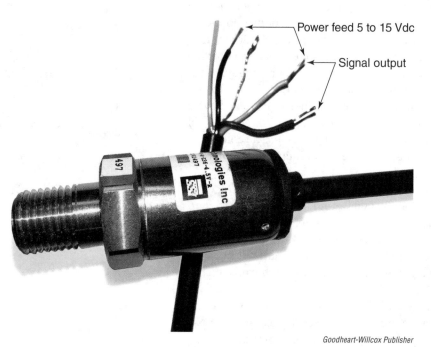

Power feed 5 to 15 Vdc

Signal output

Goodheart-Willcox Publisher

Figure 12-16. A typical transducer has three wires and supply is applied to the positive and negative leads. The output is connected to the output lead and the negative lead.

12.4 Application

This section provides circuit examples to demonstrate how switches and sensors interact. See **Figure 12-17**. The latching circuit in this example uses NO and NC momentary pushbutton switches and a relay with two NO set of contacts. When the start button is momentarily pushed, current flows through the relay coil. The normally open contacts RC1 and RC2 close. RC1 contacts provide a path to energized the coil. The start button path to feed the relay coil is not required, and therefore it can be released almost immediately. The relay coil is de-energized by pressing the stop button. This is also called a **stop/start station**.

A **lockout relay** is used to prevent automatic resetting of power after a limit switch opens and then closes when a condition returns to normal. For example, if a high-temperature switch opens due to a rising temperature and shuts down a system to prevent damage, it may not be advisable to turn the system back on when the temperature drops. The lockout relay coil is made with a higher coil resistance than typical relay and contactor coils. The lockout relay coil is wired in series with a contactor or relay coil, **Figure 12-18**. The lockout coil drops most of the voltage since it has higher resistance, and thus, it is energized. The contactor or relay coil in series with lockout drops a small amount of voltage and acts as a conductor to complete the circuit for the lockout relay. During normal operation, the lockout relay does not energize because current take the pass of least resistance through the closed pressure switch. To reset the lockout relay, power must be turned off in this example by breaking S1.

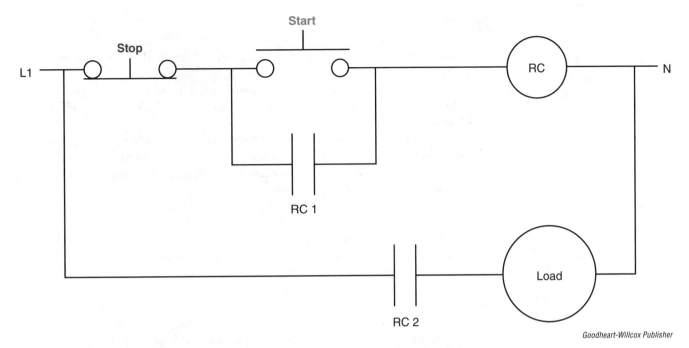

Figure 12-17. The latching circuit is turned on with a momentary push of a NO switch.

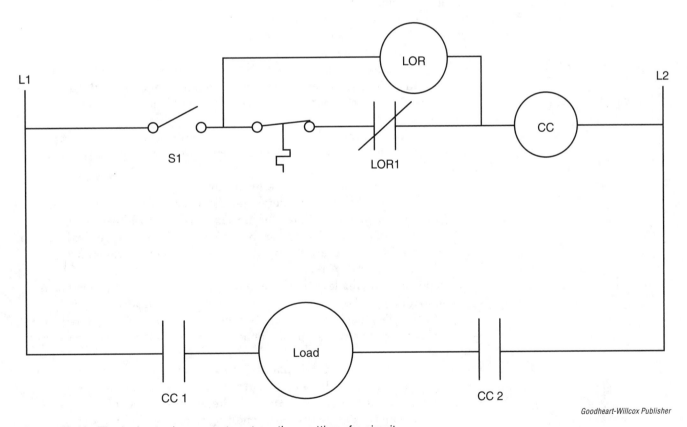

Figure 12-18. The lockout relay prevents automatic resetting of a circuit.

Summary

- Switches are available in many configurations determined by the number of poles and throws.
- A single-pole refers to one source of power feeding the switch. A single-throw indicates that power can be passed to only one terminal when the switch is closed.
- Automatically-activated switches are designed to respond to an external force.
- Switches that retain their selected state (open or close) until making another selection are called maintained. Switches that revert back to their normal state after activation are called momentary.
- Switch specifications include critical characteristics such as current and voltage capacity.
- When replacing a switch, replace it with the same or equivalent part number. Never use a new one that has a lower current and voltage capacity.
- Most solid-state devices use silicon or germanium as a base material.
- Impurities are added to change their current-conducting characteristics.
- Positive type material and a negative type material are the fundamental building blocks of solid-state electronics.
- Solid-state devices are also called semiconductors due to their ability to conduct current in varying amplitude as opposed to a fixed value.
- Electronic components are categorized as active or passive. Diodes, diacs, triacs, SCRs, and transistors are active components. Passive components include capacitors, inductors, resistors, and transformers.
- Diodes can be used to rectify ac to dc.
- Zener diodes are used for voltage regulation.
- An LED emits light when forward biased.
- Resistors are either variable or fixed. The potentiometer is a three-terminal device with an adjustable center contact called the wiper and two fixed terminals. The rheostat is a two-terminal device with one fixed terminal and an adjustable wiper.
- Resistors are rated by ohmic value, tolerance, and wattage.
- Sensors are devices that sense or detect a change in physical quantities such as temperature.
- Transducers change one form of energy to another.
- The bimetal is made from two different metals that are bonded together and warp in response to temperature. The movement is used to active a switch.
- The resistance of a positive temperature coefficient (PTC) increases with temperature rise. The negative temperature coefficient (NTC) resistance decreases as temperature rises.
- Thermocouples are made of two dissimilar metals that are joined together. A potential difference is developed across the metals when heated.
- Type K and Type J thermocouples are common thermocouples found in HVACR.
- The cadmium sulfide cell is a photoconductive device that reacts to changes in light.
- A transducer senses a physical quantity and then converts that quantity to an electrical signal that is calibrated to correspond to the sensed value.
- A lockout relay is used to prevent automatic resetting of power after a limit switch opens and then closes when a condition returns to normal.

Lab 12.1 Two-Way Lighting

The purpose of this lab is to understand how two three-way switches provide control to one bulb.

Lab Introduction

This lab allows students to practice tracing a circuit. This circuit is commonly used in hallways or stairways. For example, turning the light on when climbing a stairway to the second floor and then turning the light off/on at the second-floor switch. This is called two-way lighting. This circuit design has a flaw when both switches are activated at the same time. The HVACR technician should be aware of system's capability before looking for a non-existing problem.

Equipment

- Lab board
- Safety glasses
- Multimeter

- 2—3-way switches SPDT
- 1—SPST switch
- 1 incandescent bulb

 Procedure

1. Wire the lab board per **Figure 12.1-1**. Note that the common terminal of the first is on top. This is because the line belongs on top and the load on the bottom terminal. The second switch receives power from the first switch at the two adjacent terminals on the top and feeds the load through the common terminal that is on the bottom.

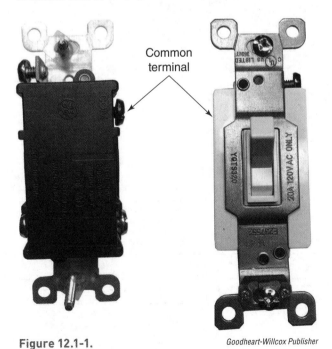

Figure 12.1-1.

Goodheart-Willcox Publisher

Continued ▶

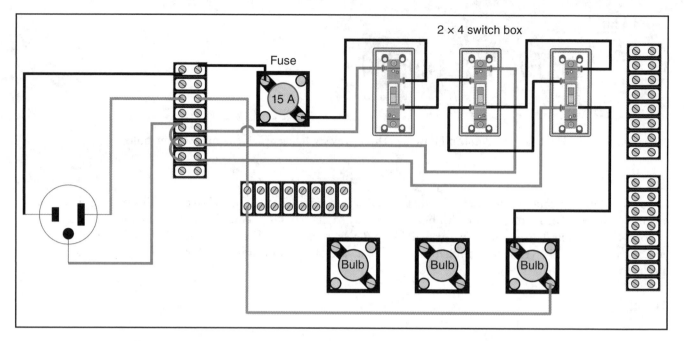

Schematic diagram

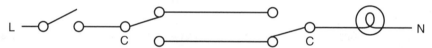

Figure 12.1-1. (Continued)

Goodheart-Willcox Publisher

2. Check the board for short circuits.
3. Turn off the SPST switch.
4. Get permission from your instructor before you plug in the board.
5. Turn on the board.
6. Activate the toggle on first SPDT switch. Observe the bulb.
7. Activate the toggle on second SPDT switch. Observe the bulb.
8. Ensure the bulb is on.
9. Activate both switches at the same time. Observe the bulb.
10. Ensure the bulb is off.
11. Activate both switches at the same time. Observe bulb.

Lab Questions

1. Did the bulb illuminate in Step 6?

_____

2. Use the schematic diagram to explain why.

_____

_____

_____

3. Did the bulb illuminate in Step 7?

_____

4. Use the schematic diagram to explain why.

_____

_____

_____

5. Did the bulb illuminate in Step 9?

_____

6. Use the schematic diagram to explain why.

_____

_____

_____

7. Did the bulb illuminate in Step 11?

_____

8. Use the schematic diagram to explain why.

_____

_____

_____

Lab 12.2 Latching Circuit

The purpose of this lab is to build a latching circuit called a stop/start station by using one relay with two normally open contacts.

Lab Introduction

The latching circuit requires only a momentary signal to energize a load and keep it running as opposed to using a maintained switch. This circuit has many uses in control systems. When this latching function is used to pass power to a load, it is called a stop/start station. This control method will correct the inherent fault with the two-way lighting in Lab 12.1.

Equipment

- Lab board
- Safety glasses
- Multimeter
- Step-down transformer 24 V
- 24 Vac coil relay or contactor with 2 NO contacts

- 1—SPST switch
- 1—momentary NO SPST switch
- 1—momentary NC SPST switch
- 1—incandescent bulb

Procedure

1. Wire the lab board per **Figure 12.2-1**. If the momentary push button switches are not available, use the SPS type to mark one as NC, and ensure it is in the ON position. The other switch should be OFF initially. Momentarily turn on and off to mimic momentary switch.

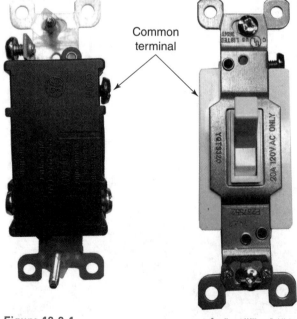

Common terminal

Figure 12.2-1.

Goodheart-Willcox Publisher

Continued▶

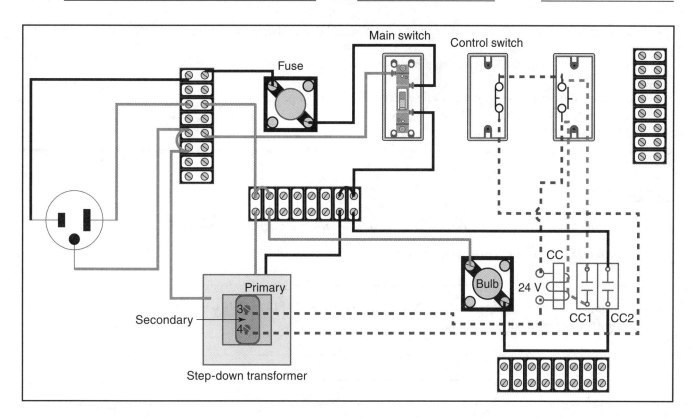

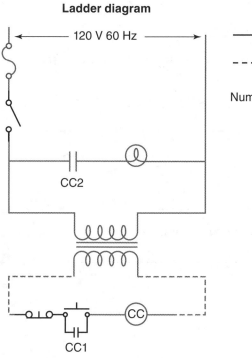

Figure 12.2-1. (Continued)

Goodheart-Willcox Publisher

2. Check the board for short circuits.
3. Turn off the SPST switch.
4. Get permission from your instructor before you plug in the board.
5. Turn on the board.
6. Turn on the main power switch. The bulb should be off.

Continued ▶

7. Momentarily press the NO push button. The relay should latch and keep the bulb on.
8. Press the NC push button. The relay should de-energized, and the bulb should turn off.
9. Unplug the board and disconnect any one wire from cc1 contacts. Cover the exposed terminal with electrical tape or a wire nut.
10. Plug board in and repeat Step 7.

Lab Questions

Use your digital multimeter or other electrical instruments to answer the following questions.

1. Did the relay latch and bulb turn on in Step 7?

_____

2. Use the ladder diagram to trace current flow in the control circuit to explain why the lamp stays on after releasing the NO momentary switch.

_____

_____

_____

3. Did the relay unlatch and the bulb turn off in Step 8?

_____

4. Use the ladder diagram to trace current flow in the control circuit to explain why the lamp turns off after pressing the NC momentary switch.

_____

_____

_____

5. After disconnecting the CC1 wire in Step 9, did the bulb stay on?

_____

6. Use the ladder diagram to trace current flow in the control circuit to explain what happened.

_____

_____

_____

Lockout Relay Operation

The purpose of this lab is to build another type of latching circuit to provide a lockout function. In addition, a pictorial diagram will be drawn from the information given in a ladder diagram.

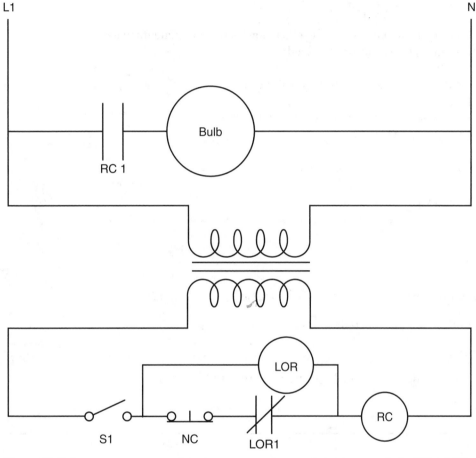

Figure 12.3-1.

Goodheart-Willcox Publisher

Lab Introduction

This is a multipurpose lab. The lockout relay operation will be examined by wiring the lockout relay in an operating circuit. However, the usual pictorial diagram for the lab board is not provided. Instead, a ladder diagram is provided in **Figure 12.3-1**. The student must first draw a pictorial diagram before attempting to wire the circuit. Instead of a sensor, a normally closed momentary SPST switch will be used. Once the circuit is broken, the lockout relay will energize.

Equipment

- Lab board
- Lockout relay 24 Vac coil (example: Mfg. # S1-02424116000)
- 1—24 Vac coil general purpose relay—require NO contacts, one set
- 1—24 Vac step-down transformer
- 1—SPST switch
- 1—SPST momentary NC push botton switch
- 1—Incandescent bulb
- Wire 14 AWG—white and black
- Wire 18 AWG—available colors

 Procedure

1. Study the ladder diagram. Understand the placement of the components in each section of the diagram (line, control, and load).
2. Draw a pictorial diagram.
3. Wire the lab board according to the pictorial diagram.
4. Ask your instructor to review the diagram.
5. Check for short circuit with ohmmeter. Turn off all switches.
6. Get instructor approval to plug in.
7. Close main switch.
8. Close the control switch—S1. Bulb should light.
9. Press the NC momentary switch and release. Bulb should turn off and not come back on when NC switch is released.

Lab Questions

Use your digital multimeter or other electrical instruments to answer the following questions.

1. Did the circuit function as expected?

_____

2. Why did the bulb not light when the NC switch was released?

_____

_____

_____

3. What had to be done to turn the lamp back on?

_____

4. Did it require turning off the main switch?

_____

5. Could turning off S1 have reset the circuit?

_____

Know and Understand

_____ 1. *True or False?* Any switch can be used as a replacement providing the contact configuration matches.

_____ 2. The DPDT switch symbol has a dotted line drawn between the two throws. What does it mean?
A. There are two separate handles.
B. Both handles must be moved at once by the operator.
C. One handle controls both sets of contacts.
D. One handle opens the contacts and the other closes them.

_____ 3. Which would be the best choice for controlling one 120 V load?
A. DPST.   C. DPDT.
B. SPST.   D. SPDT.

_____ 4. Which would be the best choice for controlling one 240 V load?
A. DPST.   C. DPDT.
B. SPST.   D. SPDT.

_____ 5. A three-phase load requires which type of switch?
A. Three SPST.   C. Two DPDT and one SPDT.
B. One SPST and one DPST.   D. 3PST.

_____ 6. A high-pressure limit switch _____.
A. outputs an analog signal
B. opens a set of contacts when a set pressure is reached
C. opens its contacts if the pressure in not high enough
D. produces a current when the pressure reaches a set value

_____ 7. Which is *not* a word or phrase associated with electronic components?
A. Solid state.   C. Active components.
B. Semiconductors.   D. Bimetal.

_____ 8. Which type of rectifier allows ac conduction in both directions?
A. Half wave.   C. Sine wave.
B. Full wave.   D. SCR.

_____ 9. Which device does *not* require a signal for it to conduct current?
A. Diac.   C. SCR.
B. Triac.   D. Transistor.

_____ 10. Which device conducts current in only one direction and requires a signal to begin conducting?
A. Diode.   C. SCR.
B. Bridge rectifier.   D. Resistor.

_____ 11. A gold band on a resistor indicates _____.
A. wattage rating   C. tolerance
B. ohmic value   D. resistance factor

_____ 12. *True or False?* The wattage rating for resistor operating under 5 V is not important due to the low supply voltage.

_____ 13. A rheostat is a _____.
A. special transistor type   C. four-wire variable resistor
B. three-terminal variable resistor   D. two-terminal variable resistor

_____ 14. Which statement is *not* true about a bimetal?
   A. It is made of two dissimilar metals.
   B. The metals have different expansion rates.
   C. It generates current when heated.
   D. Its warping motion is used to active contacts.

_____ 15. Which statement is *not* true about a thermocouple?
   A. It is made of two dissimilar metals.
   B. It has a hot junction.
   C. It warps out of shape with heat.
   D. It generates about 30 mVdc.

_____ 16. What statement is *true* about a photoconductive device?
   A. It converts light energy to electric energy.
   B. It stores light energy.
   C. It generates dc current only.
   D. Its resistance varies with light.

_____ 17. *True or False?* A transducer senses a physical quantity and then converts that quantity to an electrical signal that is calibrated to correspond to the sensed value.

_____ 18. A pressure transducer is calibrated to output 0.50 Vdc per 100 psig. What is the system pressure when the measure voltage at the sensor output is 3.0 Vdc?
   A. 200 psig   C. 1000 psig
   B. 250 psig   D. 600 psig

_____ 19. *True or False?* A lockout relay is used to prevent automatic resetting of power.

_____ 20. *True or False?* To reset the lockout relay power to the lockout relay must be turned off.

Critical Thinking

1. An LED has a current rating of 20 mA. The LED will be used to indicate that the lockout relay used in Lab 12.2 is energized by connecting it in parallel to lockout relay coil. How much resistance is needed to limit the current to of 20 A?

_____

_____

_____

2. What wattage is required for the above resistor?

_____

_____

_____

I. Pilon/Shutterstock.com

13 Electric Motors

Chapter Outline

Additional Reading

Modern Refrigeration and Air Conditioning, **21st edition**

Learning Objectives

After completing this chapter, you will be able to:

- Discuss the basic construction of ac induction motors.
- Explain the basic operation of ac induction motors.
- Describe split-phase induction motors.
- Explain how motor speed is determined.
- Describe dual voltage motors.
- Describe the shaded-pole motor.
- Describe three-phase motors.
- Describe the electrically commutated motor.
- Describe the variable frequency drive motor.
- Explain dc motors and its variations.
- Describe hermetic compressor motor application.
- Describe motor protection.

Technical Terms

bearing	dual-voltage motor	pickup voltage	slip ring
brush	electronic starting	potential relay	snap disc
brushless dc motor	relay	PTC relay	split-phase motor
capacitor-start,	electronically commutated	resistance-start,	squirrel cage rotor
capacitor-run (CSCR)	motor (ECM)	induction-run (RSIR)	start winding
capacitor-start,	end bell	revolutions per minute	stator
induction-run (CSIR)	hermetic compressor	(RPM)	stepper motor
centrifugal switch	horsepower	rotating magnetic field	synchronous speed
commutator	locked rotor amperage	rotor	torque
continuous voltage	(LRA)	run winding	variable frequency drive
current magnetic relay	magnetic pole	self-start	(VFD)
(CMR)	mechanical load	servo motor	
dc motors	permanent split	shaded-pole motor	
dropout voltage	capacitor (PSC)	slip	

Introduction

Electric motors convert electrical energy into mechanical energy, which is needed to drive various equipment that require physical movement. In the HVACR industry, motors drive components, such as compressors, fans, pumps, circulators, dampers, and valves. Because these applications require different amounts of mechanical energy to operate correctly, there are different motor designs to satisfy specific requirements. This chapter will cover the motor construction and motor controls that make it possible for the motor to deliver the required mechanical energy to motors in various applications.

Two common factors in a motor's design are its horsepower rating and applied electric power. Motors are rated by *horsepower*, which is the measure of mechanical energy. One horsepower equals 746 W, or 33,000 ft-lb of work per minute (550 ft-lb of work per second). Single-phase and three-phase ac are the main types of electric power that will be discussed in this chapter. While all motors are not identical, they operate on electromagnetic principles to generate rotational force called *torque*. Torque is applied to a mechanical load, such as a fan, to perform useful work. A *mechanical load* can be any object that requires a physical force to move it. The heavier the object is, the more torque that is required. Motors generate torque by using the repelling and attracting forces of a magnetic field.

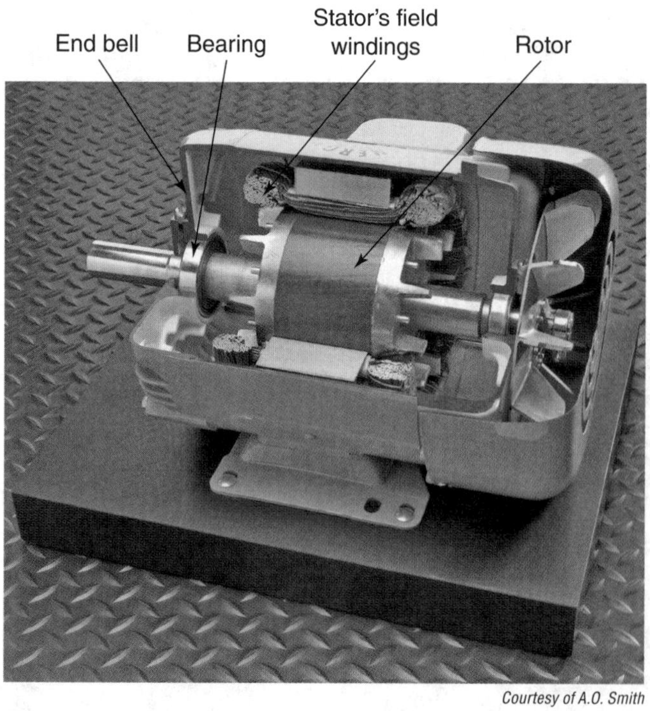

End bell   Bearing   Stator's field windings   Rotor

Courtesy of A.O. Smith

Figure 13-1. Cutaway view of an induction motor.

13.1 Basic Motor Construction

An ac induction motor is widely used in HVACR applications. It is the simplest motor design based on its construction and operation. Its construction consists of a stator, rotor, housing, and end bells. See **Figure 13-1**.

The *stator* is the stationary part of the motor and made of windings, also called field windings. The coils of wire that make up the winding are wrapped around iron cores to increase the magnetic field strength. Iron cores are made of iron laminates. Laminated iron sections produce a greater magnetic field than one solid piece of iron, **Figure 7-2**. The stator assembly is attached to the housing. Housing construction is determined by how a motor's internal parts are cooled and where the motor is located. Because motors generate heat, there must always be means for it to dissipate. The *rotor* is the rotating part and is made up of staggered iron laminates to produce diagonal bars. It is attached to the shaft. See **Figure 13-2**. These bars act as conductors.

Rings are placed at both ends of the rotor to connect the diagonal bars together and complete a circuit. The rings are called shorting rings. *End bells* are attached at both ends of the housing to support the shaft and rotor. The end bells contain *bearings* to allow rotation of the rotor. Bearing type depends on the size of the mechanical load driven by the motor.

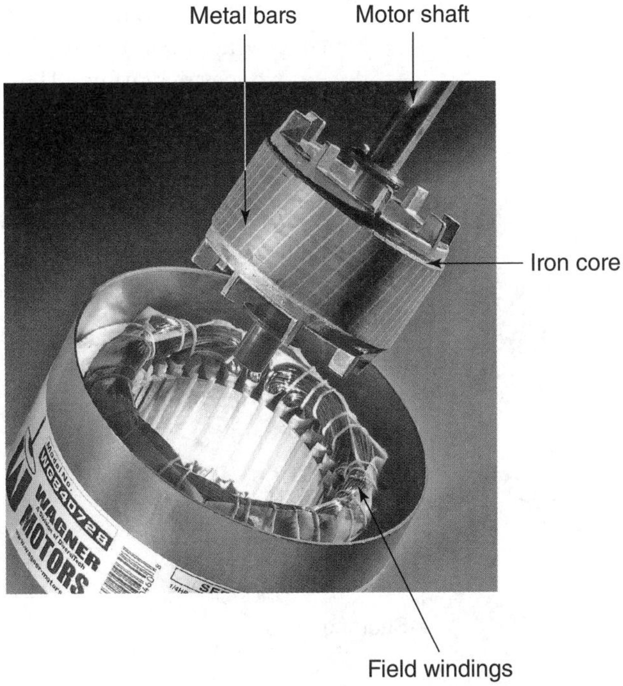

13.2 Basic Motor Operation

Electric motors require continuously changing magnetic field polarities to produce a rotating motion. Since magnets have a north and south pole, the stator must, at a minimum, produce two poles. These poles must also change polarity at some frequency. Alternating current automatically produces changing magnetic polarities as the current flow changes direction. See **Figure 13-3**.

Metal bars   Motor shaft

Iron core

Field windings

DiversiTech Corporation

Figure 13-2. Squirrel cage rotor showing the diagonal bars that are attached to the shaft.

Figure 13-3. Magnetic poles are formed by wrapping wire around poles in different direction.

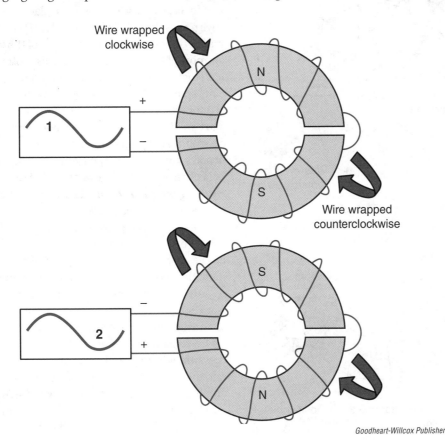

Wire wrapped clockwise

N

S

Wire wrapped counterclockwise

S

N

2

Goodheart-Willcox Publisher

Note that one wire is used to produce two *magnetic poles*. This is accomplished by wrapping the wire clockwise around one iron core and counterclockwise around an opposite iron core. These iron cores are called stator poles. Thus, the stator shown in the figure has two poles and is called a two-pole motor. The direction of the supply current during the first half of the cycle induces the north and south magnetic fields in the poles with opposite polarities. Current flows into the bottom pole and induces a magnetic field that yields in a south pole. Since the wire is wound in the opposite direction on the top pole, a north pole is produced. In the second half of the cycle, the poles change polarity due to the change in current flow direction.

For the rotor to rotate, its magnetic poles must interact with the stator poles. An ac induction motor uses a rotor called the *squirrel cage rotor*. This rotor has a similar appearance to a hamster or squirrel cage and wheel. Other types of rotors are wire wound and must be connected to power, while others are made from permanent magnets. The collapsing of the stator magnetic field induces a current in the rotor. The stator and rotor behave like a transformer—the stator is the primary, and the rotor is the secondary. This also means that the induced current from the rotor is in the opposite direction of the current in the stator. See **Figure 13-4**. A large current is induced in the rotor since there is low resistance from the bars and shorting rings. The rotor current induces a magnetic field with a polarity that depends on current flow direction.

In **Figure 13-5,** a rotor is placed within energized stator fields. This results in the rotor's magnetic poles attracting the poles of the stator for most of the sine wave cycle. This motor cannot start automatically. It would require physically spinning the rotor to get it started, which is impractical. Instead, in order for the motor to start and in the same rotor direction each time, a second set of poles is required. The second set of poles must be out-of-phase and have opposite polarity from the first set of poles. This creates a *rotating magnetic field*. The rotor then follows the rotating magnetic field.

Generating opposite polarities is determined by wrapping the stator pole in opposite directions. To create two out-of-phase magnetic field poles, two out-of-phase currents are required. To generate out-of-phase currents from single-phase power, the windings must have different dc resistance and inductive reactance resistance. The main winding is called the *run winding* and has low dc resistance but high inductive reactance resistance. The auxiliary winding is called the *start winding* and has higher dc resistance but less inductive reactance resistance. This is because the run

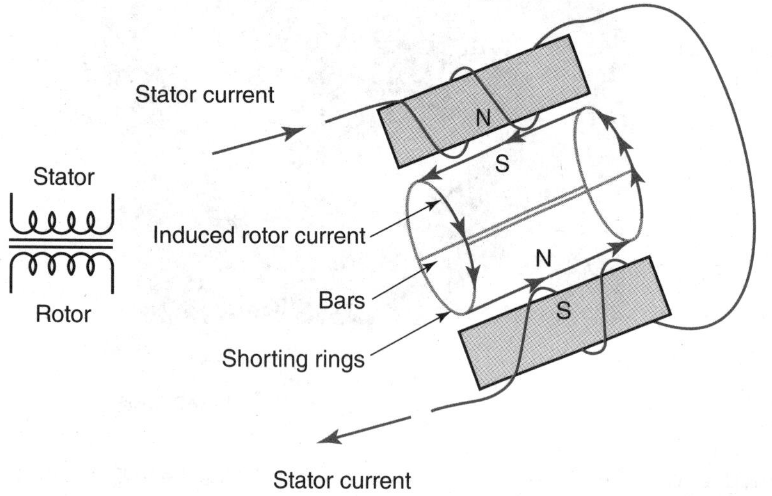

Figure 13-4. The stator induces a current in rotor similar to a transformer.

Goodheart-Willcox Publisher

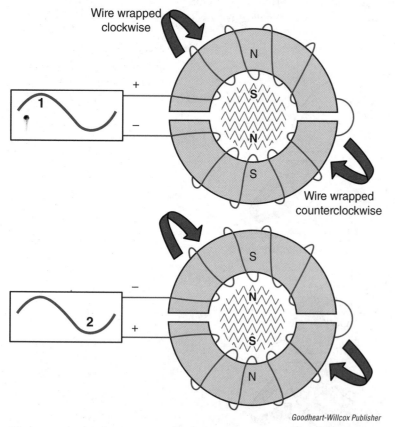

Figure 13-5. A rotor is placed within the stator and the opposite poles attract, resulting in the rotor not moving.

Goodheart-Willcox Publisher

Essential Electrical Skills for HVACR: Theory and Labs

winding has a larger wire diameter and shorter length, and the start winding has thinner wire diameter and longer length. Since the run winding has higher inductive resistance, the run winding current lags behind the start winding current. Hence, out-of-phase currents and out-of-phase magnetic polarities are created.

The two out-of-phase currents make it possible to generate a rotating magnetic field. It allows the rotor to start turning without the aid of an external force. This is called *self-start*. Note the next remaining figures in this section illustrate the automatic starting of a two-pole motor and describe a full 360° rotor rotation. Each winding requires two poles, which means there is a total of four poles. However, only the run winding poles are used to describe the number of poles a motor contains. The start winding poles are required to produce the rotating magnetic field, so there is an equal number of run and start winding poles.

In **Figure 13-6,** the leading start winding current flows from L2 to the first start winding pole, making it a south pole. Current continues onto the second pole, which becomes a north pole. The start winding poles induce currents in the rotor, so the rotor poles are opposite in polarity to the starting poles. The start winding current starts to decrease as the run winding current begins to increase. The rotor has not yet moved but has established magnetic poles.

Figure 13-7 shows the progression of a sine wave in time. It demonstrates how the run winding reaches its peak current and magnetic field. The start winding current decreases toward zero. The rotor's poles are attracted to the run winding poles, and so the rotor moves. The strong attraction allows the rotor to complete 180° of travel, **Figure 13-8.** The second half cycle of the sine begins in **Figure 13-9.** The start winding poles change polarities, causing the rotor polarities to change. As run winding current increases, the rotor is attracted to the run winding poles. The run winding current then peaks and provides the rotor with enough inertia to complete 180° rotation back to the original starting point. See **Figure 13-10.** Thus, the rotor has completed 360° of rotation. After the rotor develops enough inertia to maintain rotation, the start winding can be removed.

First Half Cycle of Sine Wave

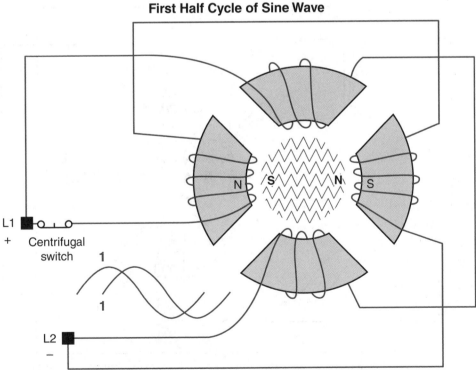

Start winding (blue lines) reaches peak current and maximum magnetic pole strength
Run winding (red lines) is a zero current and no magnetic field

Figure 13-6. The sine wave cycle begins with start winding current leading the run winding current. Magnetic poles are established by the start winding.

Figure 13-7. The run winding current peaks as the start winding current decreases. The rotor moves clockwise.

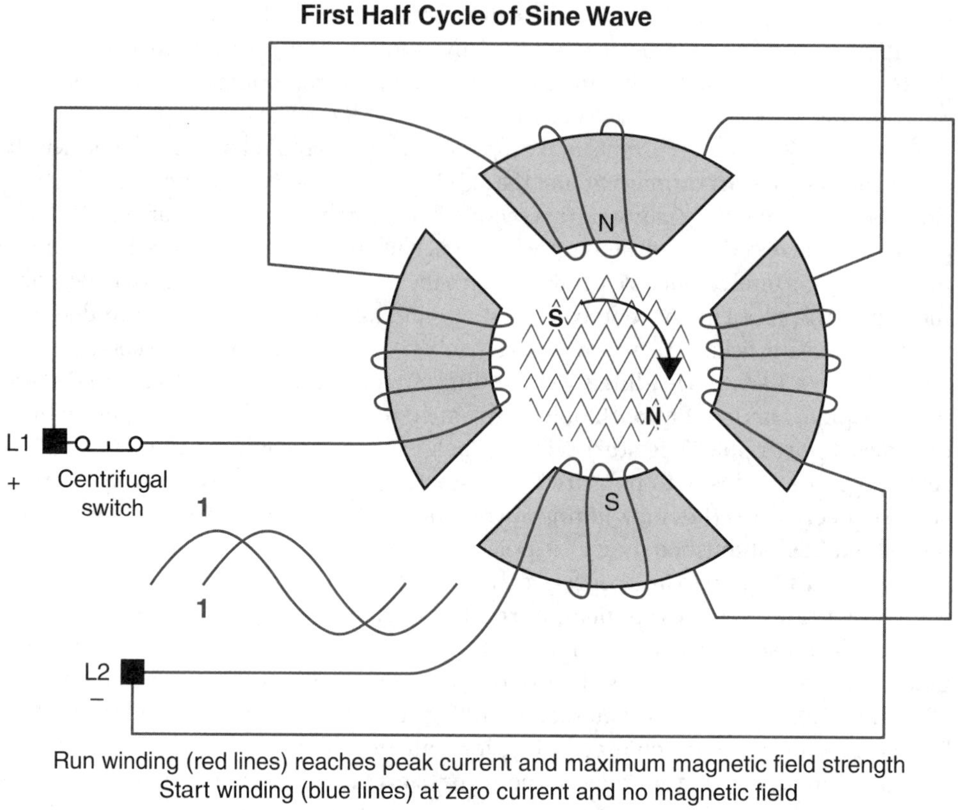

First Half Cycle of Sine Wave

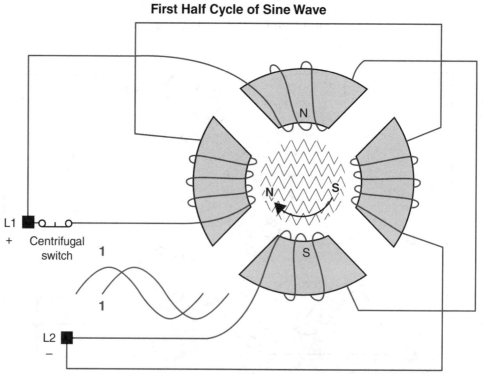

Run winding (red lines) reaches peak current and maximum magnetic field strength
Start winding (blue lines) at zero current and no magnetic field

Goodheart-Willcox Publisher

Figure 13-8. Inertia causes the rotor to complete half a rotation, 180°.

First Half Cycle of Sine Wave

Run winding (red lines) reaches peak current and maximum magnetic field strength
Start winding (blue lines) at zero current and no magnetic field
Inertia causes continued rotation to complete 180° in a half cycle

Goodheart-Willcox Publisher

256   Essential Electrical Skills for HVACR: Theory and Labs   Copyright Goodheart-Willcox Co., Inc.

Second Half Cycle of Sine Wave

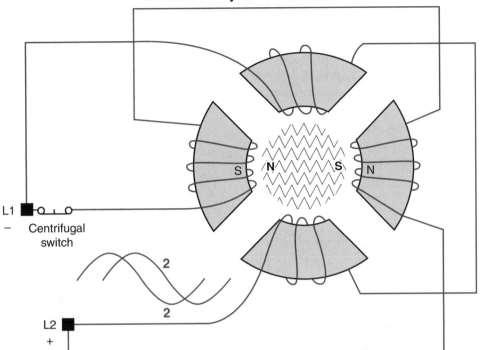

Figure 13-9. The second half sine wave cycle begins and the start winding poles change polarity.

L1

– Centrifugal switch

2

2

L2

+

Start winding (blue lines) reaches peak current and maximum magnetic pole strength
Run winding (red lines) is a zero current and no magnetic field

Goodheart-Willcox Publisher

Second Half Cycle of Sine Wave

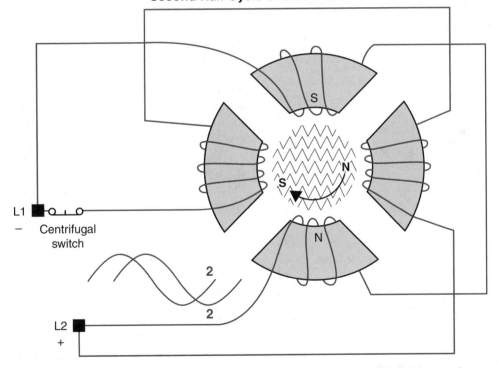

Figure 13-10. The run winding current peaks while the start winding current decreases. The rotor develops enough inertia to complete a 360° rotation.

L1

– Centrifugal switch

2

2

L2

+

Run winding (red lines) reaches peak current and maximum magnetic field strength
Start winding (blue lines) at zero current and no magnetic field
Inertia causes continued rotation to complete 180° in a half cycle
Rotor completes 360° rotation in one sine wave cycle

Goodheart-Willcox Publisher

13.3 Split-Phase Motors

Split-phase motors operate on single-phase power. This type of motor cannot self-start with only single-phase applied. Therefore, single-phase power must be split. This means that two out-of-phase currents must be created. This is illustrated in **Figure 13-7** through **Figure 13-10** of the previous section.

There are four split-phase motors currently used in HVACR applications:

- Resistance-start, induction-run
- Capacitor-start, induction-run
- Permanent split capacitor
- Capacitor-start, capacitor-run

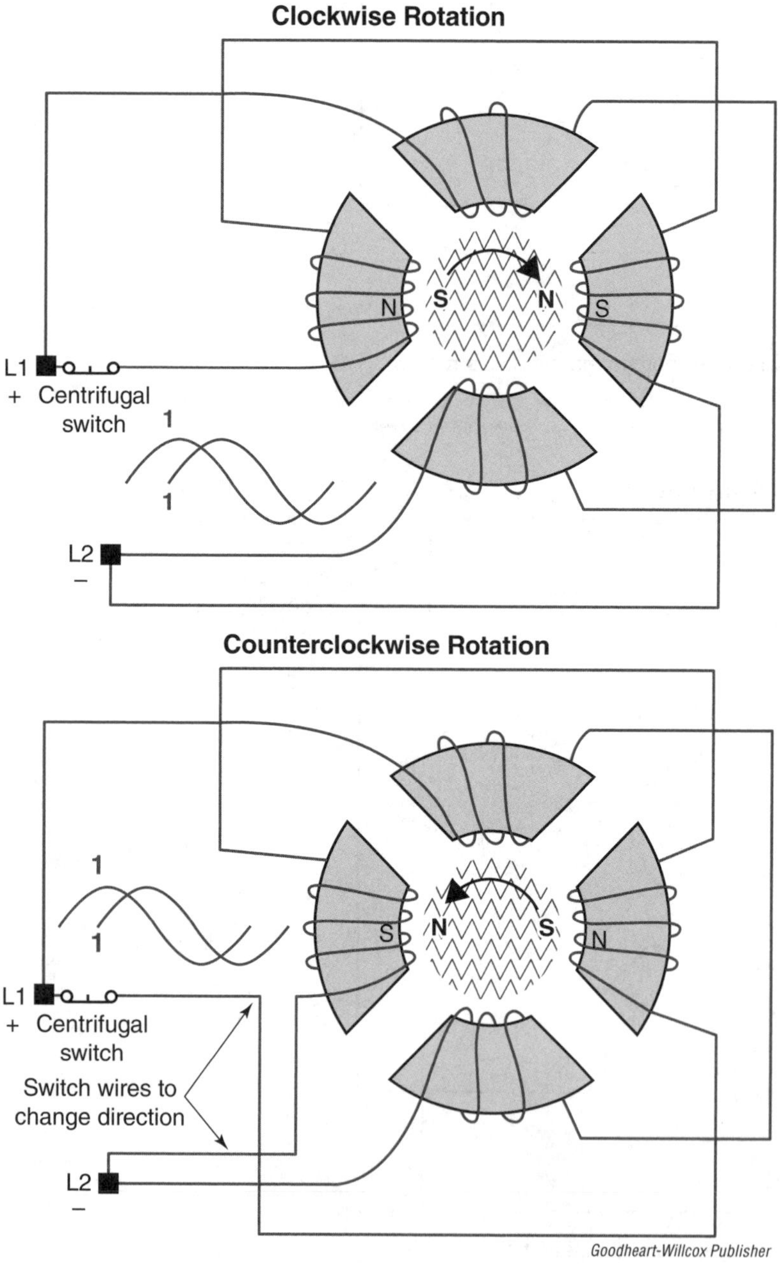

The differences in these four types are based on their starting and running torque requirements. All types require both a run and a start winding. The rotor rotation direction can be changed by switching the direction of the current in the start winding in respect to the run winding current direction. See **Figure 13-11**. However, the motor must have provision to change rotation. Manufacturers provide terminals that can be easily exchanged on select motors. These terminals are found in the motor's terminal board where external power connections are made. The start winding terminals typically identified as terminals 2 and 5 have red and black wires connected to them with quick connect mating terminals. To change the rotor direction, wires must be disconnected and exchanged when reconnecting. For example, if the red wire was on terminal 2, move it to terminal 5 with the black wire on terminal 2.

Two split-phase motors require the start winding to be disconnected after the motor starts. One method of disconnecting the start winding is using a *centrifugal switch*. Centrifugal force acts on objects moving in a circular path. When rotational speed is increased to a certain level, the object travels in a straight path and breaks away from the circular path. The typical centrifugal switch is made with weights attached by springs to the shaft. When the motor reaches from 60% to 75% of rated speed, the weights propel outward away from the shaft. The outward movement of the weights cause a set of contacts to open,

Goodheart-Willcox Publisher

Figure 13-11. Rotor rotation direction is changed by interchanging the start winding terminals with respect to the run winding terminals.

and the start winding is removed from the active circuit. See **Figure 13-12**. When the motor is turned off and it begins to slow, the weights return to the rest position and allow the contacts to close.

Split-phase motors can operate on two different voltages. They are called ***dual-voltage motors*** and have two run windings and one start winding, See **Figure 13-13**. When the run windings are wired in series, they are wired for high voltage. The first run winding drops half the applied voltage. The other half is dropped by the other run winding that is parallel to the start winding. When all three windings are in parallel, low voltage is applied.

Motors draw more current on initial start. This is because stator current is initially not impeded by inductive reactance, thus restricting current flow to only the wire's pure resistance. This also induces rotor current to the maximum value. This current is called ***locked rotor amperage (LRA)***, or also referred to as *inrush* or *starting current*. LRA can be two to six times greater than the normal operating current and lasts typically a fraction of a second.

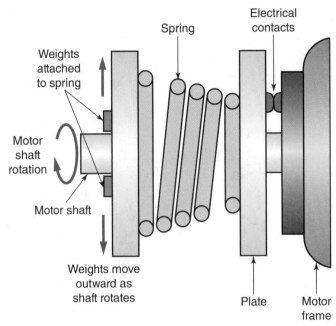

Goodheart-Willcox Publisher

Figure 13-12. The weights of a centrifugal switch move outward as the motor reaches a set speed to open the contacts.

The speed of the motor is called ***synchronous speed***, and it is determined by the motor's rotating magnetic field. The speed is measured in ***revolutions per minute (RPM)***. Synchronous speed can be calculated in the following formula:

$$Ns = 120 \times \frac{F}{p}$$

where

Ns = synchronous speed measured in revolutions per minute (RPM)

120 = constant

F = frequency measured in hertz (Hz)

p = number of stator poles (run winding)

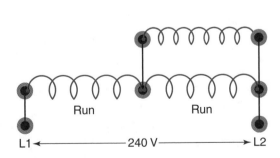

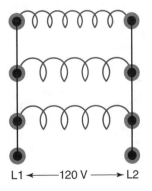

Figure 13-13. Dual voltage single-phase motor. Voltage is selected by changing the numbered connections.

Goodheart-Willcox Publisher

The following are examples of applying the formula to different pole numbers and power values:

- Two-pole motor operating at 60 Hz power:

$$Ns = \frac{120 \times 60\ \text{Hz}}{2} = 3600\ \text{RPM}$$

- Two-pole motor operating at 50 Hz power:

$$Ns = \frac{120 \times 50\ \text{Hz}}{2} = 3000\ \text{RPM}$$

- Four-pole motor operating at 60 Hz power:

$$Ns = \frac{120 \times 60\ \text{Hz}}{4} = 1800\ \text{RPM}$$

As shown in these examples, speed decreases as more poles are added. The four-pole motor rotor completes only a half revolution per ac cycle compared to motors with two poles. Lower frequency also reduces rotor speed.

Note that synchronous speed is not the actual rotor speed. The rotor lags the rotating magnetic field since a mechanical load is attached. The difference between the synchronous and the actual speed is called the *slip*. A motor's actual design speed is given as the rated full-load speed. The rated speed occurs only when the rated load is applied in addition to the rated voltage and frequency. For example, a 1/4 hp motor with a rated full-load speed 3450 RPM has a 1/4 hp load attached, and the voltage and frequency match the rated values and operates at approximately 3450 RPM. Even if the motor is running without a load, the RPM does not match the synchronous value. There are losses to the speed due to the bearings and magnetic field losses in the form of heat.

Not all motor applications require the high speed produced by a two-pole motor. Therefore, HVACR motors use between two and eight poles depending on required speeds and torque. More poles reduce speed and increase torque, or rotational force.

13.3.1 Resistance-Start, Induction-Run

A *resistance-start, induction-run (RSIR)* motor uses a high resistance (dc) to start winding to split the phase. See **Figure 13-14**. Ideally, the phase is split by 90° to gain the most starting torque. The RSIR falls short of that by creating only about half the phase shift. As a result, this motor is limited to low starting torque applications, such as fractional horsepower refrigeration compressors used in compact dorm room refrigerators.

Figure 13-14. RSIR motor schematic diagram. Note that the centrifugal switch inside the motor is shown outside the circle, representing the motor housing.

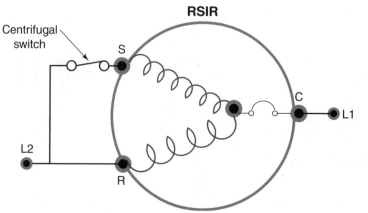

Goodheart-Willcox Publisher

13.3.2 Capacitor-Start, Induction-Run

A *capacitor-start, induction-run (CSIR)* is wired in series with the start winding and causes the phase shift between the run and start winding to approach 90°. As a result, the motor produces higher starting torque. This motor is suitable for larger refrigeration compressors, pumps, and air handlers. The start winding must be disconnected like the RSIR, **Figure 13-15**.

The start winding wire diameter is thicker than the start winding of an RSIR motor due to the increase in current flow. When the start winding is removed, the CSIR operates the same as the RSIR. The only difference between 1/2 hp RSIR and a 1/2 hp CSIR is that the CSIR requires more starting torque to move a specific load, such as a refrigeration compressor.

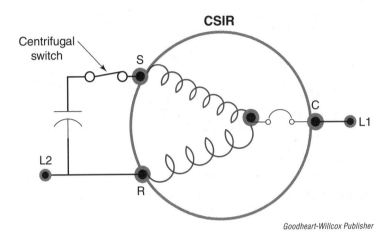

Goodheart-Willcox Publisher

Figure 13-15. CSIR motor diagram.

13.3.3 Permanent Split Capacitor

As the name implies, the phase in a *permanent split capacitor (PSC)* is always split as the motor operates because the start winding is left in the circuit. Thus, it does not require a device to remove the start winding. A run capacitor is wired in series with the start winding and improves the operating power factor. See **Figure 13-16**. A PSC is not a high-starting torque motor like a CSIR. It has a low to moderate starting torque and is used for evaporator blowers and air conditioning compressors up to three tons.

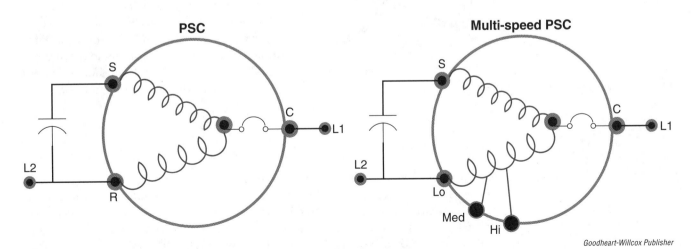

Goodheart-Willcox Publisher

Figure 13-16. PSC motor diagram single and multi-speed.

An advantage of the PSC is its slower ramping start-up. This is helpful when sound must be minimized, such as when operating an evaporator blower quietly. The PSC also draws less start-up current than the CSIR and RSIR. Air handlers need more than one speed. Forced warm air requires less speed than air conditioning. The multi-speed PSC can produce up to four speeds. Connections along the run winding, called taps, make it possible to select more or less winding resistance. Less winding resistance results in more current flow and greater speed. See **Figure 13-16**. These multi-speed motors have six to eight poles.

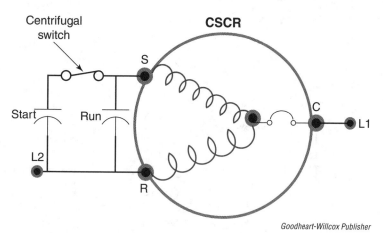

Goodheart-Willcox Publisher

Figure 13-17. CSR motor diagram using a centrifugal switch to remove the start capacitor from the circuit.

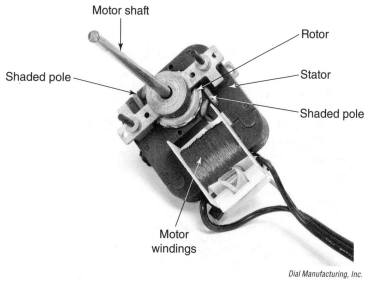

Dial Manufacturing, Inc.

Figure 13-18. Shaded-pole motor.

13.3.4 Capacitor-Start, Capacitor-Run

The *capacitor-start, capacitor-run (CSCR)* motor, also called a *CSR*, is the highest starting torque split-phase motor. When the motor is starting, both run and start capacitors are wired in parallel to each other and in series with the start winding. Capacitors in parallel increase the starting torque. The start capacitor must be removed from the circuit when the motor reaches between 60% and 75% of its rated speed. See **Figure 13-17**. After the start capacitor is removed from the circuit, the CSR operates as a PSC motor. Thus, the CSR has the same efficiency benefit of the PSC. The CSR can be used with up to 5 tons of air conditioning and many refrigeration applications. This motor is also used in large air handlers.

13.3.5 Shaded-Pole Motor

The *shaded-pole motor* is not a split-phase motor because it only has a run winding. In order to self-start, it uses a copper band that produces the secondary out-of-phase magnetic pole to create the rotating magnetic field effect. See **Figure 13-18**. These motors have minimal starting and running torque. They are used for small fans and small condensate pumps. The rotor rotation cannot be changed unless it is a special service type motor that has provisions to alter the band or a rotor-to-frame configuration. Shaded-pole motors are available from less than 1/100 hp to around 1/6 hp. This is the least efficient motor at about 50%.

13.4 Split-Phase Motor Starting Devices

A centrifugal switch cannot be used in explosive atmospheres or inside hermetic compressors. The arcing caused by switching current on and off can ignite gases. It can also create a short circuit if used in a refrigerant and oil environment inside a hermetic compressor. Four common devices are used to remove the start winding and starting capacitor. These are the current magnetic relay, PTC relay, potential relay, and electronic starting relay/timer.

The ***current magnetic relay (CMR)*** is made with a low-resistance coil and a normally open set of contacts. The coil is wired in series with the run window. The coil's low resistance does not drop significant voltage or affect the run winding operation. The coil requires a relatively large current compared to a standard relay to pull in the relay armature and close the NO contacts. The NO contacts are made of one fixed contact and a moving contact that rests on the bottom of an enclosure by gravity.

This construction makes installation critical. The open contact must be facing downward. The NO contacts are connected in series with the start winding. See **Figure 13-19**. When the motor is turned on, current flows through the run winding only due to the NO contacts. The run winding increases as it induces an increasing current in the rotor, yet the motor cannot yet start. The increased run winding current produces a strong magnetic field in the relay coil and closes the NO contacts. The start winding energizes, so the motor can start. As the motor reaches about 75% of rated speed, the run winding current decreases. The lower current reduces the relay coil magnetic field strength. The force of gravity is thus greater and pulls the contact down, and it opens the start circuit.

A ***PTC relay*** uses a PTC thermistor in series with a start winding, or both start capacitor and start winding, to control the current. When power is first turned on, the thermistor is at room temperature and exhibits low resistance. As start winding current increases, the thermistor heats up, and its resistance increases upwards of 20 kΩ. This high resistance effectively removes the winding from the circuit. The PTC relay has to be cooled down three to five minutes before the motor can be restarted. The PTC can also be used on a PSC motor as an assist start-up device. The thermistor is wired in parallel to the run capacitor, so when power is applied to the motor, current flows through the PTC and start winding for added torque. When the PTC heats up, current feeds the start winding through the run capacitor. See **Figure 13-20**.

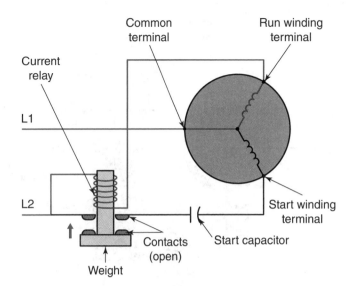

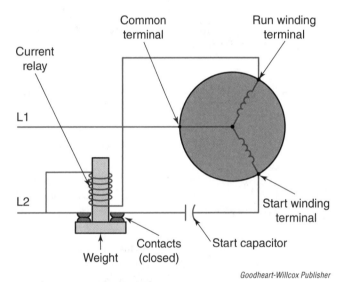

Goodheart-Willcox Publisher

Figure 13-19. Current magnetic relay.

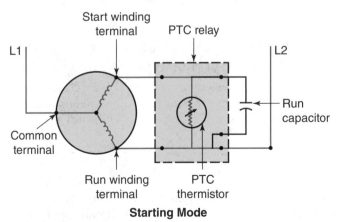

Starting Mode

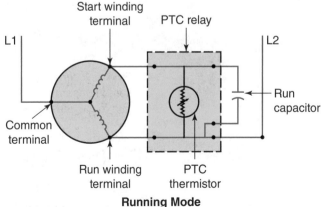

Running Mode

Goodheart-Willcox Publisher

Figure 13-20. A PTC used to assist the starting of a PSC motor.

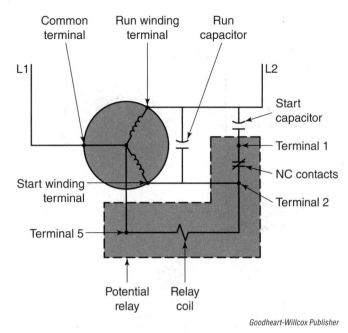

Common terminal

Run winding terminal

Run capacitor

L1

L2

Start capacitor

Terminal 1

NC contacts

Terminal 2

Start winding terminal

Terminal 5

Potential relay

Relay coil

Goodheart-Willcox Publisher

Figure 13-21. Potential relay.

A *potential relay* relies on a bemf (back electromotive force or counter electromotive force) generated by inductive reactance to operate. This relay is used mostly on high-torque CSR motors. The relay has a high resistance coil, typically 8 kΩ to 15 kΩ between terminals 2 and 5, and a normally closed set of contacts between terminals 1 and 2. See **Figure 13-21**. Terminals 3 and 4 are false terminals used for making splices and are not internally connected. The relay coil is wired in parallel to the start winding, and terminal 1 is connected to the start capacitor. When power is turned on, current feeds both the run and start winding. An insignificant amount of current flows through the relay coil because of its high resistance, leaving the majority of the current to flow in the start winding. As current increases through the start circuit, bemf is produced due to inductive reactance. The bemf rises to a level greater than the supplied voltage and causes enough current to flow through the high resistance coil and become energized. The energized coil opens the NC contacts and removes the start capacitor from the starting circuit.

Potential relays are rated by pickup voltage, dropout voltage, and continuous voltage:

- The *pickup voltage* is the required bemf applied across the coil to open the NC contacts and remove the start capacitor.
- The *dropout voltage* is the minimum bemf applied across the coil to keep the armature pulled in and keep the NC contacts open.
- The *continuous voltage* is the maximum bemf that can be applied across the coil without causing damage to the coil.

The pickup voltage occurs when the rotor reaches about 75% of rated speed. The initial LRA current that produced the bemf decreases to FLA. This lower current must produce a bemf that is above the dropout voltage. Recall that relays, solenoids, and motors require greater power to move from a standing position due to inertia.

The *electronic starting relay* has no moving parts. There are types that sense changes in current and voltage to determine when to stop current flow between terminals. For example, the transition between LRA and FLA can create high resistance between two terminals used to remove a start capacitor. Another type uses a timer to create high resistance between terminals based on the predictability that a motor starts within one second.

Pro Tip	Electronic (Timer Type) Starting Relays

Electronic timer type starting relays are not always the best replacement choice for some refrigeration compressors. A compressor can likely need more starting time due to varying cooling loads. The better choice, in this case, is the potential relay.

13.5 Three-Phase Motors

Three-phase motors have higher starting and running torque than single-phase motors. They also have higher efficiency. Since the phases are already split, these motors do not require a start winding or capacitors.

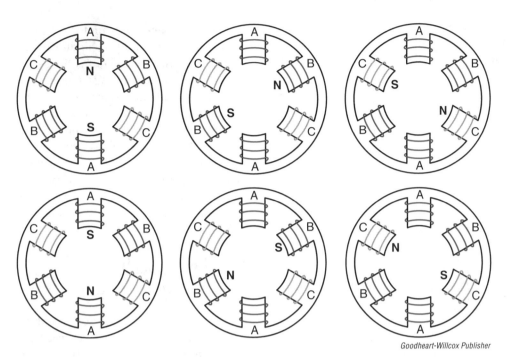

Goodheart-Willcox Publisher

Figure 13-22. A three-phase motor's rotating magnetic field.

Recall each phase is 120° apart. **Figure 13-22** shows the rotating magnetic field of a two-pole, three-phase stator for one ac cycle. Note that each phase requires two poles. The squirrel cage rotor is used as with single-phase motors to follow the rotating magnetic field. The rotor is being rotated clockwise. To change the rotation direction, any two phases can be swapped. By standard, phase one and three are normally swapped.

The ability to easily change direction in three-phase also creates a problem when the phases are accidentally switched before reaching the motor. A phase sequence and rotation tester is available to test three-phase power status. See **Figure 13-23**. Tester leads are connected to three-phase wires that are connected to the motor terminals. The tester displays the rotation based on the sequence. Three-phase supply must be labeled L1, L2, and L3, and motor terminals labeled T1, T2, and T3. However, the power source could be mislabeled or the identification not legible.

Another problem with three-phase motors is single phasing. This occurs when one phase is not supplying power. The motor is then operating with only two hot legs or a single phase. This causes the motor to burn out. Also, a line imbalance of 2% or greater can cause an increase in current through one phase and overheat the motor winding. To ensure all three phases are applied to the motor at the same time, a three-pole single-throw switch, or a three-pole contactor, is required.

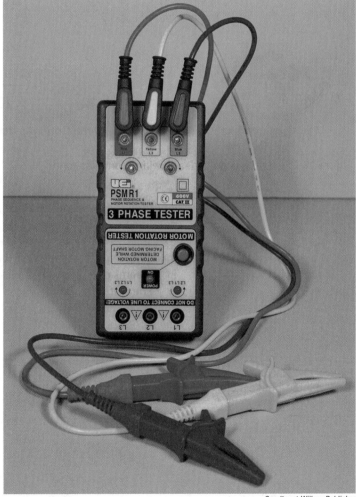

Goodheart-Willcox Publisher

Figure 13-23. Phase sequence and rotation tester.

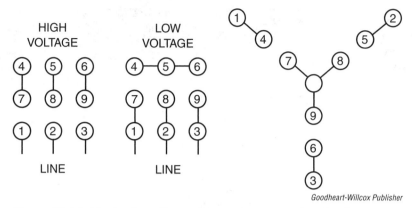

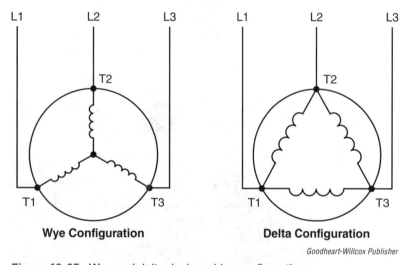

Figure 13-24. Dual voltage three-phase motor connections for the wye configuration.

Figure 13-25. Wye and delta design wiring configurations.

Goodheart-Willcox Publisher

Many three-phase motors have a dual voltage capability. A single-voltage motor has only three connections—T1, T2, and T3. A dual-voltage motor typically has nine connections, so a second set of windings can be wired in series or in parallel with the first set of windings. See **Figure 13-24**. Three-phase motors are manufactured by using the wye or delta design, **Figure 13-25**. A delta wired dual voltage motor is shown in **Figure 13-26**. Note that L1, L2, and L3 always connect to TI, T2, and T3.

13.6 Electronically Commutated Motors (ECMs)

Fractional horsepower ac induction motors, or motors that deliver less than 1 hp, are inefficient. Their efficiency is about 60% to 70% range. That means 30% to 40% of the energy is wasted. To improve efficiency, the *electronically commutated motor (ECM)* was produced. This motor uses the benefits of high-efficiency dc motors and the benefits of the distributed ac power sources.

Recall that frequency is a factor that controls motor speed. The result is precise and efficient speed control. Since the electronically produced ac signal is made from pulses that vary in amplitude, the pulse duration can be controlled. The length of the pulse determines the amount of time current is flowing. This makes it possible to control torque. Long pulses result in higher torque. The ECM is used in fractional horsepower applications where speed and torque control is required. This includes evaporator blowers, condenser fans, and flue gas inducer blowers. While the ECM seems complex to troubleshoot, there are special testers to simplify the task.

The stator is a three-phase stator that is made of three run windings. The rotor is a permanent magnet. To generate the rotating magnetic field required to turn the rotor, the stator windings are sequentially powered to produce magnetic poles.

Figure 13-26. Dual voltage three-phase motor connections for the delta configuration.

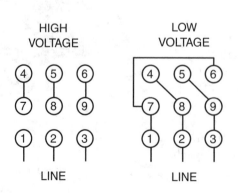

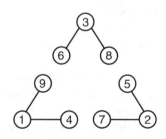

Goodheart-Willcox Publisher

The single-phase power supplied to the motor must be converted to three-phase in order to feed each of the three run windings. First, the single-phase ac is rectified to dc. The dc is processed through an inverter to produce three 120° out-of-phase signals that represent alternating current that rises and falls and changes direction. Since the three-phase signals are produced electronically, the frequency can be controlled. See **Figure 13-27**.

13.7 Variable Frequency Drive Motors (VFDs)

A *variable frequency drive (VFD)* motor provides infinite variable motor speed instead of a few fixed speed values. See **Figure 13-28**. The VFD is not limited to fractional horsepower applications. It is used in high-efficiency large blowers and refrigeration and air conditioning compressors. The VFD controller, called a drive, is applicable to single- and three-phase motors. However, the windings must be rated for VFD use since they are more robust in design. The VFD drive receives single- or three-phase power and rectifies it to dc. The dc is processed through an inverter to produce single- or three-phase signals that can vary in frequency.

> **Pro Tip**   **Inverters**
>
> High-end high-efficiency air conditioning compressors use VFD motors called inverters. Inverters allow for cooling capacity control by regulating the motor speed to maximize efficiency.

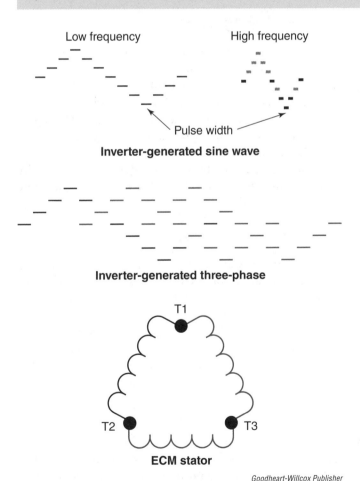

Goodheart-Willcox Publisher

Figure 13-27. Electronically commutated signal to drive an ECM motor.

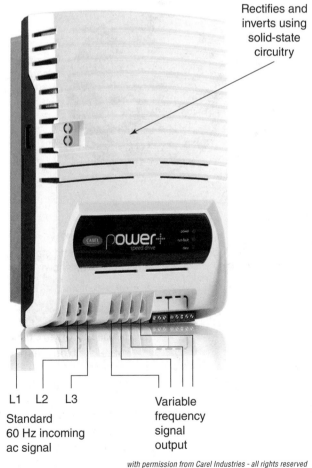

Figure 13-28. A variable frequency drive (VFD).

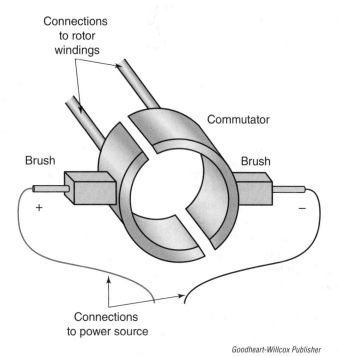

Connections to rotor windings

Commutator

Brush

Brush

+

−

Connections to power source

Figure 13-29. A dc motor commutator.

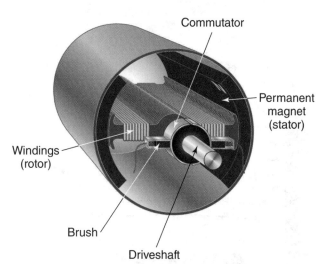

Commutator

Permanent magnet (stator)

Windings (rotor)

Brush

Driveshaft

Figure 13-30. A dc motor rotor, commutator, and stator Rotors have many windings but only one was shown for simplicity.

13.8 DC Motors

DC motors provide the highest torque and efficiency. However, they are more expensive to build as they require more parts. It is too expensive to rectify ac for the purpose of operating moderate to large torque dc motors. DC does not rise and fall or change direction. Thus, a rotating magnetic field is not possible without the use of a commutator. The ***commutator*** creates a change in current through the rotor winding. See **Figure 13-29**. The stator is a permanent magnet in a simple dc motor, and the rotor contains windings. The rotor windings are connected to ***slip rings*** that rub against the brushes that connect to the dc source. The ***brushes*** conduct current between the stationary power source to the rotating slip rings. The winding terminal slip rings contact opposite source brushes for each half-turn of the rotor. This creates alternating magnetic fields needed to rotate the rotor, **Figure 13-30**.

There are many types of dc motors designed to meet specific application requirements. Brushless dc motors, servo motors, and stepper motors are common ones used in the field. These will be discussed in the next few sections.

13.8.1 Brushless DC Motors

A ***brushless dc motor*** is a smaller version of an ECM motor that uses a permanent magnet rotor. The stator windings are energized sequentially to produce rotor movement. These motors are used in modern refrigerators for the evaporator fan and condenser fan. The motor speed can vary from a few to thousands of rpm. They are very efficient and operate at low voltages of about 12 Vdc. The motors are driven by a microprocessor and typically require four wires connecting three windings and one common terminal.

13.8.2 Servo Motors

Servo motors are similar to the brushless dc motor, but they have added feedback to the controlling microprocessor. The added signal, called feedback, provides rotor position back to the processor, so it can move the rotor in either direction. The feedback is used to position the rotor to an exact position as required by operating conditions. These motors are used for dampers and louvers in duct systems. They are also used in some electronic expansion valves.

13.8.3 Stepper Motors

The ***stepper motor*** is another variant of the brushless dc motor. It is driven by pulses that, in turn, move the rotor a certain number of degrees in either direction. The movement can be a little as 1°. Some can provide position feedback. This motor is used in some electronic expansion valves.

Essential Electrical Skills for HVACR: Theory and Labs

13.9 Hermetic Compressor Applications

Motors used inside **hermetic compressors** are surrounded by refrigerant and oil. The winding insulation must be compatible with the refrigerant and oil used. Three terminals connect the run and start windings for a single-phase motor, or the three run windings of a three-phase motor. The terminals are not insulated on the inside of the compressor and contact the refrigerant oil. The oil is not conductive, providing it is not contaminated with acid.

While the terminals should be labeled, this is not always reliable. A technician can check the terminals with an ohmmeter to determine the common, run, and start terminals. **Figure 13-31** shows a single-phase motor diagram as an example. The numbers are used to temporarily assign a name to the unknown terminals. Resistance between terminals 1 and 2, 1 and 3, and 2 and 3 are taken. The highest reading between 2 and 3 indicates that terminal 1 is the common terminal where both windings are connected. This is evident in the winding illustration since the meter is connected to the run and start windings in series, resulting in the highest reading. The second highest reading is between terminals 1 and 2, indicating that terminal 2 is the start winding. The remaining terminal, terminal 3, is the run winding. Since three-phase motors have only run windings, all three readings should be approximately the same. There should be infinite resistance between winding terminals to ground.

13.10 Motor Protection

Fuses and circuits breakers do not always protect the motor. These devices protect the wiring from overheating to protect against fire. Motors use internal and external overloads that react to heat or current. One device is the snap disc bimetal. The **snap disc** is automatically resettable when it cools down. It can be attached to motor windings, or it can be external, such as on the hermetic compressor housing. Some bimetal snap discs contain a low-resistance heating element to sense overcurrent. The overcurrent generates heat and activates the snap disc. Thus, this type protects against ambient heat and overcurrent. Thermistors can also be used to sense temperature and provide a signal to a controller.

Other motor protection methods, specifications, and motor troubleshooting will be covered in Chapter 15, *Troubleshooting Overview.*

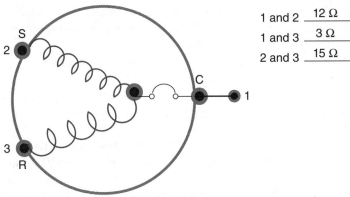

1 and 2	12 Ω
1 and 3	3 Ω
2 and 3	15 Ω

Figure 13-31. Single-phase motor diagram used to determine terminal identification.

Goodheart-Willcox Publisher

Summary

- Electric motors convert electrical energy into mechanical energy.
- Motors drive compressors, fans, blowers, pumps, circulators, dampers, and valves in the HVACR industry.
- Torque is rotational force. Motors generate torque by the repelling and attracting forces of magnetic fields.
- Basic induction motor consists of a stator, rotor and shaft, housing, end bells, and bearings.
- The stator is the stationary part of the motor and the rotor is the rotating part.
- Electric motors require continuously changing magnetic field polarities to produce rotating motion.
- The stator contains a low-resistance run winding and a high-resistance start winding.
- Split-phase motors operate on single-phase power.
- The run winding is called the main winding and the start winding is called the auxiliary winding.
- The start winding must be removed from the circuit when the motor reaches about 75% of rated speed.
- Split-phase motors have a run and start winding and include the RSIS, CSIR, PSC, and CSR motors.
- Start capacitors improve starting torque and run capacitors improve running efficiency.
- Motors draw LRA current during start-up, which can be two to five times greater than running current.
- Motor speed is given in RPM (revolutions per minute).
- A motor's speed depends upon frequency and number of poles.
- The difference between the synchronous and the actual speed is called the slip.
- HVACR motors use between two and eight poles depending on required speeds and torque.
- The shaded-pole motor is not a split-phase motor because it only has a run winding.
- The current magnetic relay or CMR is made with a low-resistance coil and a normally open set of contacts.
- The PTC relay, or PTC, uses a PTC thermistor in series with a start winding or both start capacitor and start winding to control the current.
- The PTC can also be used on a PSC motor as an assist start-up device.
- The potential relay has a high resistance coil and normally closed contacts. The coil is energized by the bemf produced by the start winding.
- The electronic relay has no moving parts. There are types that sense changes in current and voltage to determine when to stop current flow between terminals.
- Three-phase motors have higher starting and running torque and higher efficiency than single-phase motors.
- Three-phase motors do not require start windings or capacitors.
- Three-phase motors use three run windings.
- ECM motors use a three-phase stator and a permanent magnet rotor.
- The ECM is used in fractional horsepower applications where speed and torque control is required.
- The VFD provides infinite variable motor speed instead of a few fixed speed values.
- High-end, high-efficiency air conditioning compressors use VFD motors called inverters.
- DC motors provide the highest torque and efficiency. However, they are more expensive to build as they require more parts.
- Motors use internal and external overloads that react to heat or current.

Lab 13.1 Split-Phase Motor Analysis

The purpose of this lab is to observe the operation of a split-phase motor under normal and no start conditions. You will also check the state of windings with an ohmmeter.

Start capacitor
enclosed in cylinder

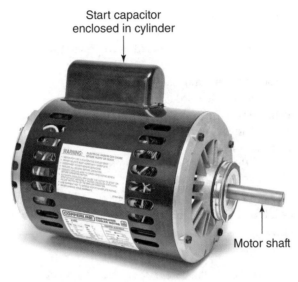

Motor shaft

Figure 13.1-1.   *Dial Manufacturing, Inc.*

Lab Introduction

The recommended motor for this lab is CSIR 1/4 hp single voltage 115 V open frame motor. Voltage, starting, and operating currents will be taken under normal conditions, and then again with the start circuit disabled. The winding resistance will be measured to check the state of windings. The windings could be open, shorted to each other, or shorted to ground. Note the motor is not connected to a mechanical load.

Equipment

- Lab board
- Safety glasses
- Multimeter with inductive clamp
- 1—SPST switch
- 1—CSIR motor 115 V 1/4 hp open frame
- 14 AWG black wire for the hot leg and 14 AWG white wire for the neutral.

Procedure—Normal Operation

1. Secure the motor to the lab board unless the motor is already attached to lab station.
2. Check the motor name for power connections. Your motor may specify the ungrounded side of line. Make sure this is the wire that is connected to the control switch.
3. Remove the connection cover from the motor, and find the connection terminals. The motor terminals typically require F-type 1/4″ crimp terminals.
4. Supply power to the motor terminals.
5. Use your ohmmeter to check the circuit for any shorts.
6. Get permission from your instructor to plug in lab board or to power up lab station.
7. Turn on the motor. Observe the sound it is making.
8. Measure the voltage applied to the motor. Record.
9. Measure the current while the motor is running. Record.
10. Turn off the motor. Listen to the sound it is making.
11. Allow the shaft to come to a complete stop.
12. Set up the clamp ammeter measure the LRA. Review the instructions for your specific meter.
13. Place the clamp around the wire, and start the motor. Record the LRA current.
14. Turn off the motor. Unplug the lab board or disable power to the lab station.

Procedure—No Start Operation

1. Remove the start capacitor cover.
2. Properly discharge the capacitor, and remove one wire from the capacitor. Attach the F-terminal to the opposite blade terminal that is still connected to the capacitor. Capacitors typically have two to four blade terminals for each connection. The intent here is to disconnect the start capacitor from the starting circuit in a safe way. Place the capacitor back into the cover and attach the cover to the motor housing.
3. Place the clamp around wire to measure current (not the LRA).
4. Have a stopwatch ready or count in seconds immediately when turning the motor on.
5. Turn on the motor. Record the current draw.
6. Observe the sound and time recorded.
7. Note any change in the sound and current draw. Record the elapsed time.
8. Disconnect the power to the lab board or lab station.
9. Remove the capacitor cover, and remove the wire from the capacitor. *Do not reconnect to the original terminal at this time.*
10. Measure the resistance between the two motor terminals where the hot and neutral wires are connected. Record resistance. This is the run winding resistance.

Continued ▶

11. To measure the start winding resistance, the start capacitor cannot be in the circuit. Connect the loose capacitor wire back on the already-connected terminal on the capacitor. The goal is to jumper out the capacitor. Both wires are attached to the same terminal. Look at the diagram on the motor to establish the terminal numbers for the start winding on the motor terminal plate. Remove either of the two wires. Place one probe on the loose wire and the other probe on the connected wire and measure resistance. Record the start winding value. By checking at this location, the centrifugal switch continuity is also checked.
12. Reconnect the start winding terminal.
13. Measure the resistance across the hot and neutral terminals again. Record value.
14. Measure the resistance between the hot to the ground terminal on the motor. Record value.
15. Repeat Step 14 for the neutral connection.
16. Reconnect the capacitor to original state.

Lab Questions

1. Describe the sound when the motor is turned on.

_____

_____

_____

2. What is the voltage supplied to the motor?

_____

3. Is the voltage within ±10% of the rated voltage found on the motor nameplate (label)?

_____

4. What is the operating current?

_____

_____

_____

5. Describe the sound when the motor is turned off?

_____

_____

_____

6. What is the LRA current?

_____

_____

_____

7. How much greater is the LRA current than the operating current?

_____

8. When the capacitor is removed, describe the sound observed.

_____

_____

_____

9. When the capacitor is removed, what is the current drawn?

_____

10. When the capacitor is removed, how long did it take for a change in sound and current to drop to zero amperes?

_____

11. What is the run winding resistance?

_____

12. What is the start winding resistance?

_____

13. What is the combined resistance of the run and start windings in parallel from Step 13?

_____

14. What is the resistance between the hot terminal and motor ground?

_____

15. What is the resistance between the neutral terminal and motor ground?

_____

Hermetic Compressor Motor

The purpose of this lab is to identify the terminals of a hermetic compressor.

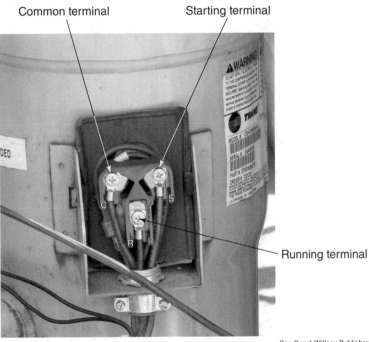

Common terminal   Starting terminal

Running terminal

Figure 13.2-1.

Goodheart-Willcox Publisher

Lab Introduction

A hermetic compressor will be used to identify the connecting terminals by measuring resistance and following a procedure. The compressor will not be powered up.

Equipment

- Hermetic compressor
- Safety glasses
- Multimeter

Procedure

1. Turn off the power to the compressor, if applicable.
2. If the compressor terminals are connected, record the wire configuration or take a picture.
3. Disconnect the compressor, if applicable.
4. Make a sketch of the terminals. Assign a number to each terminal.
5. Measure between the three combinations—1 and 2, 1 and 3, 2 and 3.
6. Review Section 13-8.
7. Measure the resistance between each terminal to the compressor housing ground.

Lab Questions

Use your digital multimeter or other electrical instruments to answer the following questions.

1. Which combination has the highest resistance?

_____

2. Which is the common terminal?

_____

3. Which combination has the second highest (middle) reading?

_____

4. Which is the start terminal?

_____

5. Which is the run terminal?

_____

6. Is there continuity between a winding terminal and ground?

_____

Lab 13.3 Three-Phase Motor Analysis

The purpose of this lab is to acknowledge the operation of a three-phase motor under normal and no start conditions and to check state of windings with an ohmmeter.

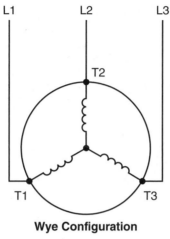

Wye Configuration

Figure 13.3-1.

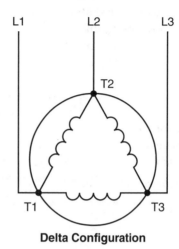

Delta Configuration

Goodheart-Willcox Publisher

Lab Introduction

This lab will require an operational three-phase motor and three-phase power. Power should be delivered to the motor through a three-pole contactor of directly from a three-pole single-throw disconnect or switch. Voltage imbalance will be calculated, and phase currents will be measured. In addition, winding resistance will be measured. Note the motor is not connected to a mechanical load.

Equipment

- 1/4 hp three-phase motor single or dual voltage (wired for low voltage)
- Safety glasses
- Three-pole disconnect, three-pole single-throw switch (rated for motor) or three-pole contactor with 120 V coil.
- Multimeter with inductive clamp

Procedure

1. Before wiring the motor, check the winding resistance between T1 and T2, T1 and T3, T2 and T3. Record the resistance values.
2. Make the power connections to the motor.
3. Get permission from your instructor to turn on motor.
4. Set up the meter to measure LRA current.
5. Start motor. Record LRA for one phase (L1 for example).
6. Measure the operating current for all 3 phases and record.
7. Turn off the motor and observe shaft rotation.
8. Disconnect the power to lab station.
9. Exchange the L1 and L3 at motor terminals.
10. Get permission from your instructor to turn on the motor.
11. Turn on the motor.
12. Perform phase imbalance by measuring the voltage to the motor. Measure the three combinations T1 and T2, T1 and T3, T2 and T3. Record values.
13. Find the average of the three values by dividing the sum of the values by 3.
14. Find the maximum deviation from average. (Consider: combination differs more from the average and by how much?) Record value.
15. Apply the formula: phase unbalance % = $\dfrac{\text{max deviation}}{\text{average}} \times 100$.

Lab Questions

Use your digital multimeter or other electrical instruments to answer the following questions.

1. Was the resistance of the three combinations nearly equal?

_____

2. What is the LRA?

_____

_____

_____

3. Were the operating phase currents nearly the same?

_____

4. Did the shaft rotation direction change when L1 and L2 were exchanged?

_____

5. Is the phase imbalance less than 2%?

_____

Know and Understand

_____ 1. Which answer is *not* equal to one horsepower?
- A. 746 Watts
- B. 550 ft-lbs per minute
- C. 33,000 ft-lbs per minute
- D. 550 ft-lbs per second

_____ 2. *True or False?* The centrifugal switch contacts are normally open when it is used to remove the start winding from the motor circuit.

_____ 3. When a dual voltage motor is wired to operate with high voltage, the windings are configured as follows:
- A. both start windings and both run windings are in series.
- B. the start winding is in series with two parallel run windings.
- C. all windings are in parallel.
- D. the run windings are in series and the start winding is in parallel with one of the run windings.

_____ 4. *True or False?* The inrush current drawn when a motor starts is called locked rotor amperage.

_____ 5. Synchronous speed describes the speed of the _____.
- A. rotor
- B. shaft
- C. bearings
- D. rotating magnetic field

_____ 6. Which factor does *not* affect rotor speed?
- A. Voltage.
- B. Frequency.
- C. Number of poles.
- D. None of the above.

_____ 7. What is *not* true about an RSIR motor?
- A. The start winding stays in the circuit after start up.
- B. The run winding has less dc resistance the start winding.
- C. The start winding is removed from the circuit after start up.
- D. It has low starting torque.

_____ 8. What is *not* true about a CSIR motor?
- A. The start winding stays in the circuit after start up.
- B. The run winding has less dc resistance the start winding.
- C. The start winding is removed from the circuit after start up.
- D. It has high starting torque.

_____ 9. Which motor has low to moderate starting torque and the start winding remains in the circuit after start up?
- A. PSC.
- B. RSIR.
- C. CSR.
- D. CSIR.

_____ 10. Which is *not* true about a PSC motor?
- A. It has high starting torque.
- B. The start winding remains in the circuit after start up.
- C. A run capacitor is in series with the start winding.
- D. It is more efficient than the RSIR motor.

_____ 11. Which split-phase motor has the highest starting torque?
- A. PSC with assist PTC relay.
- B. CSIR.
- C. CSR.
- D. Three-phase motor.

_____ 12. The current magnetic relay has _____.
   A. a high resistance coil and NC contacts
   B. a low resistance coil and NC contacts
   C. a high resistance coil and NO contacts
   D. a low resistance coil and NO contacts

_____ 13. The potential relay has _____.
   A. a high resistance coil and NC contacts
   B. a low resistance coil and NC contacts
   C. a high resistance coil and NO contacts
   D. a low resistance coil and NO contacts

_____ 14. Which is *not* a split-phase motor?
   A. PSC.   C. CSR.
   B. RSIR.   D. Three-phase.

_____ 15. What is *not* true about three-phase motors?
   A. It has three run windings.
   B. Its rotation cannot be changed.
   C. It has higher torque than single-phase motors.
   D. It can be used in hermetic compressors.

_____ 16. *True or False?* Single phasing a three-phase motor means that the motor is operating on only two hot legs.

_____ 17. What is *not* true about an ECM motor?
   A. It uses a permanent magnet rotor.
   B. The stator has three run windings.
   C. It generates three-phase power from single-phase power.
   D. It is used in applications with greater than 5 hp.

_____ 18. The stator windings generate a magnetic field that induces a current in the rotor. The rotor current is _____.
   A. in phase with the stator current
   B. 90° out of phase with stator current
   C. 180° out of phase with the stator current
   D. 360° out of phase with the stator current

_____ 19. *True or False?* Centrifugal switches can be used inside hermetic compressors to remove the start winding.

_____ 20. *True or False?* Circuit breakers are used to protect motors from overheating.

Critical Thinking

1. A potential relay coil fails (breaks) while a CSR motor is running. What happens?

   _____

   _____

   _____

2. When measuring resistance of a hermetic compressor motor, the resistance between the run terminal and the common terminal is the same as the resistance between the start terminal and the common terminal. What is wrong?

   _____

   _____

   _____

cosmopolit/Shutterstock.com

14 Troubleshooting Printed Circuit Board Control Systems

Chapter Outline

14.1 Printed Circuit Board Structure
    14.1.1 Microprocessor-based PCB (Controller) Operation
14.2 Troubleshooting the PCB
    14.2.1 Microprocessor-based PCB (Controller) Applications
14.3 Heating System PCB
14.4 Air Conditioner System PCB
14.5 Communication

Learning Objectives

After completing this chapter, you will be able to:

- Discuss the function of a printed circuit board (PCB).
- Define an input.
- Define an output.
- Explain an algorithm.
- Discuss the purpose of a lookup table.
- Define bi-directional communication.
- Discuss system annunciation and diagnostics.
- Describe the operation of a microprocessor-based PCB.
- Explain how to troubleshoot a PCB.
- Describe microprocessor-based PCB applications.
- Discuss the process of a heating system PCB.
- Discuss the process of an ac system PCB.
- Explain microprocessor-based PCB communication.

Additional Reading

Modern Refrigeration and Air Conditioning, 21st edition

14.3 Circuit Boards and Microprocessors

18.3 Servicing Fan Motors

38.5 Blower Controls

45.5 Building Control Protocols

Technical Terms

algorithm	direct digital control (DDC)	lookup table	programmable logic controller (PLC)
bi-directional communication	event	open-loop system	
	input	output	signal
closed-loop system	integrated furnace controller (IFC)	printed circuit board (PCB)	trace
diagnostics			varistor

Introduction

Modern HVACR equipment uses microprocessor-based controls. The microprocessor and its supporting electronic components are mounted on a ***printed circuit board (PCB)***. A PCB receives input signals from system sensors and switches. The microprocessor's algorithm then uses this information to produce output signals to control an HVACR system.

In order for the technician to troubleshoot PCB controls, they must understand the basic structure of the PCB as it applies to most controllers. A technician can then apply this knowledge to specific HVACR system PCB controllers.

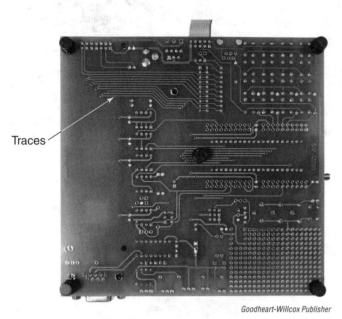

Traces

Figure 14-1. Copper conducting paths called traces.

14.1 Printed Circuit Board Structure

Most components mounted on a PCB are connected with strips of copper called traces, **Figure 14-1**. *Traces* provide electrical connectivity between electronic components. They vary in thickness based on their current capacity. The microprocessor, electronic components, and electromechanical devices are mounted on the PCB and soldered to the traces. See **Figure 14-2**. The only serviceable device on most PCB boards, as shown, is the fuse because it can be replaced. Miniature relays are soldered in place and not meant to be field serviceable. Some PCBs may use plug-in relays that can be replaced. PCBs can use jumper connectors or DIP switches to configure system-operating profiles. See **Figure 14-3**.

Regardless of the equipment controlled, the microprocessor-based PCB can be represented in block form, as shown in **Figure 14-4**. A block represents all of the circuits and components that make up a section of the board that performs a specific function.

Figure 14-2. Electronic and electromechanical devices are soldered to the traces.

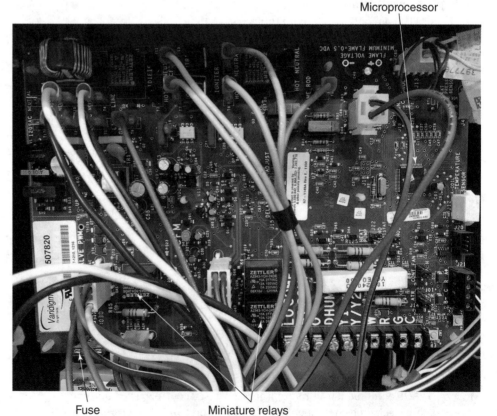

Microprocessor

Fuse   Miniature relays

Essential Electrical Skills for HVACR: Theory and Labs

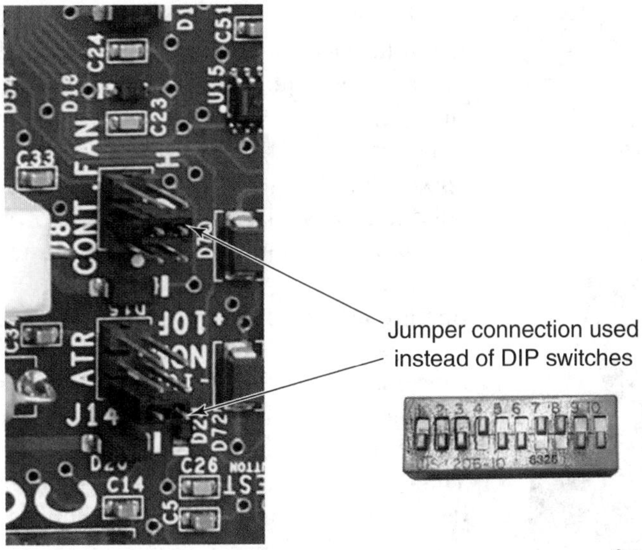

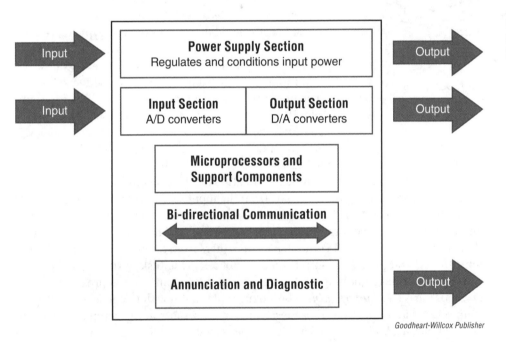

Figure 14-3. Jumpers connectors are used on this board in place of DIP switches.

Jumper connection used instead of DIP switches

Goodheart-Willcox Publisher

Input

Input

Power Supply Section
Regulates and conditions input power

Output

Input Section
A/D converters

Output Section
D/A converters

Output

Microprocessors and Support Components

Bi-directional Communication

Annunciation and Diagnostic

Output

Figure 14-4. Microprocessor-based PCB block diagram broken down by sections.

Goodheart-Willcox Publisher

The circuit details are not useful to the field technician. The six blocks shown perform the following functions that receive power and signals, process information, communicate information, and send power and signals out. ***Input*** is the term used for signals and power fed to the PCB from the system. ***Output*** refers to the processed signals and power sent to the system loads and other elements of the system.

Pro Tip

PCB Terminology

The terms PCB, board, microprocessor, control, and controller are often used interchangeably in the HVACR industry. A technician must be able to use the terms correctly. Note a microprocessor is mounted onto a PCB, which is a board. It is used to control the operation of an entire system or a single motor.

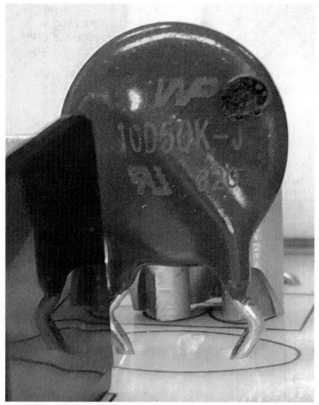

Goodheart-Willcox Publisher

Figure 14-5. A varistor protects against voltage spikes.

The power supply section supplies the processor and other electronic components with stable, noise-free power at the required dc voltage levels. This section may include replaceable fuses or a nonserviceable varistor. A ***varistor*** is a one-time-use resistor that reacts to voltage spikes. See **Figure 14-5**. The power section receives line voltage or low 24 Vac from a step-down transformer. The power section supplies transducers, sensors, and other parts of the system with power. The output to operate sensors is typically 5 to 15 Vdc and 24 Vac.

The input section receives inputs from system sensors, transducers, limit switches, system selector switches, and profile selection DIP switches or jumpers. The input voltage or current is referred to as a ***signal***. The signal supplies the PCB with information to make decisions based on the algorithm running the system's program. The type of input signal that is supplied depends on the signal source. The signal source can be digital or analog. A simple selector switch provides 24 Vac or 0 Vac. Because it has only two states, on and off, it is considered digital. However, a typical pressure transducer input signal is between 0 and 5 Vdc. This is an analog signal since it is within a range of values.

Inputs are labeled on the PCB for their specific use. The pressure transducer may be labeled HPin, for example. HPin stands for high-pressure input signal and is found in the diagram legend. The incoming analog signal to HPin is converted to a digital equivalent signal and fed to the microprocessor. The signal is expected to be within a specified range. An input has a dedicated terminal identification for the expected signal.

A microprocessor receives signals from the input section and then sends signals out to the output section. The microprocessor uses a set of instructions, known as an algorithm, to process the input signals. The ***algorithm*** is a computer program that is specific to a piece of equipment. This differs from a word processor and other multi-functional programs on personal computers. One task of the algorithm is to compare the input signals against the expected values found in the program's memory. This is called a ***lookup table***. The lookup table contains the electrical signal value from sensors and the physical property value it represents in the computer memory. For example, a 3 Vdc signal on the HPin input equates to 300 psig. Lookup tables are mostly used for variable signal inputs and conditions.

A condition refers to the relationship between input signals. It can help explain whether two or more signals have to occur at the same time or if there is a required sequence for the signals to occur. Another part of the algorithm is making decisions based on input conditions. If the pressure signal reaches a specific value, then the compressor is turned off. This is known as an example of decision making.

The output section sends the signal required to operate the load. The signal can be line voltage to drive the load directly, or it can be low voltage to drive a relay or contactor. For a low-voltage output, the processor sends a signal to the output section, where it is converted to an analog signal that is then sent to the compressor contactor coil. Thus, the output to a load depends on the status of input signals. The relationship between inputs and respective outputs are written in the algorithm and lookup tables. For an output to exist, there may be several input conditions that have to be met first. These conditions may include timing sequences of the inputs. The output may be delayed after all the input conditions are met.

Some PCB microprocessor controllers have bi-directional (receive and send) communication capability. ***Bi-directional communication*** means the board may communicate to other system boards and space temperature control. For example, the furnace/air handler, condensing unit, thermostat, and zone control module can communicate digitally among each other. These systems can also connect to the internet through a modem or Wi-Fi. In addition, the PCB may have a port for troubleshooting and updating system software with a laptop or specialized human interface device. An example of a human interface is a tablet, pc, or custom display that can be read by a human being.

The annunciation and diagnostics section uses one or more LEDs to indicate system status. The ***diagnostic*** section receives status and specific failure information from the microprocessor and uses the annunciation section to announce the information to the technician through LEDs or a human interface. The LED can be flashed a certain number of times to announce system faults or normal operation. As an example, the LED may rapidly flash on and off three times, pause, and then repeat the cycle. This repeated flash rate indicates Code #3. The technician must lookup in the supplied code table for the description of the problem. These code definitions and methods of annunciation vary among manufacturers. Some systems may use a numerical display, which is easier to read. Systems with bi-directional communication can provide the technician with historical failure data and the inputs and outputs (referred to as I/O) status.

14.1.1 Microprocessor-based PCB (Controller) Operation

The PCB is also known as a controller since its function is to control a system. The operation of microprocessor-based systems is similar, but they differ in equipment grade level, application, and algorithms with diagnostics. When the controller is first powered up, the processor can check voltage level requirements and internal connections between the processor and support components, such as the A/D and D/A converters. Once internal operation is established, the algorithm may begin checking the inputs for in-range values when the system is off. This is called polling. The outputs may be polled for short circuit conditions.

After the initial self-check is complete, the algorithm begins executing the system operation instructions sequentially. Execution time computes thousands of instructions per second. The system is constantly polling the inputs for a change in status that can cause the processor to execute an output signal. When an input change causes the processor to execute a command, it is called an ***event***. For example, when a switch is closed, a 5 Vdc signal is sent to an input terminal. The event results in the program executing an output signal. The system operation instructions are repeated from the start to end of a program unless there is a special cause to interrupt the execution of the algorithm instructions. The interrupt may cause equipment shutdown.

Pro Tip

System Startup

When main power is shut off and turned back on to a microprocessor controller, the system takes longer to start than when the system is turned off by a thermostat. This is because the processor sees initial power and must perform self-check. This process can take several minutes.

14.2 Troubleshooting the PCB

An example will be used in this section to illustrate the troubleshooting process for a system that includes dependent, sequential, and timing conditions for the system. Use **Figure 14-6** as a guide. Dependent means two or more signals are required before an output is generated. Sequential means a specific order must be followed. Timing indicates that a signal must be received within a specified time period. Also, the signal has to be active for a specified length of time.

The input consists of two single-pole single-throw (SPST) maintained and two SPST momentary switches that supply the input with 5 Vdc when activated. The outputs consist of a green and a red LED supplied by 5 Vdc and a 24 Vac relay coil. The green LED annunciates the board is powered up and ready to operate.

Figure 14-6. Inputs and outputs to describe the troubleshooting process.

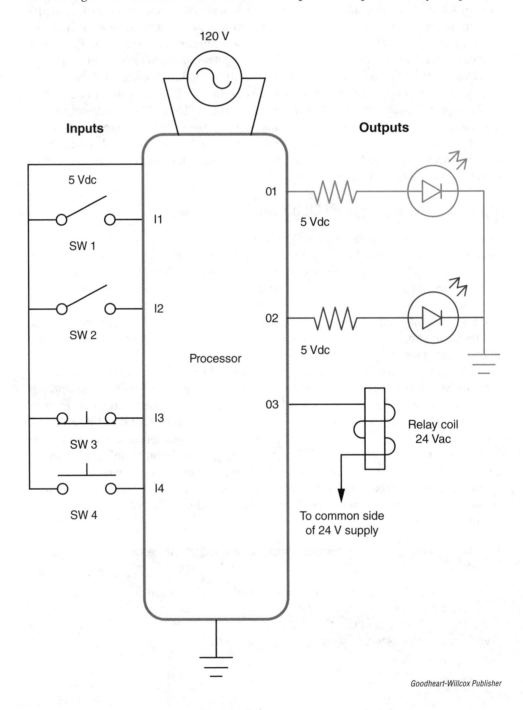

The red LED indicates system fault. To start the process, the green LED must be on. For the relay coil to energize, inputs from SW1, SW2, and SW4 must initially be in the open state. This then should show 0 V at the corresponding terminals I1, I2, and I4. Input from SW3 must be in the closed state initially and during normal operation. This means 5 Vdc should be at terminal I3.

In addition, output O3 must be at 0 V. If these conditions are met when SW1 and SW2 are both closed and sending 5 V signals to I1 and I2, then output O3 supplies 24 Vac to the relay coil. However, if SW1 and SW2 are both closed, but the initial conditions are not met, the red LED signals a system fault and the relay does not become energized. Assuming that the relay is energized, the next event that must occur is having SW4 closed for at least one second within the first 30 seconds of coil activation. If the event occurs within the timeframe, then the coil is allowed to operate until either SW1, SW2, or SW3 is opened. If the SW4 event does not occur, the coil does not become energized, and the red LED is flashed four times.

Assume that the coil does not stay energized. It energizes for about 30 seconds and then shuts off. The red LED is flashing four times per cycle. The system's troubleshooting legend states that SW4 must close for one second within 30 seconds of energizing the coil. Thus, the SW4 signal is not reaching terminal I4 within the required time. If this occurs, the technician must investigate the SW circuit and understand how the circuit works. Use a voltmeter to check for voltage at I4. The meter should be set up and the probes connected before closing SW1 and SW2 because this is a timed event. This is performed to rule out that the problem is not within the PCB, as would be the case if terminal I4 received the required signal. Once the PCB is ruled out as the cause, the SW4 circuit is examined to find the root cause of the problem by using the methods discussed in Chapter 11, *Introduction to Practical Circuits*.

The I/O should be checked with a voltmeter. An in-line ammeter is needed to measure the board input when a transducer outputs current instead of voltage (for example, 4 to 20 mA). An inductive clamp can be used for the output in some situations. Never use an ohmmeter to check any location on the board. The meter's battery voltage can be destructive to the board.

PCBs are not repaired by the HVACR technician. They are replaced. Some applications have generic boards that can be used for replacement. However, others require the exact genuine replacement. PCBs are sometimes replaced when the root cause of the problem is not with the controller PCB, but instead, there is an external intermittent problem that exists. This is the result of not understanding the function and operation of the board or the sequence of operation of the system under control.

Pro Tip

Replacing a PCB

Before installing a new replacement PCB, the technician must know if a fault in the system caused the existing board to fail.

Pro Tip

Electrostatic Discharge

Electrostatic discharge can damage a PCB. Use caution and do not touch terminals or unprotected points on the board as you can store a static charge. To protect the board, a technician can wear a wrist grounding strap connected to a ground while working with a PCB.

14.2.1 Microprocessor-based PCB (Controller) Applications

The microprocessor PCB is designed for many applications. The PCB is dedicated to a specific piece of equipment, such as a gas-fired furnace. The PCB is used in ***direct digital controls (DDC)***, where it controls an entire HVAC system in a building and can include building automation. The PCB is used in ***programmable logic controllers (PLC)***, where the controller can be programmed for a wide variety of applications. The PLC allows the user to write a program using a special high-level language. Other PCB applications include motor control, such as an ECM and VFD.

14.3 Heating System PCB

The dedicated equipment algorithm is not accessible to the user, which means the user cannot make changes to the program. Instructions are set according to safety standards and codes. Safety is critical because combustible fuels are used, and potential carbon monoxide poisoning can occur. The only allowed changes are system profile information that is set through DIP switches, jumpers, or human interface. This information affects only the lookup tables and allows the system to operate at maximum performance and efficiency.

The PCB is typically known as the ***integrated furnace controller (IFC)***. These boards can also control the ac portion of a system. While there are various manufacturers and models, the function is relatively the same. A generic board illustration is shown in **Figure 14-7**.

Figure 14-7. A generic gas-fired furnace PCB.

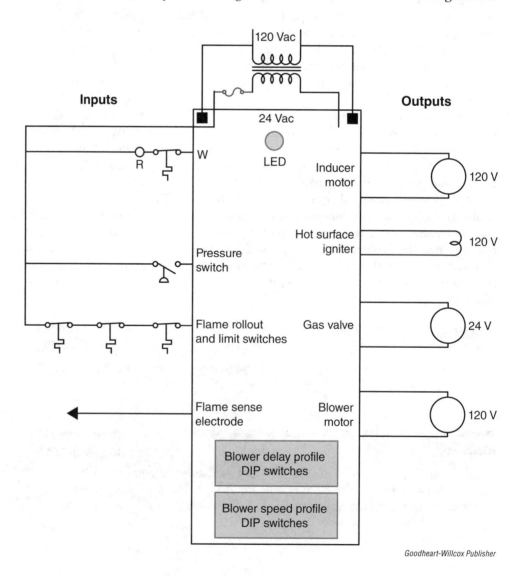

Goodheart-Willcox Publisher

Assuming that the board has been powered up and the initial board diagnostics successfully completed, the LED is illuminated, and the system is ready to operate upon command. The sequence of operation begins when terminal W receives a 24 V signal from the thermostat.

Next, the processor checks the state of the inputs. If the inputs are not in the required state, the process ends, and a trouble code is displayed. The flame rollout and limit switch must be in the closed position to indicate that the switches are in the circuit. The pressure switch must be in the open position initially. This indicates that the contacts are not closed due to component failure. It could also indicate a jumper wire was purposely attached to bypass the safety switch or that the inducer fan is already operating. The inducer fan should not be operating at this time. If the pressure switch is in the open position, the next step is to energize the inducer fan. The pressure switch must close within a predetermined time, thereby applying 24 V to the pressure input terminal. The inducer fan then continues to operate for a predetermined time known as pre-purge before the next event.

The hot surface ignitor is next energized. Since it takes time to heat up, the gas valve becomes energized after a predetermined time. Next, the gas valve is energized, and the gas is ignited by the heated hot surface igniter. The flame must be proven for the gas valve to stay energized. The flame rectification signal must be received at the flame sense terminal within a specified time. If the signal is not received, the gas valve shuts off. The system then attempts to reignite for a set number of trials depending on system design, or it may shut off completely until the system is manually reset. The blower motor becomes energized after the plenum in the duct system heats up to a satisfactory temperature. Older systems used temperature sensors, but the function has evolved to a program timer that can be selected through the delay profile DIP switches.

When the call for heat ends, the gas valve shuts off, and the inducer fan may continue to operate for a predetermined time. This is called post-purge. The blower motor continues to run for a timed period determined by the blower profile settings. Blower speed is also selectable through profile DIP switches or equivalent.

The above example is for a basic single-stage noncondensing furnace that uses a PSC blower motor and shaded-pole inducer motor. However, the troubleshooting strategy remains the same for more complex boards used for dual-stage and modulating condensing furnaces. These systems likely use ECM motors for the blower and inducer and then a stepper motor to control the gas valve. Differential pressure and multiple single pressure switches are used due to the variable speed inducer motor. The technician must know the required input signals and the correct sequence that generate the expected output from the board.

14.4 Air Conditioning System PCB

Generic high-efficiency split-system central air conditioner boards are illustrated in **Figure 14-8**. The system consists of an outdoor unit PCB, indoor unit PCB, thermostat PCB, and zone controller PCB. Digital bi-directional communication is used to share information between the four boards. The thermostat sends a call for cooling to the indoor board and passes the required signals to the outdoor board and zone controller boards. The boards can send information back to the thermostat. This information is recorded and used for built-in performance and error logging reporting for system troubleshooting and efficiency evaluation.

When the boards are initially powered, they perform a self-test and report status to the indoor board. When a call for cooling is sent to the indoor board, it signals the outdoor unit to start. The zone controller is also signaled to start its algorithm. Inputs to the indoor board are the temperature transducers and communication from the thermostat and the zone controller. The input information, together with the algorithm, determines the blower speed and expansion valve position.

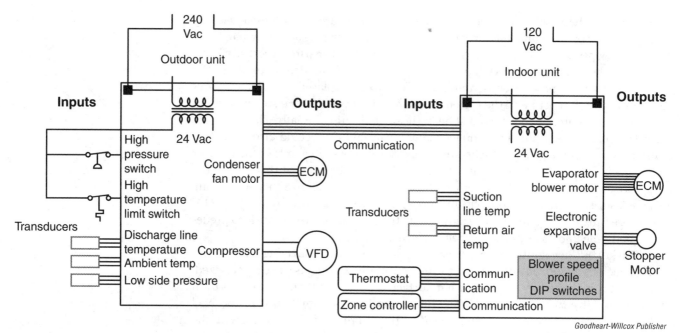

Figure 14-8. A generic ac split system PCB.

Connections for 120 V power

Low-voltage control circuit connections

Emerson Climate Technologies

Figure 14-9. An ECM with attached controller uses a four-terminal and a five-terminal connector.

The number of wires to connect the blower ECM depends on where the ECM controller is located. Some designs have a dedication controller microprocessor in the PCB, while others use a controller housed in the motor. The PCB controller-type has three wires to connect the motor. Recall that the ECM uses a three-phase stator, and thus it requires a simulated three-phase power to operate. The phases vary in frequency and amplitude to control motor speed. When the controller is attached to the motor, a four-terminal and a five-terminal connector are required. See **Figure 14-9**. The four-terminal connector is used for line voltage, and the five-terminal connector is used for profile and speed control.

There are testers on the market to test this type of ECM controller. The electronic expansion valve (EEV) uses a stepper motor that typically has a four-wire connection. When the EEV commands to rotate and open or close the valve, it affects the evaporator temperature. The sensed temperature is sent by the transducers to the controller, and it adjusts the EEV as needed to produce the desired temperature. This is called a *closed-loop system* since there was information feedback to the activation of the EEV. An *open-loop system* does not provide feedback.

If the above EEV operated without feedback, the controller would not be able to accurately position the valve. The outdoor board used its input sensors and communicated information from the indoor board to control the compressor motor speed. The compressor uses a variable frequency three-phase stator motor. The three-phase power varies in frequency and amplitude. The condenser fan uses an ECM motor with the controller built-in to board, resulting in only three wires connecting to the motor and uses feedback from the transducers to regulate its speed.

The indoor PCB can include a forced-air gas-fired furnace control to provide central ac and heating. The PCB can also contain control for a heat pump by adding outdoor EEV, reversing valve, and defrost logic.

14.5 Communication

The above ac board uses digital communication instead of the traditional thermostat wiring. Instead of using multiple wires to communicate between boards, only two wires are required to transmit and receive digital data. This is called multiplexing. Recall that digital refers to two states on and off. The off state is represented by 0 and is the equivalent of 0 Vdc. The on state is represented by 1 and is equivalent to 5 Vdc. Digital information is made up of a series of sequential signals that have only two states. See **Figure 14-10**. The illustration shows eight pieces of information called an 8-bit word. Each bit occurs in one clock cycle in serial communication. Serial means that 1 bit is processed at a time. A bit is one signal with a value of 0 or 1 and is represented by 0 V or 5 Vdc.

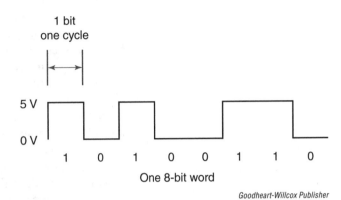

Figure 14-10. An 8-bit word sample of digital communication.

All microprocessor activity occurs in precise intervals that are timed by a system clock. The value represented by the word can take on different meanings depending on the proprietary system software. Typical word length can be up to 64 bits. The bi-directional digital data not only contains system-operating information that needs to be communicated between the boards, but it also addresses a validity check of the information. Each board must have a unique address, so it can receive and send data to another particular board.

Pro Tip

Plug-in Devices

Many manufacturers use a plug-in device that the PCB microprocessor reads that contains the board's address. When a board is changed, this plug-in device must be installed to the replacement board.

Pro Tip

Talking

The term used when communication is taking place between systems or boards is talking. For example, the indoor board is talking to the thermostat.

In order to interpret the digital data, a human interface device is required. This could be a specific troubleshooting tool for the specific equipment or the system's communicating thermostat. In some cases, a communication program can be used in a laptop computer or tablet. Since the communication data is a string of pulsating dc, the presence of a signal can be determined with a voltmeter set on dc. If a communication signal is present, a fluctuating 1 to 4 Vdc is displayed. This is not meaningful information, but it demonstrates that communication is taking place, and the connectivity to the board under test should be good.

Communicating boards can usually connect to the internet via Wi-Fi. Commercial and industrial systems that use equipment from different manufacturers use a standard communication protocol such as BACnet.

Summary

- Modern HVACR equipment uses microprocessor-based controls.
- The microprocessor and its supporting electronic components are mounted on a printed circuit board (PCB).
- The microprocessor-based PBC can be represented in block form. Components include power supply, inputs, outputs, microprocessor and support devices, bi-directional communication, and annunciation/diagnostic sections.
- The input section receives inputs from system sensors, transducers, limit switches, system selector switches, and profile selection DIP switches or jumpers.
- The microprocessor receives signals from the input section and sends signals out to the output section.
- The microprocessor uses a set of instructions known as an algorithm to process the input signals. The algorithm is a computer program that is specific to a piece of equipment.
- One task of the algorithm is to compare the input signals against the expected values found in its program's memory.
- The output section sends the required signal to operate the load.
- Some PCB microprocessor controllers have bi-directional (receive and send) communication capability and can connect to the internet via Wi-Fi.
- The annunciation and diagnostic section uses one or more LEDs to indicate system status.
- The operation of microprocessor-based systems is similar but differ in equipment grade level, application, and algorithms that include diagnostics.
- The inputs and outputs (I/O) should be checked with a voltmeter.
- Never use an ohmmeter to check anywhere on the board.
- PCBs are not repaired by the HVACR technician – they are replaced.
- Some applications have generic boards that may be used for replacement.
- Electrostatic discharge can damage a PCB.
- The PCB is used in direct digital controls (DDC) where it controls an entire HVAC system in a building and can include building automation.
- The PCB is used in programmable logic controllers (PLC) where the controller can be programmed for a wide variety of applications.
- PCBs may use digital communication instead of the traditional thermostat wiring.
- A human interface device is required to interpret digital data.

Lab 14.1 Wiring an Integrated Furnace Controller PCB

The purpose of this lab is to set up an IFC board on the lab board. This will be used for the next two labs. Wiring the board improves your understanding and learning how an IFC works.

Lab Introduction

An IFC board will be wired on the lab to simulate the operation of a gas-fired warm-air furnace. This will require connecting inputs and outputs to the IFC. The inputs, with the exception of the flame sensor, are digital on and off signals. Jumper wires will be used to supply the input terminal with the required signal. The outputs will be represented by light bulbs. A circuit will be constructed to simulate the presence of a flame. The inputs will be labeled so that the correct conditions are applied to the board to simulate normal operation. The outputs will be labeled to observe the sequence of operation.

Equipment

- Lab board
- Safety glasses
- Multimeter
- Grounding wrist band
- IFC # 1165-83-301A or ICM286 or equivalent single stage universal IFC
- Step-down 224 V, 20 or 40 VA
- 24 Vac wired pilot lamp or 100 W incandescent (filament will illuminate adequately for observation) an extra bulb fixture is required
- Molex 0.062 inch round female terminals for 12 and 2 pin board connectors or harness is available – sku # 0259f00007p
- 1–1N1001 diode
- 1–8.5MΩ 1/4 W 5% resistor
- 1 additional SPST to the main power lab board switch for the flame sensor detection circuit
- 5–alligator clip jumper wires
- 3–low wattage 120 V bulbs incandescent or LED
- 1/4″ F-terminals
- 14 AWG black and white wire
- 18 AWG multiple colors for low-voltage inputs
- Nonconductive rubber pad to place PCB on the Lab board or static mat

1. Protect the PCB from static electricity. Use a wrist grounding strap to ensure you are free of static electricity.
2. Read the electrostatic discharge precautions and safety considerations.
3. Unpack the PCB from static shield bag. Compare your board to **Figure 14.1-1** and **Figure 14.1-2**.

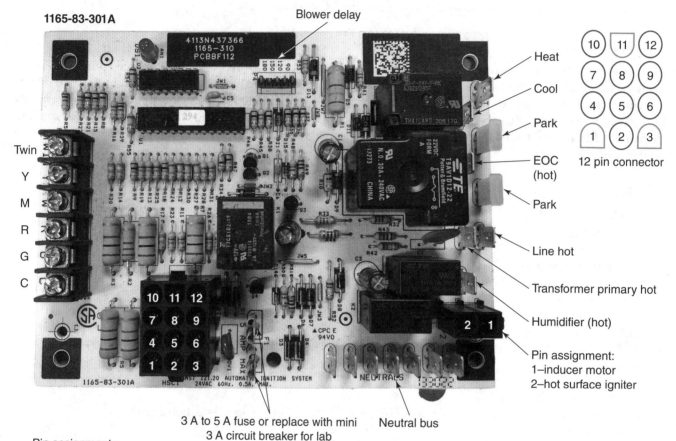

Pin assignments:
1 and 7–Primary limit switch(es)
2–Flame rectification signal
3–24 Vac hot side
4 and 10–Pressure switch
5 and 11–Rollout switch(es)
6–24 Vac common side
8–Chassis ground (used for flame rectification)
9 and 12–Gas valve

Note: The flame sensor line voltage potential is developed across terminals 2 and 8.

Figure 14.1-1.

Continued ▶

ICM286

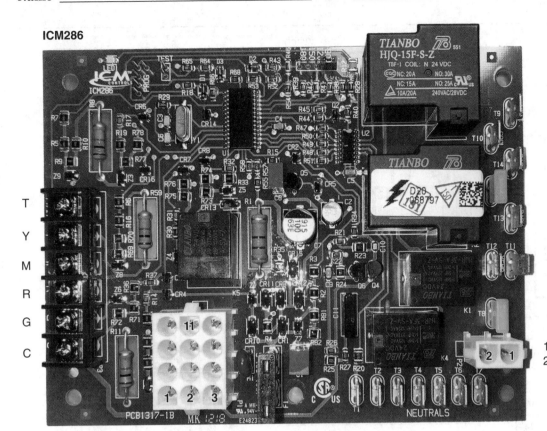

1–Inducer motor hot
2–Hot surface ignitor hot

Pin assignments:
1–Limit switch input
2–Flame sense input
3–24 V (hot) input
4–Pressure switch output
5–Rollout switch input
6–24 Vac common input
7–Limit switch output
8–Gas valve common
9–Common 24 Vac
10–Pressure switch input
11–Rollout switch output
12–Gas valve output

Note: Terminals 6, 8, and 9 are connected together to the 24 Vac common side internally. Terminal 8 is not used for the flame sensor circuit. Terminal 8 is not connected to the ground. The flame sensor line voltage potential is developed across terminal 2 and neutral.

Goodheart-Willcox Publisher

Figure 14.1-2

4. Read the Main Operation section.
5. Review the 12 pin connection assignments for your board.
6. Confirm if terminal 8 is grounded. This will determine how the flame sense circuit is wired. See **Figure 14.1-3**.

Continued▶

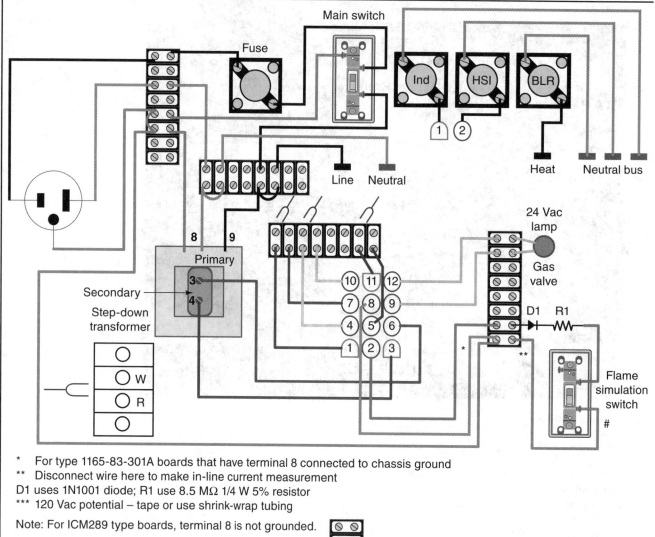

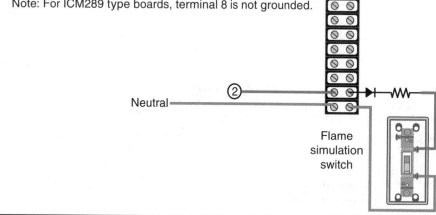

*   For type 1165-83-301A boards that have terminal 8 connected to chassis ground

**   Disconnect wire here to make in-line current measurement

D1 uses 1N1001 diode; R1 use 8.5 MΩ 1/4 W 5% resistor

*** 120 Vac potential – tape or use shrink-wrap tubing

Note: For ICM289 type boards, terminal 8 is not grounded.

Figure 14.1-3.

Goodheart-Willcox Publisher

Continued ▶

7. Connect the resistor to the cathode end of the diode using a butt connector or small wire nut. Tape over the connection. Connect the anode side to the terminal strip. Connect a section of wire to the free end of the resistor. Connect the wire to the SPST switch.

8. Place the switch in the off position.

9. Prepare the 12 pin and 2 pin connector wires. Use 18 AWG wires (multiple colors if available), and connect to the round terminals. Follow the diagram in **Figure 14.1-3**. Label each of the following pairs of wires to be jumped together: 1 and 7 limit switch, 4 and 10 pressure switch, and 5 and 11 rollout switch. Label the outputs (bulbs), which are the inducer, hot surface ignitor, blower, and gas valve. Before connecting the hot terminals for the loads, ensure there is continuity for each load circuit.

10. Assemble the jumpers, or use purchased parts. Ensure that connections can be made easily and quickly. Small toggle switches may be used instead of jumpers.

11. Review all the notes in **Figure 14.1-3** before proceeding.

12. Prepare to use the in-line ammeter by using one jumper wire to connect to the terminal strip and loose wire.

13. Review the features and specifications for your IFC.

14. Get your instructor's permission to plug in the lab board.

15. Turn on the main switch.

16. Check for 24 V at the transformer secondary.

17. Check for 120 V between the line and neutral terminals on the IFC.

Lab Questions

1. Is terminal 8 grounded?

_____

2. Is the flame sensor terminal 2 potential reference to ground terminal 8 or neutral?

_____

3. What is the inducer pre-purge time?

_____

4. What is the heat blower on delay time?

_____

5. What is the heat blower off delay time?

_____

6. What is the ignitor "on" time?

_____

7. What is the inducer post-purge time?

_____

8. How many trials for ignition?

_____

9. What is the Auto reset time?

_____

10. What does a steady LED "on" indicate?

_____

11. Is there 24 V on the transformer secondary?

_____

12. Is there line voltage across line and neutral IFC terminals?

_____

Lab
14.2 Analyzing the Heat Cycle

The purpose of this lab is to execute a heating cycle using the completed setup from Lab 14.1.

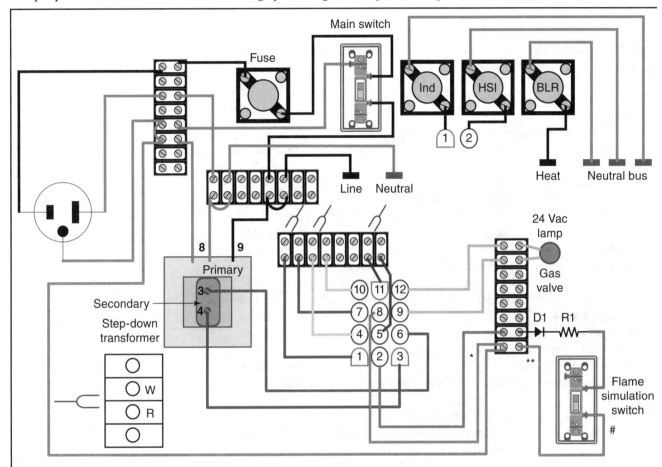

*   For type 1165-83-301A boards that have terminal 8 connected to chassis ground
** Disconnect wire here to make in-line current measurement
D1 uses 1N1001 diode; R1 use 8.5 MΩ 1/4 W 5% resistor
*** 120 Vac potential – tape or use shrink-wrap tubing

Note: For ICM289 type boards, terminal 8 is not grounded.

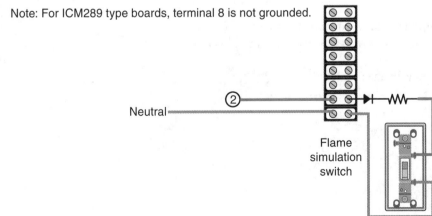

Figure 14.2-1.

Goodheart-Willcox Publisher

Lab Introduction

Inputs required for a normal heat cycle operation will be administered manually by using jumper wires to send the correct signals to the IFC. Providing that the signals are sent in the correct sequence and within the required timing, the IFC will produce the correct outputs.

Equipment

- Lab board with wired IFC from Lab 14.1
- Safety glasses
- Multimeter
- Jumper wires

Procedure

1. Turn off all switches.
2. Prepare the IFC for initial start.
3. Ensure that thermostat jumper R to W is disconnected.
4. Ensure the pressure switch is open. This means jumper 4 to 10 should be disconnected.
5. The rollout switch should be closed, so jumper 5 to 11 should be connected.
6. The limit switch should be closed, so jumper 1 to 7 should be connected.
7. Ensure the flame simulation switch is off. The IFC must see zero current flow in the flame sensor circuit. Initially, the circuit is incomplete. When the flame is lit, it will complete the circuit between the flame sensor probe and ground or neutral. The current will be rectified by the flame. Only about 7 microamperes should flow. The flame is simulated by the diode that will rectify the 120 V ac to about 60 Vdc, and the resistor will limit the current to about 7 microamperes.
8. Timing is critical. Read the following steps through before proceeding.
9. Call for heat by connecting jumper R to W.
10. Observe the inducer motor that will soon start.
11. Connect pressure jumper 4 to 10 within 3 to 5 seconds after the inducer starts.
12. After the inducer pre-purge ends (about 15 seconds), the hot surface ignitor turns on.
13. Observe the gas valve turn on after around 7 seconds.
14. Within 2 to 3 seconds after the gas valve turns on, turn on the flame switch.
15. The heater blower should turn on after 30 seconds.
16. Disconnect the thermostat jumper R to W, and quickly turn off the flame switch. If the flame switch is detected to be off before the jumper is disconnected, a failure code will be set. The IFC acknowledges it as a flame failure.
17. Check as the gas valve shuts off.
18. The post-purge feature should keep the inducer running for about 15 seconds.
19. The heat blower will run for 90 seconds after the gas valve shuts off. This is a selectable time.
20. Take note as the normal cycle has now been completed.

Lab Questions

Use your digital multimeter or other electrical instruments to answer the following questions.

1. When the board was initially powered, was a LED on steadily?

_____

2. When the call for heat was made, did the inducer fan turn on?

_____

3. Did the hot surface ignitor turn on indicating the pressure switch closed within the expected time?

_____

4. Did the hot surface ignitor turn on indicating the pressure switch closed within the expected time?

_____

5. Did the gas valve turn on next?

_____

6. Did you execute the flame switch in time to not set a trouble code?

_____

7. Did the heat blower turn on?

_____

8. How long did it take to turn on?

_____

9. After shutting of the thermostat, how long did the inducer motor run?

_____

10. After shutting off the thermostat, how long did the heat blower run?

_____

The purpose of this lab is to investigate the IFC diagnostic capabilities by introducing faults before and during system operation.

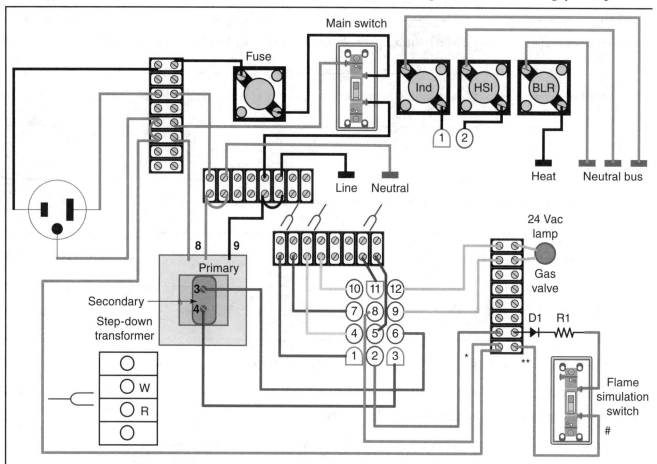

* For type 1165-83-301A boards that have terminal 8 connected to chassis ground
** Disconnect wire here to make in-line current measurement
D1 uses 1N1001 diode; R1 use 8.5 MΩ 1/4 W 5% resistor
*** 120 Vac potential – tape or use shrink-wrap tubing

Note: For ICM289 type boards, terminal 8 is not grounded.

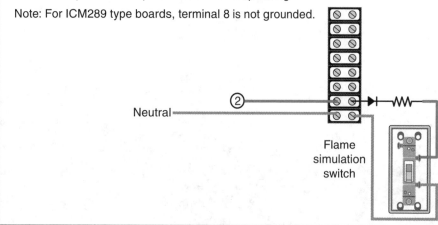

Figure 14.2-1.

Lab Introduction

Faults will be observed in this lab through the failure to provide signals within the required timing. This will allow you to observe the board's diagnostic capabilities and resulting system operation. Blower motor operation profile selection and the ac portion of the board will also be examined in the lab.

Equipment

- Lab board with wired IFC from Lab 14.1
- Safety glasses
- Multimeter
- Jumper wires

Procedure—Setup

1. Prepare the board settings for initial start. This means the flame switch is off, limits and rollout are closed, the pressure switch is open, and the thermostat is not calling for heat.
2. Send a call for heat.
3. After the inducer starts, do not jumper the pressure switch.

Procedure—Testing Flue Air Flow

1. Observe the operation. Note a trouble code by counting LED flashes. Record the information.
2. Reset the system by disconnecting 24 V at the transformer.
3. Restart the system. Perform normal operation.
4. When the heat blower is running, turn off the flame switch.

Procedure—Testing Flame Failure during Operation

1. Observe the operation. Note a trouble code by counting LED flashes. Record the information.
2. Reset the system by disconnecting 24 V at the transformer.
3. Open the rollout switch, and then restart the system.

Procedure—Testing Abnormal Flame

1. Observe the operation. Note a trouble code by counting LED flashes. Record the information.
2. Reset the system by disconnecting 24 V at transformer.
3. Restart the system. Perform normal operation.
4. After the heat blower is operating, disconnect the limit switch.

Procedure—Testing Overheated Heat Exchanger

1. Observe the operation. Note a trouble code by counting LED flashes. Record the information.
2. Reset the system by disconnecting 24 V at transformer.
3. Turn on the flame switch, and restart the system.

Procedure—Testing for Abnormal Startup (Flame Detected before System Start)

1. Observe operation. Note a trouble code by counting LED flashes. Record the information.

Procedure—Adjusting Heat Blower Turn Off Time

1. Change the heat blower off delay to 180 seconds by moving the jumper or setting the DIP switch on the board. See **Figure 14.3-2**.

Jumper is set for 90 seconds

Figure 14.3-2. *Goodheart-Willcox Publisher*

2. Run the normal heat cycle, and then disconnect the thermostat.
3. Observe the gas valve and heat blower shut off time difference.

Procedure—Testing AC Operation

1. Set the board for ac operation. Disconnect the blower from the heat terminal on the IFC. Connect the wire to the cool terminal.
2. Jumper the R to the G terminal.
3. Observe the operation of the blower now set for cooling.

Procedure—Testing Blower Off Time

1. Remove the thermostat jumper, and observe blower operation.

Lab Questions

Use your digital multimeter or other electrical instruments to answer the following questions.

1. Describe the operation in testing flue air flow, including the trouble code.

   _____
   _____
   _____
   _____

2. Describe the operation in testing flame failure, including the trouble code.

   _____
   _____
   _____
   _____

3. Describe the operation in testing the abnormal flame, including the trouble code.

   _____
   _____
   _____
   _____

4. Describe the operation in testing the overheated heat exchanger, including the trouble code.

   _____
   _____
   _____
   _____

5. Describe the operation while testing abnormal setup, including the trouble code.

_____

_____

_____

_____

6. Describe the operation in adjusting the heat blower turn off time.

_____

_____

_____

_____

7. Did the blower start immediately upon connecting the jumper?

_____

8. How long did the blower run after removing the R to G jumper?

_____

Know and Understand

_____ 1. What are the conductive paths to connect components in a PCB called?
- A. Inputs.
- B. Outputs.
- C. Varistors.
- D. Traces.

_____ 2. Which of the following is a serviceable PCB component?
- A. Fuse.
- B. Relay.
- C. Input.
- D. Transistor.

_____ 3. Which of the following is *not* an input?
- A. Fan motor.
- B. Temperature sensor.
- C. Pressure transducer.
- D. Limit switch.

_____ 4. Which is *not* a typical voltage to operate input devices?
- A. 5 Vac
- B. 5 Vdc
- C. 24 Vac
- D. 15 Vdc

_____ 5. *True or False?* A digital signal voltage level is the range of voltages between 1 and 5 V.

_____ 6. Which best describes an algorithm?
- A. Power output.
- B. Digital input.
- C. Computer program.
- D. Relays and contactors.

_____ 7. In a bi-directional communication system, which board receives initial status information from the other system boards?
- A. Outdoor unit.
- B. Indoor unit.
- C. Thermostat.
- D. Zoning control board.

_____ 8. *True or False?* The annunciation and diagnostic section uses LEDs to indicate system status.

_____ 9. *True or False?* When a microprocessor-based PCB is initially powered up, it performs self-diagnostics.

_____ 10. The PCB diagnostics feature aids in troubleshooting by annunciating which of the following conditions?
   A. A loose wire on a sensor output.
   B. A defective sensor.
   C. The wrong sensor is being used.
   D. The input signal is outside the expected operating range.

_____ 11. Which meter should *not* be used to troubleshoot a PCB?
   A. Voltmeter.
   B. In-Line Ammeter.
   C. Ohmmeter.
   D. Inductive clamp ammeter.

_____ 12. Which is *not* a correct method to troubleshoot a microprocessor-based PCB?
   A. Measure input voltage.
   B. Measure output voltage.
   C. Measure resistance between input and output.
   D. Measure input current.

_____ 13. Which event occurs first in a heating algorithm?
   A. Gas valve energized.
   B. Ignition energized.
   C. Inducer motor energized.
   D. Blower motor energized.

_____ 14. Which one of the following is *not* a required initial heating sequence condition?
   A. The pressure switch must be closed.
   B. The flame current must be zero.
   C. The rollout switch must be closed.
   D. The high-limit switch must be closed.

_____ 15. Which event is directly dependent on the correct operation of the inducer motor function?
   A. Roll ut switch.
   B. Pressure switch.
   C. Flame sensor.
   D. High-limit switch.

_____ 16. The gas valve is energized after _____.
   A. the high-limit switch closes
   B. the pressure switch closes
   C. the hot surface igniter has been energized for a predetermined time
   D. the flame sensor sends the proper signal to the PCB

_____ 17. *True or False?* If the flame sensor does not detect a flame, the gas valve is shut off.

_____ 18. A stepper motor is typically used for which device?
   A. Electronic expansion valve.
   B. Inducer motor.
   C. Compressor.
   D. Blower motor.

_____ 19. Which is true about digital communication type boards?
   A. Use terminal Y for air conditioning.
   B. Use terminal W for heating.
   C. Requires 7 wires.
   D. Requires 2 or 4 wires.

_____ 20. *True or False?* Digital communication information can be interpreted with a voltmeter.

Critical Thinking

1. In addition to receiving an input signal within the required voltage or current range, what else may be necessary for an output to be executed?

_____

_____

_____

2. What should be checked before condemning a microprocessor PCB?

_____

_____

_____

15 Troubleshooting Overview

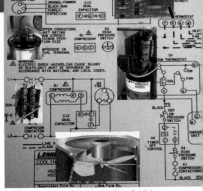

Learning Objectives

After completing this chapter, you will be able to:

- Describe wiring diagrams.
- Explain a split central ac and heating system.
- Describe the operation of a condensing unit.
- Analyze a condensing unit wiring diagram.
- Describe the operation of a furnace and blower motor.
- Evaluate a furnace and blower motor wiring diagram.
- Trace a circuit using a wiring diagram.

Technical Terms

full-load amperage (FLA)	indoor unit outdoor unit	rated load amperage (RLA)	service factor wiring diagram

Introduction

This final chapter will expand on troubleshooting an entire HVACR system, while building on and applying all the concepts learned throughout this textbook. Effective troubleshooting requires an understanding of circuit structure, equipment and component operation, equipment specifications, test equipment, and wiring diagrams. *Wiring diagrams* show wire terminal and connector details, which makes them different from schematic, ladder, and pictorial diagrams. A technician also must be familiar with new and older equipment from the past thirty years, by various manufacturers. This starts with understanding the fundamental function of the equipment and translating equipment diagrams and specifications.

15.1 AC and Heating System

A split central ac and heating system will be examined for troubleshooting purposes in this section. A split system has its condensing and evaporator units installed in separate locations. A condensing unit is typically installed outdoors and referred to as the *outdoor unit*. See **Figure 15-1**. The evaporator and blower is referred to as the *indoor unit*.

Figure 15-1. A condensing unit.

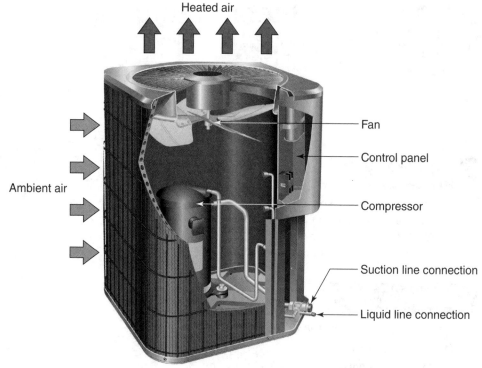

Heated air

Fan

Control panel

Compressor

Ambient air

Suction line connection

Liquid line connection

Lennox Industries Inc.

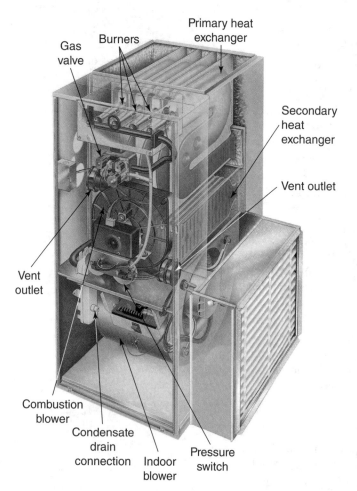

Gas valve

Burners

Primary heat exchanger

Secondary heat exchanger

Vent outlet

Vent outlet

Combustion blower

Condensate drain connection

Indoor blower

Pressure switch

Carrier Corporation, Subsidiary of United Technologies Corp.

Figure 15-2. A furnace with blower assembly.

See **Figure 15-2**. The indoor unit also contains a furnace when central forced warm air and air conditioning are both provided. A split system differs from a packaged unit that contains both units.

15.1.1 The Condensing Unit—Outdoor Unit

A typical condensing unit wiring diagram is shown in **Figure 15-3**. The top section of the diagram shows a pictorial view of the contactor, capacitor, and time delay board. A few things can be determined from the diagram. The contact configuration of the contactor is double-pole single-throw (DPST) and has a 24 Vac coil. The line side of the contacts is supplied with L1 and L2 through a disconnect box. The load side contacts feed the compressor and condenser fan. A dual-run capacitor feeds the start windings of the PSC motors. This unit uses a time delay board to prevent short-cycling that can be harmful to the compressor. When the thermostat sends a cooling signal on terminal Y to the timer board, it starts a five-minute timer. After the elapsed time, 24 Vac is sent to the contactor coil. Other components shown in the condenser unit are the compressor motor, condenser motor, low-pressure switch, and high-pressure switch.

> **Safety Note**
> ## Grounding the Run Capacitor
> The run capacitor must be grounded to the chassis with the supplied strap. The unit must be grounded in accordance with national and local codes.

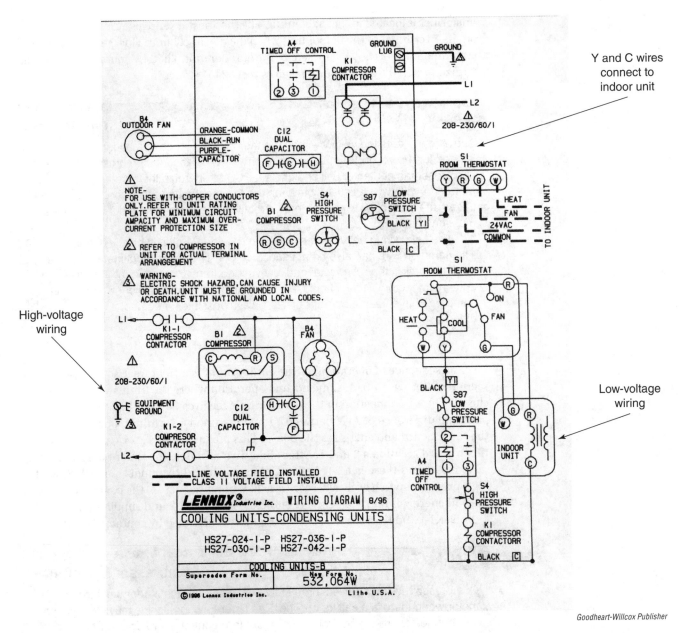

Figure 15-3. A condensing unit wiring diagram.

The high-voltage section of the diagram includes the contactor contacts, motors, a dual capacitor, and a ground lug. For troubleshooting, the high-voltage section can be broken down into line, control, and load sections. The line originates at the circuit breaker panel that feeds a disconnect box. The disconnect box then feeds the load side of contacts K1-1 and K1-2. These contracts control the power to the motors. The contacts are controlled by the low-voltage contactor coil. Both motors are PSC and require a run capacitor wired in series with the start winding. When K1-1 and K1-2 close, L1 is fed to the compressor run winding, capacitor common terminal, and the condenser fan run winding. The compressor start winding is fed from capacitor terminal H (hermetic compressor), and the fan start winding is fed from terminal F (fan motor). The C (common) terminals of both motors are connected to the opposite potential L2 to complete the circuit.

The low-voltage section of the diagram includes the contactor coil, the low- and high-pressure switches, and the time delay board. Like similar systems, the low-voltage transformer is located in the indoor unit. The thermostat supplies a 24 V hot side signal from its Y terminal to the indoor unit when it calls for cooling. This cooling signal and the common side of the 24 V are sent to the outdoor unit. The 24 V contactor coil is then energized when the outdoor unit receives the Y 24 V signal.

Pro Tip

Yellow Wire

The Y represents the yellow wire in the thermostat wire color standard, which is used for the cooling signal. This wire either energizes the compressor and condenser fan contactor coil or signals a PCB to energize the compressor and fan motors. A yellow wire is used between the thermostat and the indoor unit. In some cases, the Y terminal wire may route directly to the outdoor unit, but local codes must be observed.

The field wiring between the indoor and outdoor unit is typically different. The common wire can also be of various colors. Therefore, a voltmeter must be used to confirm the thermostat and low-voltage transformer secondary wires. The equipment factory wiring should follow the diagram colors. Although wire color standards have been established, it is always possible for someone to use a different color wire.

The sequence of operation for the outdoor unit is as follows. Use the low-voltage section of **Figure 15-3**. Assume normal operation where the limit switches are both closed. The Y1 connection in the outdoor unit receives 24 V. At this point, there is a potential difference of 24 V between Y1 and common C terminal. Current passes to the timer board and energizes the board. This means there is a 24 V potential across timer board terminals 2 and 1. After about five minutes, the timer board relay contacts close, and there is a 24 V potential across board terminals 3 and 1. Since the high-pressure switch is closed, there is a 24 V potential across the contact coil. Thus, the coil energizes and then closes contacts K1-1 and K1-2 and supplies power to the motors. K1-1 and K1-2 are the contactor contacts actuated by contactor coil K1.

Pro Tip

Compressor Start Delay

Compressor start is delayed three to five minutes after the indoor unit blower motor when there is a call for cooling. The timer function device may be built into the thermostat, indoor control board, or outdoor control board. Remember to wait for the elapsed time before troubleshooting.

The operation sequence, as discussed above, is critical to understand for troubleshooting. A technician also must understand what occurs when the controlling limit switches open and close. For example, a low-pressure switch may be in an open state because there is low pressure in the system, or the switch has failed electrically. If actual system pressure is 70 psig and the low-pressure switch is designed to open at 25 psig, then there is a fault with the switch or connecting wires. This information can be found in the manufacturer's service literature.

Troubleshooting a system begins when power is established and delivered to the outdoor unit, and there is a call for cooling. If there is a call for cooling and the indoor blower is operating, but outdoor unit is not working, the technician must isolate the problem. Ask: is it the line, load or control section, or a combination of sections? If neither motor is operating, manually pressing the contactor armature can indicate the location of the problem. If the motors turn on, it indicates a control circuit problem.

This means there is a break in the control circuit. It could be one of the two pressure switches, contactor coil, timer board, or a connectivity problem.

Remember, there can be more than one problem. If the motors do not operate when the armature is pressed manually, check the voltage across the line side of the contacts. Check also for the power feeding the motors by measuring the voltage across the load side of the contacts. If power is reaching the motor, but it is not starting, turn off the power, disconnect the motor, and check the windings with an ohmmeter. If the motor is attempting to start, check the run capacitor. Recall when there is a starting circuit problem, the motor makes a humming sound but does not start.

15.1.2 Furnace, Blower, and Evaporator—Indoor Unit

An indoor unit contains a furnace and blower assembly that houses the integrated furnace control board, ignition component, and blower. An evaporator may also be connected to the furnace assembly to provide cooling. If an evaporator is added to the indoor unit, the furnace blower is used to circulate air through the evaporator when cooling is required. This is possible because the furnace blower has multiple speeds to supply different speeds for heating and cooling.

Figure 15-4 provides an example of a wiring diagram for a condensing gas-fired furnace. The top portion is a pictorial diagram, which shows the relative location of the components. They are shown separately in **Figure 15-5**. The furnace is separated into two compartments, each with a door panel. The top compartment houses the burner components and heat exchanger. The electrical burner components include the inducer motor, gas valve, hot surface ignitor, flame detector, roll out switch, primary gas limit, and pressure switch. Power is fed into this compartment by a SPST switch mounted on the side of the furnace.

There are three (plug and jack) connectors used to connect the electrical components in the top compartment to the lower blower and control board. Each connection requires a jack that attaches to the plug. The lower compartment contains the indoor blower and motor, step-down transformer, and control board. The thermostat wires and wires leading to the outdoor unit enter in the blower compartment. This compartment door must be closed for the unit to operate, so it has an interlocking door switch. This is a SPST momentary NO switch. When the door is in place, the door pushes the switch plunger and closes the contacts. When the door is opened, the switch contacts open. The blower motor is a PSC motor, thus it uses a run capacitor. Board A92 has two round pin connectors and a screw-on terminal connector for the thermostat wires that include terminals C and Y for the outdoor unit.

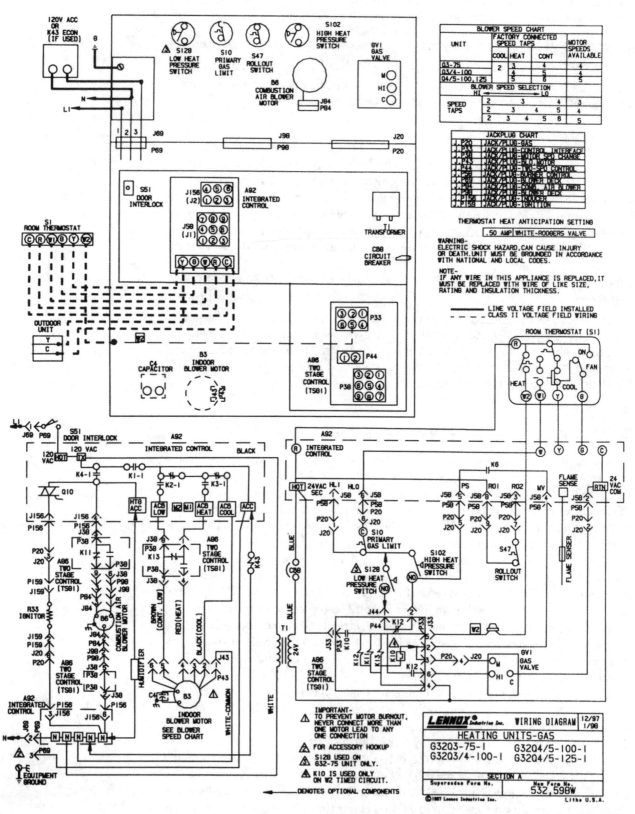

Figure 15-4. A condensing furnace wiring diagram.

Goodheart-Willcox Publisher

Pictorial Diagram

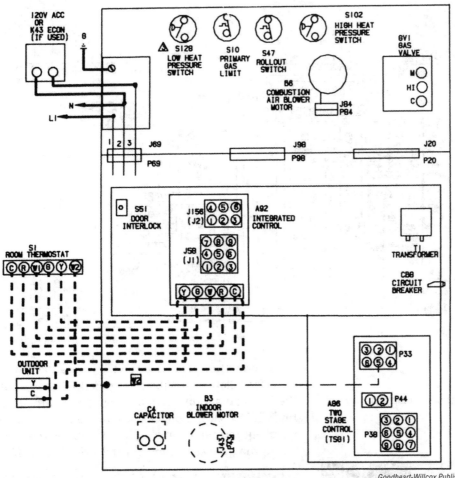

Figure 15-5. Pictorial section of the furnace wiring diagram.

Goodheart-Willcox Publisher

Board A86 contains three round pin connectors. The connector numbers and function name are shown in the jack/plug chart, **Figure 15-6**. Each jack and plug number is assigned a specific function. For example, J.P69 (jack and plug number 69) is for the blower deck connection.

Figure 15-6. A jack/plug chart.

	Jack/Plug Chart	
1	J.P20	Jack/Plug - Gas
2	J.P33	Jack/Plug - Control Interface
3	J.P38	Jack/Plug - Motor SPD Change
4	J.P43	Jack/Plug - Blo, Motor
5	J.P44	Jack/Plug - Two-Spd Control
6	J.P58	Jack/Plug - Burner Control
7	J.P69	Jack/Plug - Blower Deck
8	J.P84	Jack/Plug - Comb Air Blower
9	J.P98	Jack/Plug - Blower Deck
10	J.P156	Jack/Plug - Inducer
11	J.P159	Jack/Plug - Ignition

Goodheart-Willcox Publisher

Figure 15-7. Wiring section of the wiring diagram.

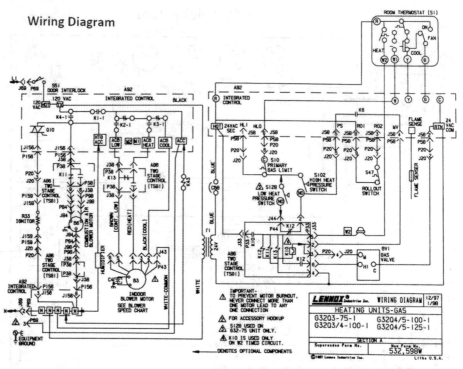

Goodheart-Willcox Publisher

In order to trace this circuit, a wiring diagram is provided below the pictorial. **Figure 15-7** shows only the wiring diagram portion. It can prove difficult to find where the component leads from the burner compartment terminate on the circuit board without the wiring diagram and connector chart. See **Figure 15-8.**

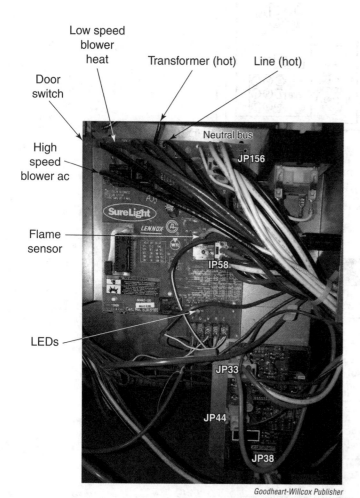

Goodheart-Willcox Publisher

Figure 15-8. The jack/plug numbers shown on board picture.

> ### Pro Tip
> #### Tracing a Wire without a Wiring Diagram
>
> It is time-consuming to trace a wire without a wiring diagram. The wire may be routed within a group of wires that are bundled together, known as a harness. The wire may not be visible inside the harness and may require separating the wires apart in order to find it. In addition, the wire may change colors as it leaves the other side of a connection. There could be several connection interfaces. An alternative and better choice when the wire's path is not visible is to turn off power and unplug the connectors from the control board and check continuity between points.

The general sequence of operation for a gas-fired furnace begins with a call for heat, so the thermostat sends a 24 V signal to the IFC. The board performs diagnostics that include checking that the high-temperature limit and roll out switches are in the closed position and the pressure switch is open. If these tests pass, the inducer is turned on for a pre-purge period. During this period, the pressure switch should close to indicate flue gases can flow freely. The hot surface ignitor is energized, and after a warm-up period, the gas valve is energized. The hot surface ignitor is de-energized. The flame rectification circuit must provide a flame within one to four seconds. If it does not, the gas valve shuts off. About 30 seconds after ignition, the blower motor is energized. When the thermostat is satisfied, the gas valve is shut off, and the inducer runs until the post-purge times out. The blower shuts down after the off-delay time period elapses.

The wiring diagram shows connector terminal numbers for tracing out a circuit from the board to a system component. Use **Figure 15-9** as an example of tracing the inducer neutral wire connection to the neutral bus on the A92 board. Starting at the inducer motor neutral wire, the wire is connected to terminal 2 in J84. J84 is mated to P84 and to terminal 2 in J98. J98 mates to P98 terminal and connects to terminal 5 in J38. J38 mates to P38 terminal 5 on board A86. The wire comes out of the A86 board J38 terminal 2 to terminal 6 of P156. P156 mates to J156 on the A92 board that contains the neutral bus. The same approach is used to trace any wire in the diagram.

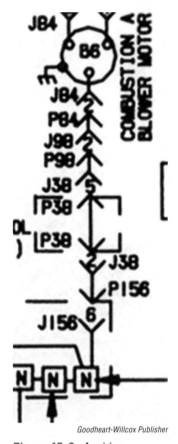

Goodheart-Willcox Publisher

Figure 15-9. A wiring diagram cutaway showing connector and terminal numbers.

Pro Tip	Manufacturers and Wiring Diagrams

Manufacturers use one wiring diagram to cover several models. Be aware of components and circuits that are used only on some models and special accessory features.

15.2 Specifications in Troubleshooting

Knowing a system's limitations and power requirements are necessary when troubleshooting HVACR equipment. The following paragraphs highlight the information that is referenced to the equipment nameplate and literature.

The minimum and maximum voltages must be considered. Most open motors are allowed to operate at ±10% of rated line voltage. For example, a 115 V rated motor may have line voltage between 103.5 V and 126.5 V. However, a compressor rated for a 208 V to 230 V range has a minimum that is only 5% below 208 V and a maximum of 10% above 230 V. This means the range is 197 V to 253 V. The lower the supply voltage, the greater the current, which causes the motor to run hotter and less efficient.

Minimum circuit ampacity is also a factor that must be determined and understood. It defines the minimum amount of current that must be available to the system to operate per design. In addition, the wires, switches, contactor contacts, and relay contacts must be able to safely pass this current to the load. Motor current draw is further described by the FLA, RLA, and LRA. *Full-load amperage (FLA)* for non-compressor motors is current drawn when a motor is operating at rated RPM, and the rated mechanical load horsepower has the rated voltage and frequency applied to it. *Rated load amperage (RLA)*, for an ac or refrigeration compressor, is the manufacturer's established compressor current draw when operating under normal design conditions. The voltage and frequency must also be at design values. Lock rotor amperage (LRA) is the motor startup current. This value is typically two to six times greater than the FLA depending on the mechanical load and type of motor.

Motor startup current is grouped into LRA code letters that specify LRA in terms of kVA per horsepower. For example, a 0.5 hp motor rated for 230 V, 4.2 FLA and code letter M draws a starting current (LRA) between 21.7 and 24.3 A. Code M represents 10 – 11.19 kVA or 10,000 VA through 11,190 VA. The calculation for the above maximum value of the range is as follows:

$$LRA = \frac{\text{Rated hp} \times \text{max VA}}{V}$$

$$= \frac{0.5\ hp \times 11,190\ VA}{230\ V}$$

$$= 24.3\ A$$

Motors may operate above their rated horsepower without causing internal damage. Small fractional horsepower typically provides between 25% to 35% reserve horsepower. Larger motors of 1 hp and greater are limited to about 10% reserve horsepower. The percentage of horsepower is given as a decimal value and is called the *service factor*. This way the rated horsepower can be multiplied by the service factor to yield the maximum motor horsepower that includes the reserve capacity. A service factor of 1.0 means that the motor does not provide reserve horsepower.

The following are also included in the specifications and manufacturer literature:

- **Frequency and phase.** Frequency is a factor in speed.
- **Maximum circuit protection.** The largest fuse or circuit breaker that can be used and still provide protection against dangerous current.
- **Horsepower.** The motor's mechanical power output.
- **Class ratings.** The safe operating motor winding temperature level.
- **Ambient temperature.** The maximum air temperature surrounding the operating motor.
- **Voltage and current ratings.** Used for contacts used in relays, contactors, sensors, and switches.
- **Contact configuration.** Used for relays, contactors, and switches.
- **Temperature and pressure set points.** The values at which the contacts open and close.
- **Conductor current capacity and insulation dielectric strength.** The amount of current that can safely flow through the conductor and how much potential difference the insulation can restrain to prevent current flow through the insulation.

> **Safety Note**
>
> ### Manufacturer Safety Warnings
>
> Read and adhere to the manufacturer's safety warnings and notes.

Summary

- Effective troubleshooting requires understanding of circuit structure, equipment and component operation, equipment specifications, test equipment, and wiring diagrams.
- A technician must be knowledgeable of newer and older equipment.
- The evaporator and blower are housed in a different location than the condensing unit in a split system.
- Packaged units contain the evaporator and blower and the condensing units inside one housing.
- In addition to the operation sequence, the technician must know when the controlling limit switches open and close.
- Compressor start is delayed three to five minutes after the indoor unit blower motor when there is a call for cooling to prevent compressor short cycling.
- Allow a hot motor to cool down to allow the internal overload that could have tripped to reset.
- A wiring diagram shows connectors and terminal numbers for tracing out a circuit from the control board to a system component.
- Knowing system limitations and power requirements are necessary when troubleshooting HVACR equipment.
- Read and adhere to the manufacturer's safety warnings and notes.

Lab 15.1

Heating System Evaluation

The purpose of this lab is to evaluate a gas-fired furnace by recording and measuring key operating characteristics.

Lab Introduction

The lab will require the use of a condensing furnace. A noncondensing furnace can be used if a condensing type is not available. The unit will be run to ensure it operates normally. You will also record the operating values of key components. In order to record the necessary information, the unit may have to cycle a few times.

Equipment

- Condensing furnace (or noncondensing type if necessary)
- Safety glasses
- Multimeter
- Furnace literature – wiring diagram, diagnostic code chart, and service literature if available

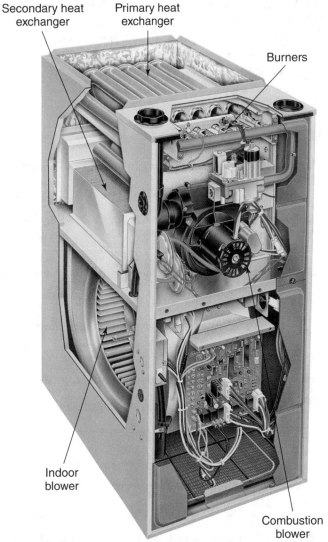

Rheem Manufacturing Company

Figure 15.1-1.

 Procedure

1. Inspect for proper power and ground connections.
2. Have your instructor check the unit for safe operation.
3. Close blower door.
4. Get your instructor's approval to operate the system.
5. Have a time watch, paper, and pen to record event timing.
6. Send a call for heat by setting thermostat to a higher temperature than the ambient.
7. Observe the LEDs through blower door window. Record status. Count the time it takes for the board self-diagnostic to complete.

Continued ▶

8. Observe the inductor motor turning on and time.
9. Observe and record when the hot surface igniter turns on.
10. Observe and record when the gas valve opens and the flame lights.
11. Start time count for the blower to turn on. Record.
12. Lower the thermostat to shut off the gas valve.
13. Record how long the inducer motor runs after the flame cuts out.
14. Record how long the blower runs after the inducer shuts off.
15. Turn off the unit, and set up the multimeter to measure in-line current. Set the meter to microamperes, and connect it in series with the flame probe wire.
16. Run the furnace measure. Record the flame rectification current when the flame is lit.

Lab Questions

1. How long did the board diagnostics take to complete when the system was first powered?

_____

2. Describe the annunciation LED or LEDS after the initial power up.

_____

3. How long did the inducer motor run before the hot surface igniter was energized?

_____

4. How long was the hot surface igniter on before the gas valve opened and the flame lit?

_____

5. Did the flame continue to operate after four seconds?

_____

6. Did the hot surface igniter shut off?

_____

7. How much time elapsed between the flame lighting and the blower motor turning on?

_____

8. When the thermostat is turned off, how much longer did the inducer motor run?

_____

9. How much longer did the blower motor run after the inducer motor shut off?

_____

10. What was the total run time for the blower after the flame shut off?

_____

11. What is the flame rectification current?

_____

12. Is there line voltage across line and neutral IFC terminals?

_____

Lab

15.2 Evaluating Heating System Diagnostics

The purpose of this lab is to introduce faults into the furnace used in Lab 15.1. The unit's diagnostics will be used to aid in troubleshooting the fault.

Lab Introduction

Using the same furnace as in Lab 15.1, faults are introduced by disconnecting wires at different times during the operating cycle. These faults are identified in the system diagnostics chart.

Equipment

- Condensing furnace (or noncondensing type if necessary)
- Safety glasses
- Multimeter
- Furnace literature – wiring diagram, diagnostic code chart, and service literature if available
- Digital manometer

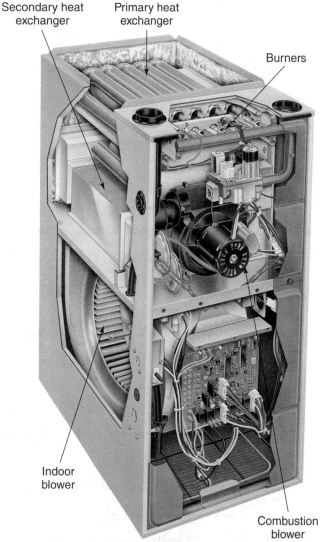

Secondary heat exchanger

Primary heat exchanger

Burners

Indoor blower

Combustion blower

Rheem Manufacturing Company

Figure 15.2-1.

 ## Procedure

1. Get permission from your instructor to operate the furnace.
2. Use the operating parameters from Lab 15.1.
3. Select a diagnostic code failure from the chart. Start with a limit switch open.
4. Disconnect the switch before a call for heat.
5. Observe the operation and diagnostic LED(s).
6. Reset system and reconnect the limit switch.
7. Call for heat. Observe diagnostic LED(s).
8. Reset the unit.

Continued ▶

9. Call for heat, and allow system to run. After blower turns on, disconnect the limit switch.
10. Observe the diagnostic LED(s).
11. Run the unit. When the blower starts, disconnect the pressure switch.
12. Observe the diagnostic LED(s).
13. Disrupt the circuit pertaining to each diagnostic code. Observe the system operation and the diagnostic code displayed.

Lab Questions

Use your digital multimeter or other electrical instruments to answer the following questions.

1. Describe what happened when the limit switch was disconnected before a call for heat occurred.

_____

_____

_____

2. Describe what happened when the limit switch was disconnected after a call for heat occurred.

_____

_____

_____

3. Describe what happened when the limit switch was disconnected after the blower turned on?

_____

_____

_____

4. Describe how the system operates and what diagnostic code is displayed for each circuit disrupted in Step 13.

_____

_____

_____

5. Did the diagnostic display respond according to expectations when a circuit was disrupted?

_____

_____

_____

Lab 15.3 AC Unit Evaluation

The purpose of this lab is to evaluate the outdoor ac unit that receives control signals and from the indoor unit in Lab 15.1.

Lab Introduction

The outdoor ac condensing unit receives signals from the indoor unit to control the compressor and condenser fan. The furnace blower will be used to circulate air through the attached evaporator coil.

Equipment

- Condensing unit standard efficiency with PSC compressor and fan motors
- Safety glasses
- Multimeter

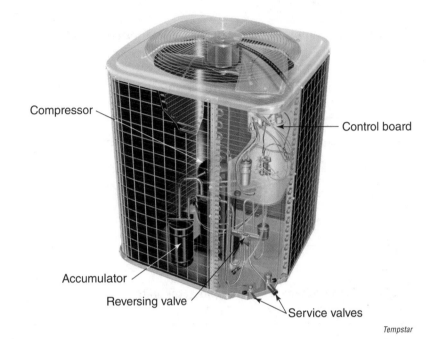

Compressor

Control board

Accumulator

Reversing valve

Service valves

Tempstar

Figure 15.3-1.

Procedure

1. Ensure the condensing unit is properly powered and safe to operate.
2. Get your instructor's permission to start the condensing unit.
3. Turn on the disconnect.
4. Set the thermostat to call for cooling by lowering the temperature setting below ambient.
5. Observe the blower motor operation.
6. Record the time between the blower turning on and compressor and condenser turning on.
7. Turn off the thermostat and condenser disconnect.
8. Remove the condenser unit control cover. Replace the disconnect plug or turn on the disconnect.
9. Check the voltage across the line side of the contactor. Record.
10. Check the voltage across the contactor coil.
11. Disconnect the contactor coil on one side.
12. Carefully manually push in the contactor armature with nonconductive tool.
13. Replace the coil wire, and set the thermostat to call for cooling.
14. Check the voltage across the contactor coil.
15. Check the RLA of the compressor. Record.
16. Check the FLA of the condenser fan. Record.
17. Turn off the thermostat.
18. Prepare the meter to check LRA of the compressor. Call for cooling, so the unit starts.
19. Record the LRA.
20. While the compressor is running, check the voltage across the line and load side of each contactor contact.

Lab Questions

Use your digital multimeter or other electrical instruments to answer the following questions.

1. What was the compressor on delay time?

   _____

2. What is the supply voltage to the line side of the contacts?

   _____

3. Is the supply voltage within the operating limits?

   _____

4. With the coil disconnected, what happens when the contactor armature was manually pressed in?

   _____

   _____

   _____

   _____

   _____

5. What is the voltage across the contactor coil?

   _____

6. What is the compressor RLA?

   _____

7. What is the condenser fan FLA?

   _____

8. What is the compressor LRA?

   _____

9. What is the voltage drop across the L1 and T1 contacts from step 20?

   _____

10. What is the voltage drop across the L2 and T2 contacts from step 20?

   _____

Know and Understand

_____ 1. What is *not* correct about a split system?
   A. The evaporator and blower are in one housing.
   B. The condensing unit is in one housing.
   C. The thermostat signal goes to the indoor unit.
   D. The thermostat signal goes directly to the condensing unit in all cases.

_____ 2. Which is *not* included in the high-voltage section of condensing unit?
   A. Condenser and compressor contacts.
   B. Line side of contactor contacts.
   C. Load side of contactor contacts.
   D. Contactor coil.

_____ 3. *True or False?* The Y terminal wire color is yellow throughout the entire circuit.

_____ 4. Which one is *not* included in the condensing unit low-voltage side?
   A. Time delay board input.   C. High pressure switch.
   B. Run capacitor common terminal.   D. Time delay board output.

_____ 5. For the cooling operation, what event happens after a call for cooling?
   A. Only the compressor turns on.
   B. The condenser fan only turns on.
   C. The indoor blower turns on.
   D. The indoor blower and condenser fan turn on.

_____ 6. Which thermostat terminals control the condensing unit contactor?
   A. R and C.   C. Y and R.
   B. C and Y.   D. W and R.

_____ 7. A time delay is applied to the compressor to _____.
   A. prevent short cycling
   B. allow the indoor fan to speed up
   C. allow the hot surface igniter to heat up
   D. start the compressor before the fan

_____ 8. *True or False?* All outdoor units contain a step down transformer.

_____ 9. If a condensing unit contactor armature is manually pressed in and the compressor and condenser motors operate, what is the fault?
   A. Open-start capacitor.   C. Open-run capacitor.
   B. Open-control circuit.   D. Failed contacts.

_____ 10. Which diagram shows all connector and terminal identification?
   A. Ladder.   C. Wiring.
   B. Schematic.   D. Pictorial.

_____ 11. *True or False?* The blower motor in a furnace will operate with the blower compartment door open.

_____ 12. In a gas-fired furnace, which event happens first?
   A. The hot surface ignitor is energized.   C. Flame sensor current is produced.
   B. The heat blower is energized.   D. The inducer motor is energized.

13. The gas valve is energized _____.
    A. before the pressure switch closes
    B. after the post purge
    C. after the hot surface igniter heats up
    D. immediately after the pressure switch closes

14. *True or False?* Knowing system limitations and power requirements are necessary when troubleshooting HVACR equipment.

15. What is the supply voltage tolerance for a single nominal voltage rated motor?
    A. ± 20%.   C. ± 10%.
    B. ± 5%.   D. – 5% / + 10%.

16. What is the supply voltage tolerance for a 208 to 230 v motor?
    A. ± 10%.   C. – 10% / + 5%.
    B. – 5% / + 10%.   D. ± 5%.

17. *True or False?* Minimum circuit ampacity is the minimum amount of current that must be available to the system to operate per design.

18. Which current draw is based on a compressor's operating parameters and supply voltage and frequency?
    A. LRA.   C. RLA.
    B. FLA   D. KVA.

19. Which current is drawn by a motor at start-up?
    A. FLA.   C. RLA.
    B. KVA.   D. LRA.

20. What is the service factor of a motor?
    A. Reserve capacity in decimal form.
    B. Motor efficiency rating.
    C. How often the motor needs to be serviced.
    D. A form of wattage rating.

Critical Thinking

1. How can the internal overload of a compressor be checked with an ohmmeter?

_____

_____

_____

2. Why is it important to know a system's operating limitations and specifications?

_____

_____

_____

Glossary

A

ac generator. A generator that generates ac by using slip rings and brushes to keep the end loop wires from twisting together as it turns within the magnetic flux. (6)

active component. An electronic component that regulates current flow based on an external signal or special condition applied. (12)

algorithm. A computer program that is specific to a piece of equipment. (14)

alternating current (ac). Current that changes direction at a periodic rate. (6)

alternator. A device that only produces alternating current. An alternator has fixed wire loops to form a stator. (6)

American Society of Heating, Refrigerating, and Air-Conditioning Engineers (ASHRAE). An organization that develops standards for the HVACR trade, including electrical applications. (1)

American Wire Gauge (AWG). The standard for wire size in the United States. It is based on the number of extrusion details used to extrude, or stretch, the wire to its final diameter. (5)

analog meter. An electromechanical device that measures current flow to determine the basic electrical quantities of voltage, current, and resistance. (10)

analog signal. The signal is converted from analog information. The signal is expressed in a long string of zeros and ones. (10)

analog to digital converter (A/D). A device that converts the analog signal into digital information. (10)

apparent power. The power applied to a circuit. Calculated in volt-amperes (VA). (9)

armature. A movable plate used in electromagnetic devices. (7)

atom. The makeup of any natural element; comprised of three subatomic parts: protons, neutrons, and electrons. (2)

audible continuity test. A method to verify continuity between two points in a circuit. (10)

automatic ranging. A meter feature that analyzes the measured current and function used and then selects the range to be displayed. (10)

B

back emf. A reverse flowing current in a magnetic field. Also known as *counter emf.* (7)

battery. A type of dc power source that consists of one or more cells. (6)

bearing. A part used to allow rotation inside a rotor. (13)

bi-directional communication. A PCB function that allows the board to communicate to other system boards and space temperature control. (14)

bimetal. A material made from two different metals that are bonded together. (12)

bipolar transistor. A three-layer device that consists of a collector, emitter, and base. (12)

branch. Each individual path that makes up a circuit. (5)

break. A technical word used to describe an open contact. (12)

break-before-make (BBM). A switch designed to break a circuit when a slide is moved away from the first position. (12)

brush. A dc motor component that conducts current between the stationary power source to the rotating slip rings. (13)

brushless dc motor. A smaller version of an ECM motor that uses a permanent magnet rotor. (13)

C

cadmium sulfide cell. A photoconductive device that reacts to changes in light. (12)

capacitance. The measure, in ohms, of a capacitor's ability to store a charge. (8)

Note: The number in parentheses following each definition indicates the chapter in which the term can be found.

capacitive reactance (Xc). The opposition that results in resistance to current flow. (8)

capacitor. A component that has the ability to store energy in the form of electrical charge. (8)

capacitor-start, capacitor-run (CSCR). A split-phase motor with the highest starting torque due to the run and start capacitors wired in parallel to each other and in series with the start winding. (13)

capacitor-start, induction-run (CSIR). A split-phase motor wired in series with the start winding and causes the phase shift between the run and start winding to approach 90°. (13)

center-tapped secondary. A connection point in the middle of a winding, which creates two windings in a series. (9)

centrifugal switch. A switch that disconnects the start winding in a motor by using centrifugal force to remove the winding from the circuit. (13)

circuit breaker. A manually resettable circuit protection device. (9)

closed-loop system. A system in an ECM that provides information feedback to the activation of an EEV. (14)

coil. Current-carrying wires designed to produce an effective magnetic field. (7)

commutator. A component of a dc motor that creates a change in current through the rotor winding. (13)

complex circuit. A combination of series and parallel circuits. Also known as *combination circuits* and *series-parallel circuits*. (5)

conductor. A material that allows current to flow through them easily. A conductor is typically made of metal. (2)

conductor. Any material that allows a current to flow through it. (3)

contacts. Components of a relay that are configured in normally open or normally closed combinations. (7)

contactor. An electrically operated switch used for larger loads and three-phase powered equipment. (7)

continuous voltage. The maximum bemf applied across a coil without causing damage to the coil. (13)

cross-sectional area. The area of a flat slice of the wire, or the area of the circle you see when you look straight into a wire. (5)

current electricity. A source of direct current that provides power to a circuit, so current can flow through it. The current can be ac or dc. (6)

current-interrupting capacity. The maximum current that can be stopped when contacts open. (12)

current magnetic relay (CMR). A relay made with a low-resistance coil and a normally open set of contacts. (13)

current transducer. A transducer that senses current by magnetic induction like the inductive clamp on a multimeter. (12)

D

dc generator. A generator that generates dc by using a special slip ring that is split in half, allowing current to flow in one direction only. (6)

dc motor. A motor that provides the highest torque and efficiency. (13)

delta (Δ). A three-phase configuration that has its three phases connected like a triangle. (9)

diac. A thyristor that conducts ac in both directions. (12)

diagnostic. The section of a PCB that receives status and specific failure information from the microprocessor and uses the annunciation section to announce information through LEDs or a human interface. (14)

dielectric. An insulator located in between two conducting plates of a capacitor. (8)

dielectric strength. A measure of a substance's ability to block electric flow. (1)

digital meter. A device that displays electrical quantities as a numerical value rather than a position on a meter. (10)

diode. A two-layer device made of one layer of positive and one layer of negative doped material. (12)

direct current (dc). Current that always flows in the same direction. (6)

direct digital control (DDC). A function that controls an entire HVAC system in a building and can include building automation. (14)

directly proportional. A direct relationship between variables that states as one variable changes, the other changes in the same manner. (4)

doping. The process of adding impurities to change a device's current-conducting characteristics. (12)

dropout voltage. The minimum bemf applied across a coil to keep the armature pulled in and keep the NC contacts open. (13)

dual capacitor. A capacitor comprised of two capacitors with different capacity values that share one common plate. (8)

dual-voltage motor. A split-phase motor that operates on two different voltages. (13)

duty cycle. A meter function expressed as a percentage of on-versus-off time for a square waveform used in pulse width modulation. (10)

E

earth ground. The wire connected to the center tap. (9)

electric current. The flow or movement of electrons. (2)

electric shock. A discharge of electric current that passes through the human body. (1)

electrical power. A measurable quantity that exists when current flows due to a potential difference (voltage). (4)

electrode. A copper rod inserted into the ground to conduct current into the ground. (9)

electrolytic. A type of large-value capacitor that can achieve very high capacitance. (8)

electromagnet. A temporary magnet controlled by turning power on and off. (7)

electromagnetic induction. A process that occurs when two coils are placed in close proximity and one coil becomes energized, which then induces an emf and current into the second coil. (7)

electromagnetism. A type of magnetism generated by electric current. (7)

electronic starting relay. A relay without moving parts, and instead senses changes in current and voltage to determine when to stop current flow between terminals. (13)

electronically commutated motor (ECM). A motor that uses high-efficiency dc motors and distributed ac power sources. (13)

end bell. Bells with bearings attached at both ends of a motor housing to support the shaft and rotor. (13)

event. When an input change causes the processor to execute a command. (14)

F

farad (F). The unit for capacitance. (8)

float switch. A switch activated by liquid level. (12)

flow switch. A switch activated by air or liquid movement. (12)

frequency. The number of cycles that occur per section in an alternating current. (6)

full-load amperage (FLA). Current drawn when a motor is operating at rated RPM and the rated mechanical load horsepower has the rated voltage and frequency applied to it. (15)

full-wave rectifier. A rectifier that uses two diodes and conducts during the entire cycle but with reduced voltage. (12)

fuse. A one-time device that must be replaced after an overcurrent event. (9)

fusible disconnect. Disrupts power to the circuit either manually (by the operator) or automatically (by a fuse). (11)

G

generator. A device with a wire loop arrangement where the loop rotates within the magnetic poles. (6)

ground-fault circuit interrupter (GFCI). A device used to trip a circuit when there is a current imbalance between the hot and neutral feeds of a power source. (1, 9)

ground wire. A wire connected to the center tap and to the electrode. (9)

H

half-wave rectifier. A rectifier with only one diode and conducts only for half a cycle of the sine wave. (12)

hermetic compressor. A compressor surrounded by refrigerant and oil. (13)

Hertz (Hz). The unit of measurement for frequency. (6)

high leg delta. A configuration that produces an unstable high voltage in delta three-phase power sources. (9)

horsepower. The measure of mechanical energy. (13)

hot wire. Ungrounded phase wires. (9)

I

impedance (Z). The overall opposition to flow, or total resistance in an ac circuit. (9)

indoor unit. Any system unit installed indoors. (15)

inductive load. A load made of coils of wire and powered by an alternating current. (7)

inductive reactance (X_L). The opposing resistance to current changing flow. (7)

infinity. Indicates that a meter is out of range. (10)

input. The signals and power fed to a printed circuit board from the system. (14)

insulator. A material that has an atomic structure that does not readily release electrons, thus restricts the flow of electrons through it. (2)

integrated furnace controller (IFC). A board that controls all PCB-related functions along with the ac portion of the system. (14)

inversely proportional. An inverse relationship between variables that states as one variable changes, the other changes in an opposing manner. (4)

iron core. A core designed to concentrate magnetic flux around a coil producing an electromagnet. (7)

K

Kirchhoff's current law (KCL). The law that states the current through each individual branch adds up to the total current supplied and returned to the power source. (5)

Kirchhoff's voltage law (KVL). The law that states the sum of the voltage drops in a series circuit equals the source voltage. (5)

L

ladder diagram. A circuit schematic that consists of vertical side rails that deliver power and horizontal rungs that make up the parallel load branches. (11)

Law of Charges. The law that states like charges repel one another and unlike charges attract. (2)

light emitting diode (LED). A diode made from different materials that emits light when forward biased. (12)

limit switch. A switch that closes or opens in response to movement. (12)

line side. The side of a switch connected toward the power source. (11)

line-voltage control. A control that can directly operate the load if it is in series with the load. (11)

live-dead-live (LDL) method. A method used when working on equipment to verify that power is disabled and the circuit is de-energized. (1)

load. An electric device that requires electric energy to operate, using the power from the source to perform work. (3)

load side. The side facing toward the load. (11)

locked rotor amperage (LRA). A rotor's current maximum value. Also referred to as *inrush* or *starting current.* (13, 15)

lockout relay. A relay used to prevent automatic resetting of power after a limit switch opens and then closes when condition returns to normal. (12)

lockout/tagout (LOTO). A procedure that requires a physical lock and a tag to be filled out by a technician to ensure maintenance or service on equipment can be completed in a safe manner. (1)

lookup table. A computer algorithm function that contains the electrical signal value from sensors and the physical property value it represents in the computer memory. (14)

low-voltage control. A control that can only directly operate a load. (11)

M

magnetic flux. Lines of force that connect the north and south poles of a magnet. (6)

magnetic induction. The principle used to generate electricity in many devices, resulting from the generation of electromagnetic force (emf). (6)

magnetic pole. Points on the end of each magnet produced by alternating current. (13)

maintained. The status of a switch that retains the selected state, open or closed, until making another selection. (12)

make. A term used to describe a closed contact. (12)

make-before-break (MBB). A switch that connects both positions one and two to the source terminal during the transition between the two positions. (12)

mechanical load. Any object that requires physical force to move it. (13)

megohmmeter. Tests the resistance of electrical insulators. Also called *megger.* (10)

meter movement. The movement of the magnetic coil and pointer in an analog meter. (10)

microfarad (mfd). One millionth of a farad, or 1/1,000,000, as denoted by the prefix *micro.* (8)

microprocessor. The central processing unit and contains all the internal functions. (10)

momentary. The status of a switch that reverts to the normal state after activation. (12)

multimeter. An instrument with various uses in electrical work, such as testing voltage or checking continuity. (1)

N

National Electric Code (NEC). A standard published by the NFPA that addresses electrical systems and safe installation. Also known as *NFPA 70*. (1)

National Electrical Manufacturers Association (NEMA). An organization that develops standards for electrical component specifications. (1)

National Fire Protection Association (NFPA). An organization that develops codes and standards to protect the public from hazards. (1)

neutral wire. A current-carrying conductor that is at the same electrical potential as the ground wire. (9)

noncontact voltage (NCV). A meter feature that checks for the presence of voltage without using the test probes. (10)

nonpolarized. Capacitors without positive, or negative, polarity. (8)

NPN transistor. A bipolar transistor that requires a forward bias current to flow from the emitter to the base. (12)

N-type material. An electronic material that has been doped to have a surplus of electrons. (12)

O

Occupational Safety and Health Administration (OSHA). An organization created to mandate workplace safety requirements. OSHA provides training, standard development, and enforcement in the work place. (1)

Ohm's law. A law that describes the relationship between the three electrical components of voltage, current, and resistance. (4)

open-loop system. A system in an ECM that does not provide information feedback to the activation of an EEV. (14)

outdoor unit. Any system unit installed outdoors. (15)

output. The processed signals and power sent to the system loads and other elements of the PCB system. (14)

overcurrent. A process that occurs when a circuit is overloaded beyond its rated current capacity. (9)

overload. A process that takes place when too many loads are connected to a single source. (9)

P

packaged ac unit. Contains the entire system in one housing. (11)

parallel circuit. A simple circuit that has more than one path for current flow. (5)

passive component. An electronic component that does not require an external signal for operation. (12)

peak emf. The maximum amount of electromagnetic force (emf) in a loop. (6)

peak value. The maximum value when calculating rms voltage or current of alternating current. (6)

permanent split capacitor (PSC). A capacitor with a permanently split phase during motor operation due to the start winding left in the circuit. (13)

personal protective equipment (PPE). A type of equipment worn to protect an individual from a number of safety hazards. (1)

phantom voltage. Voltage measured in a circuit that is high enough to cause confusion of the validity of a meter. (10)

phase conductor. The top and bottom wires connected to the secondary winding. (9)

pickup voltage. The required bemf applied across a coil to keep the armature pulled in and keep the NC contacts open. (13)

pictorial diagram. Shows the relative location of specific components and physical connections within a unit. (11)

PNP transistor. A bipolar transistor that operates the same as the NPN, except that polarities are changed. (12)

polarized. Capacitors with positive polarity. (8)

pole. The source of power that feeds a switch. (12)

positive temperature coefficient thermistor (PTC). A temperature sensitive resistor. Also known as a *resettable fuse*. (9)

potential relay. A relay that relies on a back electromotive force (bemf) or counter electromotive force generated by inductive reactance to operate. (13)

potentiometer. A three-terminal device with an adjustable center contact, called the wiper, and two fixed terminals. (12)

power factor. The ratio of real power absorbed by a circuit's load to the apparent power flowing in the circuit. (9)

power source. A circuit component, generally a battery or a power grid, that supplies power to the circuit. (3)

primary coil winding. The first coil of wire in a transformer's laminated metal core. (7)

printed circuit board (PCB). A device that receives input signals from a microprocessor system's sensors and switches. (14)

programmable logic controller (PLC). A computer that allows the user to write a program using a special high-level language. (14)

PTC relay. A relay that uses a PTC thermistor in series with a start winding, or both the start capacitor and start winding, to control current. (13)

P-type material. An electronic material that has been doped to have a deficit of free electrons. (12)

R

rated load amperage (RLA). The manufacturer's established compressor current draw when operating under normal design conditions. (15)

relay. An electrically operated switch controlled by an external signal. (7)

resistance. The restriction, or opposition to, current flow. (2)

resistance-start, induction-run (RSIR). A motor that uses a high resistance (dc) start winding to split the phase. (13)

resistive load. Heating elements for refrigeration evaporator defrost, hermetic compressor crankcase, and resistors. (7)

resistor. An electronic component that limits current flow. (12)

revolutions per minute (RPM). Unit of measure for a rotating magnetic field. (13)

rheostat. A two-terminal device with one fixed terminal and an adjustable wiper. (12)

root mean square (rms). The actual value of electricity performing work. (6)

rotating magnetic field. A magnetic field created by two sets of poles with one set out-of-phase and with opposite polarity to the other set of poles. (13)

rotor. The rotating component of a motor made up of staggered iron laminates that produce conducting diagonal bars. (13)

run capacitor. A single-phase motor capacitor that is energized while the motor is operating and dissipates the heat generated by current flow. (8)

run winding. A motor's main winding that has low dc resistance and high inductive reactance resistance. (13)

S

schematic. A diagram that shows the complete path that makes up a circuit and its related components. (3)

schematic diagram. A diagram that shows detailed information, such as wire color, AWG wire size, and terminals and connectors. (11)

scientific notation. Used to represent a large or small number. Scientific notation expresses a number by a power of ten in increments of three. (10)

secondary winding coil. The second coil of wire in a transformer's laminated metal core. (7)

self-induction. The process of back emf. (7)

self-start. The process of a rotor starting to turn without the aid of an external force. (13)

sensor. A device that detects a change in physical quantities, such as temperature, and sends a signal to an electronic device or activates a mechanical switch. (12)

series circuit. A simple circuit that has only one path for current flow, and the current travels from one load to the next. (5)

service factor. The percentage of horsepower given as a decimal value. (15)

servo motor. A motor, similar to the brushless dc motor, but with added feedback to the controlling microprocessor. (13)

shaded-pole motor. A motor with only a run winding and a copper band to produce the magnetic pole needed to create the rotating magnetic field effect. (13)

short circuit. A process that occurs when all available current is allowed to flow. (9)

shunt. A resistor that produces a low-resistance path for current to pass through another part of the circuit. (10)

signal. The input voltage or current in a printed circuit board. (14)

silicone-controlled rectifier (SCR). A rectifier that conducts current in one direction by using a gate to control current flow. (12)

sine wave. A graphical representation of one complete cycle of alternating current. (6)

single-phase power. Power that delivers one current and voltage within one cycle. (9)

slip ring. Rings that rub against the brushes that connect to the dc source in a dc motor. (13)

slip. The difference between the synchronous and actual speed of a motor. (13)

snap disc. A device used to protect wire overheating and is automatically resettable when it cools down. (13)

solenoid. An electromagnetic device used in many trade industries and varies based on its intended purpose. (7)

solid state. The name used to describe modern electronic components or devices. (12)

solid wire. Small diameter 18 AWG wires that are generally used in home and light commercial wiring. (5)

split ac unit. Includes an indoor unit that houses the evaporator and blower assembly and an outdoor unit that consists of the compressor, condenser, and fan. (11)

split-phase motor. A motor that operates on single-phase power. (13)

squirrel cage rotor. A special rotor used in an ac induction motor. (13)

start capacitor. A single-phase motor capacitor that assists a motor in starting up. (8)

start winding. A motor's auxiliary winding that has a higher dc resistance but lower reactance resistance than the run winding. (13)

start-capacitor, induction run (SCIR). A motor wired in series with the start winding and causes the phase shift between the run and start winding to approach 90°. (13)

static electricity. A source of direct current that builds up an electrical charge on an object or surface. (6)

stator. A stationary component of a motor composed of windings. (13)

step-down transformer. A transformer with a higher primary voltage than the secondary voltage. (7)

stepper motor. A variant of the brushless dc motor that is driven by pulses which move the rotor a certain number of degrees in either direction. (13)

step-up transformer. A transformer with a higher secondary voltage than the primary voltage. (7)

stop/start station. The part of a circuit where contacts are provided a path to energize a coil when the start button is pushed and can also be de-energized when pressing a stop button. (12)

stranded wire. A type of wire that provides flexibility and is used in applications where there is vibration or movement. (5)

switch. A device that can change, or control, a connection in an electrical circuit. (3)

synchronous speed. The speed of a motor. (13)

T

temperature transducer. A transducer with output voltage or current that corresponds to a specific temperature. (12)

test lead. A conductor connected to the meter to make measurements. (10)

thermistor. A resistor that responds to temperature changes by altering its ohmic value. (12)

thermocouple. An electronic device made of two dissimilar metals that are joined together. (12)

three-phase power. A system that consists of three currents separated by 120 electrical degrees within one cycle. (9)

throw. The indication of how many terminals power can be passed to in a switch. (12)

torque. Rotational force generated by electromagnetic principles. (13)

trace. Strips of copper that provide electrical connectivity between electronic components in a printed circuit board. (14)

transducer. A device that changes one form of energy to another. (12)

transformer. A device that uses coils of wire to produce magnetic fields and transfers ac from one coil to the other. (7)

triac. A thyristor similar to a diac, but has a control gate that allows for a low-voltage control signal to turn on conduction in both directions without reaching breakover voltage. (12)

true power. The actual power consumed by a circuit load. (9)

turns-ratio. A ratio that determines the amount of secondary voltage produced in a transformer by the number of turns used to form the coils. (7)

V

variable frequency drive (VFD). A motor that provides infinite variable motor speed instead of a few fixed speed values. (13)

varistor. A one-time-use resistor that reacts to voltage spikes. (14)

voltage. The force behind the movement of electrons. Also known as *electromotive force (EMF).* (2)

volt-ampere (VA) rating. The apparent power rating of a transformer. (7)

volt-ohm-milliammeter (VOM). An analog meter that combines the three basic functions in one unit. (10)

W

watt. A unit used to measure electrical power. (4)

wattmeter. An instrument used to measure the true power consumed by a circuit and display power factor. (9, 10)

wetting current. The minimum current that can break through the contact's surface film. (12)

winding. Two coils wrapped around a laminated metal core in a transformer. (7)

wiring diagram. A diagram that shows wire terminal and connector details of a circuit. (15)

wye (Y). A three-phase configuration that has three hot wires connected to a neutral point. (9)

Z

Zener diode. A diode that is heavily doped and used for voltage regulation to produce different values of PIV. (12)

Index

electronically commutated motors (ECM), 266–267
electronic components, 230–235
    bipolar transistor, 232–233
    diacs, tracs, SCRs, 232
    diodes, 230–231
    resistors, 233–235
electronic starting relay, 264
electronic timer type starting relays, 264
electrostatic discharge, 287
EMF, 18
end bells, 253
error rate for meters, 187
event, 285

F

farad (F), 130
feeding power, 203
FLA, 317
float switch, 228
flow switch, 228
frequency, 84, 183
full-load amperage (FLA), 317
full-wave rectifier, 230
fundamentals. *See* electrical fundamentals
fuse, 152
fusible disconnect, 202

G

generator, 82
GFCI, 6–7, 154
GFI, 7
ground fault circuit interrupter (GFCI), 6–7, 154
ground integrity, 149–150
ground wire, 148
ground, safety, 149

H

half-wave rectifier, 230
heat cycle analysis, 299–301
heating system
    circuits, 210
    evaluating diagnostics, 321–322
    evaluation, 319–320
    printed circuit boards (PCBs), 288–289
    troubleshooting, 309–317
hermetic compressors, 269, 275–276
Hertz (Hz), 84
high leg delta, 157
horsepower, 252

I

impedance (Z), 151
indoor unit, 309
inductive reactance, 103, 115–118
infinity, 176
input, 283
insulation ratings, 5
insulators, 19–20
integrated furnace controller (IFC), 288
inversely proportional, 48
iron core, 98

K

Kirchhoff's current law (KCL), 64
Kirchhoff's voltage law (KVL), 63

L

lab board, 40–41
ladder diagram, 202
latching circuits, 244–246
Law of Charges, 16
LDL, 3
LED, 230
light emitting diode (LED), 230
limit switch, 228
line side, 203
line-voltage control, 202, 204–206
live circuits, 4
live-dead-live (LDL) method, 3
load, 35
load side, 203
locked rotor amperage (LRA), 259
lockout relay, 238
lockout relay operation, 247–248
lockout/tagout (LOTO), 3, 10–12
lookup table, 284
LOTO, 3, 10–12
low-voltage control, 202, 207–208
LRA, 259

M

magnetic flux, 82
magnetic induction, 82
magnetic poles, 254
maintained switch, 228
make, 226
make-before-break (MBB), 226
mechanical load, 252